Rules of Thumb for Maintenance and Reliability Engineers

Rules of Thumb for Maintenance and Reliability Engineers

Ricky Smith

R. Keith Mobley

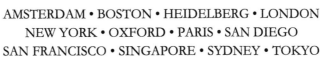

AMSTERDAM • BOSTON • HEIDELBERG • LONDON
NEW YORK • OXFORD • PARIS • SAN DIEGO
SAN FRANCISCO • SINGAPORE • SYDNEY • TOKYO

Butterworth-Heinemann is an imprint of Elsevier

ELSEVIER

Butterworth–Heinemann is an imprint of Elsevier
30 Corporate Drive, Suite 400, Burlington, MA 01803, USA
Linacre House, Jordan Hill, Oxford OX2 8DP, UK

Library of Congress Cataloging-in-Publication Data
Smith, Ricky.
 Rules of thumb for maintenance and reliability engineers / by Ricky Smith and R. Keith Mobley.
  p. cm.
 Includes bibliographical references and index.
 ISBN-13: 978-0-7506-7862-9 (pbk. : alk. paper) 1. Systems engineering—Management.
 2. Reliability. I. Mobley, Keith. II. Title.
 TA168.S57 2007
 620'.00452–dc22   2007019635

British Library Cataloguing-in-Publication Data
A catalogue record for this book is available from the British Library.

ISBN: 978-0-7506-7862-9

For information on all Butterworth–Heinemann publications
visit our Web site at www.books.elsevier.com

Transferred to Digital Printing in 2012

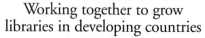

I would like to thank my wife, B. J., for providing me the environment to succeed, my Children, Amy, Ernest, and Christian who endured the long hours and days away from home traveling around the world in quest of reliability knowledge.

And to my mom and dad who believed in me.

—Ricky Smith

To my mother and father for teaching me the first lesson of project engineering—integrity,

To Sigrid Melle, my wife and best friend, who got me started on this project, graciously gave me the freedom to pursue it, and always believed in it,

To my daughter and three sons for their inspiration and the confidence they have given me in the next generation.

To the memory of Lesley Summerhayes, editor, colleague, and friend.

—Keith Mobley

Contents

P A R T

I

THE BASICS OF MAINTENANCE AND RELIABILITY

C H A P T E R

1

Understanding Maintenance and Reliability

C H A P T E R

2

The Functional Maintenance Organization and Its People

CHAPTER

11

Bearings

CHAPTER

12

Compressors

CHAPTER

13

Gears and Gearboxes

CHAPTER

14

Packing and Seals

CHAPTER

15

Electric Motors

PART

III

ADDITIONAL READINGS ON MAINTENANCE AND RELIABILITY

CHAPTER

16

Reliability Articles

CHAPTER

17

MTBF Users Guide

APPENDIX

A

Workflow for Planning

APPENDIX

B

Checklists and Forms

Introduction—The Recommended First Step to Rules of Thumb in Reliability Engineering

Begin your journey with this introduction to reliability, making this book a great tool for you to be successful. We came up with the idea of the book so someone whose sole purpose in life is reliability can go to a simple book to quickly find answers to issues facing his or her organization. The answer my not be simple; however, the book provides direction for anyone needing an answer to most reliability issues. The first recommendation is to follow these steps:

Step 1. Find some education for yourself:
- Attend a one- or two-week RCM training workshop. If you can, RCM training in your plant would be even better, so that part of the workshop could be applied to an asset in your plant.
- Attend a workshop on Maintenance Best Practices and Key Performance Indicators.
- Attend training in Six Sigma.
- This sounds like a lot of training but it is not. A true reliability engineer must have the tools required to accomplish the job, and very few universities offer real-world training and education.

Step 2. Educate management at your site in what truly is reliability and how it affects plant capacity, asset availability, and utilization.

Step 3. Read the article in Chapter 16 "Put a Plant-wide Focus on Functional Failures."

Step 4. Take the maintenance/reliability assessment in the book (Chapter 1) and identify the gaps. Be honest with your answers.

Step 5. Rank the plant's assets based on consequence and risk to the business (see Chapter 5).

Step 6. Develop a business case (see Chapter 1) and present it to executive leadership. This business case should include the cost of change, return on investment, project plan, and so forth. You want an executive engaged in your reliability initiative. This is not a journey with an end. Reliability must become a way of life for the plant.

Step 7. Execute your plan. Be sure key performance indicators (see Chapter 6) are in place before you begin this journey in order to measure and manage the project and thus the results.

A few certifications are also recommended:

1. CPMM (certified plant maintenance manager). Go to www.afe.org for more information. This certification is an open book and can be given by your plant HR manager. This certification is a great education rather than a great certification. The book they send you is a very good reference book for the future.
2. RCM certified as an RCM facilitator. There are many sources for this training certification.
3. CMRP (certified maintenance and reliability professional). Go to www.smrp.org for more information. This certification provides credibility to your position, and joining the Society for Maintenance and Reliability Professionals provides access to some great information.
4. Six Sigma black belt. This certification can be provided by many sources.

Be aware most companies try to put a quick fix on a sometimes complex problem, asset reliability. Over 80% of companies try to implement a good reliability strategy but fail to reach their ultimate goals. The reliability assessment and book will help your organization become successful. If you have questions concerning reliability contact the authors any time. Their email addresses are: askrickysmith@gmail.com and kmobley@LCE.com

THE BASICS OF MAINTENANCE AND RELIABILITY

1

Understanding Maintenance and Reliability

1.1. THE MAINTENANCE FUNCTION

The ultimate goal of maintenance is to provide optimal reliability that meets the business needs of the company, where reliability is defined as "the probability or duration of failure-free performance under stated conditions." Although many organizations view maintenance as adding little value to the firm, when properly developed and managed, it preserves the company's assets to meet the need for reliability at an optimal cost.

John Day, formerly the maintenance and engineering manager for Alumax Mt. Holly, was one of the best-known proactive maintenance management advocates and my mentor and manager for a number of years. John spoke all over the world about his model of proactive maintenance. His insight into what a successful plant considers "maintenance" provides us the section that follows.

1.2. STRATEGY TO ACHIEVE WORLD-CLASS PRODUCTION THROUGH RELIABILITY*

Alumax of South Carolina is an aluminum smelter that produces in excess of 180,000 MT of primary aluminum each year. It began operation in 1980 after a two-year construction phase. The plant is the last greenfield aluminum smelter constructed in the U.S. Alumax of

SC is a part of Alumax, Inc., which has headquarters in Norcross, Georgia; a suburb of Atlanta, Georgia. Alumax, Inc. is the third largest producer of primary aluminum in the U.S. and the fourth largest in North America.

The vision of general management was that the new smelter located on the Mt. Holly Plantation near Charleston, SC, would begin operations with a planned maintenance system that could be developed into a total proactive system. At the time in 1978–79, there were no maintenance computer systems available on the market with the capability to support and accomplish the desired objectives. Thus TSW of Atlanta, Georgia was brought on site to take not only the Alumax of SC maintenance concepts and develop a computer system, but they were to integrate all the plant business functions into one on-line common data base system available to all employees in their normal performance of duties.

Since the development and initial operation of the Alumax of SC maintenance management system, it has matured and rendered impressive results. These results have received extensive recognition on a national and international level. The first major recognition came in 1984 when *Plant Engineering* magazine published a feature article about the system. Then in 1987 A. T. Kearney, an international management consultant headquartered in Chicago, performed a study to find the best maintenance operations in North America. Alumax of SC was selected as one of the seven "Best of the Best." And in 1989, *Maintenance Technology* magazine recognized Alumax of SC as the best maintenance operation in the U.S. within its category and also as the best overall maintenance operation in any category. Mt. Holly's proactive model is shown in Figure 1.1.

*Section 1.2 is taken from John Day, "Strategy to Achieve World-Class Production through Reliability," portions also appeared in Ricky Smith, "Using Leading KPIs to Spot Trouble, *Plant Services Management* (August 2006). Used by permission of the author and publisher.

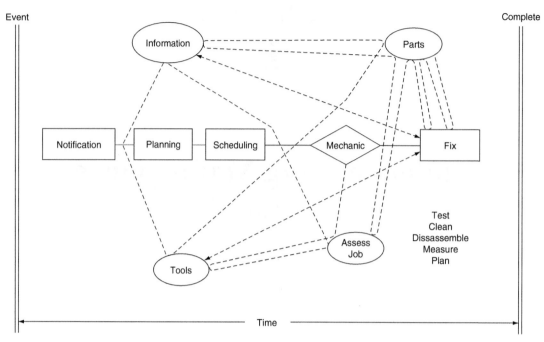

FIGURE 1.1.   Reactive maintenance model.

1.2.1. Maintenance Approaches

From a basic point of view there are two maintenance approaches. One approach is reactive and the other is proactive. In practice there are many combinations of the basic approaches. The reactive system (see Figure 1.1) responds to a work request or identified need, usually production identified, and depends on rapid response measures if effective. The goals of this approach are to reduce response time to a minimum (the computer helps) and to reduce equipment down time to an acceptable level. This is the approach used by most operations today. It may well incorporate what is termed as a preventative maintenance program and may use proactive technologies.

The proactive approach (see Figure 1.2) responds primarily to equipment assessment and predictive procedures. The overwhelming majority of corrective, preventative, and modification work is generated internally in the maintenance function as a result of inspections and predictive procedures. The goals of this method are continuous equipment performance to established specifications, maintenance of productive capacity, and continuous improvement. Alumax of SC practices the proactive method. The comments which follow are based upon the experience and results of pursuing this vision of maintenance.

1.2.2. Maintenance Management Philosophy

Alumax of SC began development of the maintenance management concept with the idea that maintenance would be planned and managed in a way that provides an efficient continuous operating facility at all times. Add to this that maintenance would also be treated as an investment rather than a cost, and you have the comprehensive philosophy on which the maintenance management system was built. An investment is expected to show a positive return, and so should maintenance be expected to improve the profitability of an operation. The management philosophy for maintenance is just as important as the philosophy established for any business operation. For most industry, maintenance is a supervised function at best, with little real cost control. But it must be a managed function employing the best methods and systems available to produce profitable results that have a positive effect on profitability.

The development of a philosophy to support the concept of proactive planned maintenance is important. It is believed that many maintenance management deficiencies or failures have resulted from having poorly constructed philosophies or the reliance upon procedures, systems, or popular programs that have no real philosophical basis.

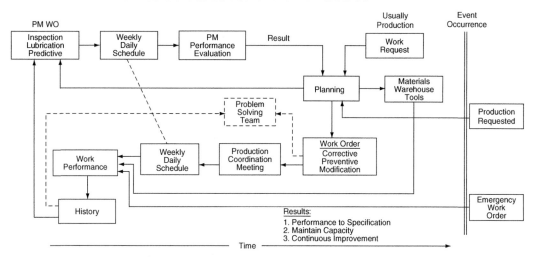

FIGURE 1.2. Mt. Holly's proactive maintenance model.

1.2.3. The Function and Control System

Today there is little disagreement that the function and control system of a good maintenance management program must be computer based.

Using the philosophy that maintenance management is to be considered in the same way that all other business functions are considered, it is difficult to justify any other approach other than complete integration of maintenance management functions with total organizational management functions. The computer is the tool to use to accomplish this difficult and complex task.

The computer, in an integrated operation, must be available for use by every member of the maintenance organization as well as all other plant employees who have a need. It is an essential part of the maintenance employee's resources for accomplishing his work. It is just as important to a mechanic or electrician as the tools in his toolbox or the analysis and measurement instruments that he uses daily.

The computer must supply meaningful and useful information to the user as opposed to normal computer data.

A successful integration of data systems will tie together maintenance, warehouse, purchasing, accounting, engineering, and production in such a way that all parties must work together and have the use of each other's information. This is part of the answer to the question being asked almost universally, how do you break down the barriers between departments and get them to work as part of the whole or as a team?

The computer system must be on line, available, and time responsive. A batch system or semi-batch system will not provide the support needed for a dynamic, integrated, maintenance management system.

In the integrated system with a common data base, data is entered only once and immediately updates all other files so that its use is immediately available to all functional areas. This means that anyone in any functional area can use or look at data in any other area, unless it is restricted. Some have referred to this effect as the "fish bowl effect" since everything is visible to all. This stimulates cooperation, in fact, it dictates cooperation.

1.2.4. What Is Maintenance?

Everyone knows what maintenance is; or at least they have their own customized definition of maintenance. If the question is asked, words like fix, restore, replace, recondition, patch, rebuild, and rejuvenate will be repeated. And to some extent there is a place for these words or functions in defining maintenance. However, to key the definition of maintenance to these words or functions is to miss the mark in understanding maintenance, especially if you wish to explore the philosophical nature of the subject. Maintenance is the act of maintaining. The basis for maintaining is to keep, preserve, and protect. That is to keep in an existing state or preserve from failure or decline. There is a lot of difference between the thoughts contained in this definition and the words and functions normally recalled by most people who are "knowledgeable" of the maintenance function; i.e., fix restore, replace, recondition, etc.

1.2.5. Specification

If we shift our defining thoughts to maintenance in the pure sense, we force ourselves to deal with keeping, preserving, and protecting. But what are we to keep, protect, or preserve? You may think that it is the machine, equipment, or plant, and that is true. But how are you to define the level to which the machine, equipment, or plant is to be kept. One way would be to say—"keep it like new." At face value the concept sounds good, but it is more subjective than objective. The answer to maintenance levels must be defined by a *specification*.

A specification is a detailed precise presentation of that which is required. We must have a specification for the maintenance of equipment and plant. In actual usage today the specification, if it exists, is not detailed or precise. A specification usually does exist informally in the mind of the mechanic or management member even though they may be unable to recite it. This means that at best, it is a variable, general-type specification. This kind of specification is defined in terms of and is dependent upon time available, personnel training level, pressure to produce a current order now, money allocated or available, or management opinion. Obviously, a specification like this will not qualify as a true specification, nor will it qualify as a supporting component of the act of maintaining. The true maintenance specification may be a vendor specification, a design specification, or an internally developed specification. The specification must be precise and objective in its requirements. The maintenance system and organization must be designed to support a concept based on rational specifications. Detailed work plans and schedules may be constructed to provide the specification requirement at the maintenance level. In the maintaining context, the specification is not a goal. It is a requirement that must be met. The maintenance system must be designed to meet this requirement. The specification must be accepted as the "floor" or minimum acceptable maintenance level. Variation that does occur should be above the specification level or floor. The specifications will probably be stated in terms of attributes and capacity.

In reference to maintenance specifications, included are individual equipment specifications, process specifications, and plant performance specifications.

1.2.6. The Maintenance Function

The maintenance department is *responsible* and *accountable* for maintenance. It is responsible for the way equipment runs and looks and for the costs to achieve the required level of performance. This is not to say that the operator has no responsibility for the use of equipment when in his hands—he does. The point is that *responsibility* and *accountability* must be assigned to a single function or person whether it be a mechanic or operator. To split responsibility between maintenance or any other department where overlapping responsibility occurs is to establish an operation where no one is accountable. Alumax of SC considers this a fundamental principle for effective operation of maintenance.

The maintenance function is responsible for the frequency and level of maintenance. They are responsible for the costs to maintain, which requires development of detailed budgets and control of costs to these budgets.

Just as the quality function in an organization should report to the top manager, so does the maintenance function for the same obvious reasons. This allows maintenance problems to be dealt with in the best interest of the plant or company as a whole. Maintenance efforts and costs must not be manipulated as a means for another department to achieve its desired costs results.

Where the maintenance department or group is held responsible and accountable for maintenance, the relationship with other departments takes on new meaning. The maintenance department can't afford to have adversary relationships with others. They must have credibility and trust as the basis of interdepartmental relationships. This is an essential element for the successful operation of a maintenance management system.

The organizational chart or better yet the organizational graphic is constructed on the basis that the central functional element for core maintenance is the Technical team. The relational (syntax) aspects of the organization are shown with concentric bands of teams. The nearer band of teams represents the tighter relationship to the core teams. Radial connecting lines show a direct relationship to a team or band of teams. Concentric connecting lines show a more indirect relationship between teams. The outer band of teams requires a Relational Organizational Chart similar to the maintenance teams chart to define their close relationships and full relationship to other plant teams. This particular chart is predicated on the relationship of all teams to central core maintenance teams.

Technical Teams—Core Maintenance—These teams perform core maintenance for the plant. They are composed of qualified electricians, mechanics, and technicians. The teams are assigned based on a functional requirement plant wide or on the basis of a geographic area of responsibility. The focus, direction of the team, and individual team member needs are provided by an assigned member of the facilitator and directional control team.

Facilitator and Directional Control Team—Members of this team have been trained and qualified to provide team organizational dynamics and traditional supervisory functions as required. With the facilitator, the team must address work performance by categories, administrating, training/safety/housekeeping, budgeting and cost control and information reporting as well as the technical requirements of the team. These members perform the necessary traditional supervisory functions, especially related to personnel functions, for the technical teams.

Work Distribution and Project Coordination Team—This team works with the Facilitator, Planning and Engineering teams to staff technical teams to meet work load requests, inventory requirements, contractor support, and field superintendence of engineering projects.

Job Planning Team—This team works closely with the Technical teams and the Facilitator team to plan and schedule maintenance, overhaul, and contractor work. Where operators are doing maintenance functions, the same applies. In addition, information and reports are prepared by this team for all other teams as required or requested. Quality control of the data input is a responsibility of this team. Coordination of production requirements must also be performed.

Technical Assistance Team—This team is a resource to the Technical teams and Facilitator team for continuous improvements, modifications, trouble shooting, and corrective action.

Materials Support Team—This team works with the Planning team, Facilitator team, and the Technical teams to meet planned job requirements and emergency material requirements.

Maintenance Management Team—This team provides overall coordination of maintenance and material functions to meet the plant capacity requirement. Overview of budget and cost control is also provided.

User/Operator Maintenance Team—This is a team of designated operators who perform assigned and scheduled maintenance work. They must be selected, trained and qualified prior to being assigned to this team.

Plant Engineering Team—This team provides projected management for the Plant capital budget program. They provide consulting and trouble shooting to the Technical Teams on an as requested basis.

Other teams in the outer band of the organizational chart must be specifically defined by individual relational organization charts.

For each of the above teams, a detailed performance requirement document must be developed. Individual team members are guided by a specific job performance document. These documents detail the vision, mission, processes used, and strategies employed.

Does the maintenance function provide a service or produce a product? Again, definition is important in the development of this part of the philosophy. Service is defined as a useful labor that does not produce a tangible commodity. A product is something that is produced, usually tangible, but definitely measurable. In the case of the maintenance function and the development of this philosophy, both a service and a product are considered as an output of maintenance. The current thinking which is related to traditional maintenance (reactive maintenance) suggests that the maintenance function is for the most part a service function. But the philosophy being developed here considers the maintenance function as the provider of a product with a small but limited service component. Consider the product produced by maintenance to be capacity (Production/Plant capacity). Writers on the subject of maintenance have suggested this concept in the past, but little has been made of developing the idea to date. A predominate service approach to maintenance, as is currently practiced, is a reactive mode of operation, and is typical of most maintenance operations today. React means response to stimulus. Most maintenance operations today are designed to respond to the stimulus of breakdown and the work order request, except for small efforts related to preventative maintenance and predictive maintenance, usually less than 25% of work hours worked. This simply means that the maintenance function must be notified (stimulated) of a problem or service requirement by some means, usually by someone outside of the maintenance organization, then maintenance reacts. Rapid response is the "score card" of this system.

It is being suggested by this proactive philosophy that the maintenance function be addressed as the producer of the product—*capacity*. Capacity is measured in units of production or output (or up time). A total proactive system must specifically be designed to produce capacity (product). If the maintenance function is to be classified as proactive, it cannot stand by and wait for someone to call or make a request. In a total proactive approach, maintenance must be responsible and accountable for the capacity and capability of all

TABLE 1.1.   Benchmarks at Alumax, Mt. Holly

	Mt. Holly	Typical
Planned/scheduled	91.5%	30–50%
Breakdowns	1.8%	15–50%
Overtime	0.9%	10–25%
Inventory level	½ normal	Normal
Call-ins	1/month	Routine
Off-shift work	5 people	Full crew
Backlog	5.5 weeks	Unknown
Budget performance	Varies, 1–3%	Highly variable
Capital replacement	Low	High
Stock outs	Minor	Routine

equipment and facilities. The function must provide a facility and equipment that performs to specification and produces the product (capacity). Stated again, the maintenance function is a process that produces capacity which is the product.

The results of this model created a benchmark that hundreds of companies followed and many continue to adopt. Table 1.1 shows the "world-class benchmarks" of Alumax, Mt. Holly.

Companies that adopted John Day's philosophy and strategy achieved results beyond what was known within the company. One company was a large manufacturing company. Once senior management understood and adopted Day's philosophy and approach, it resulted in the following:

1. Plant capacity increased by $12 million in the first year.
2. A large capital project was deferred when the capacity it was to provide was found to exist already.
3. The need to hire a projected 12 additional maintenance staff members was eliminated.
4. The plant maintenance staff was reduced by 20% over the following three years because of attrition.

The approach to proactive maintenance is not magic; implementing the process is very difficult but the results are worth the effort. To develop a true proactive maintenance process, a company must have commitment at all levels to follow known "best practices."

1.3. WHAT IS RELIABILITY?

Most maintenance professionals are intimidated by the word *reliability*, because they associate reliability

with RCM (reliability centered maintenance) and are unclear on what it actually means. Reliability is the ability of an item to perform a required function under a stated set of conditions for a stated period of time. However, many companies focus on fixing equipment when it has already failed rather than ensuring reliability and avoiding failure.

A common reason for this finding is the lack of time to investigate what is needed to ensure the reliability of equipment. Yet, a growing awareness among these reactive maintenance organizations is that the consequences of poor equipment performance include higher maintenance costs, increased equipment failure, asset availability problems, and safety and environmental impacts. There is no simple solution to the complex problem of poor equipment performance. The traditional lean manufacturing or world-class manufacturing is not the answer. These strategies do not address the true target; but if we focus on asset reliability, the results will follow.

1.3.1. Companies That Get It

Imagine a corporation fighting an uphill battle to survive despite foreign competition, an aging workshop, and many other issues. The chief executive officer (CEO) decides to focus on reliability because maintenance is the largest controllable cost in an organization and, without sound asset reliability, losses multiply in many areas. Over a two year period, a dedicated team of over 50 key employees researched the world's best maintenance organizations, assimilating the "best practices" they found and implementing them in a disciplined, structured environment. Focusing on reliability was found to offer the biggest return with the longest lasting results.

Corporations that truly understand reliability typically have the best performing plants. Some common characteristics of a "reliability focused" organization are

- Their goal is optimal asset health at an optimal cost.
- They focus on processes—what people are doing to achieve results.
- They measure the effectiveness of each step in the process, in addition to the results.
- Their preventive maintenance programs focus mainly on monitoring and managing asset health.
- Their preventive maintenance programs are technically sound, with each task linked to a specific failure mode. Formal practices and tools are used to identify the work required to ensure reliability.

1.3.2. Why Move Toward Proactive Work?

Many companies focus their entire maintenance efforts on a preventive maintenance (PM) program that does not meet the actual reliability needs of the equipment, often because "this is the way we've always done it." Others use statistical analysis to improve reliability rather than statistical analysis techniques, such as Weibull analysis, to identify assets where reliability is a problem. Here are some sobering facts that will make you think twice about the effectiveness of a time-based PM program:

1. Less than 20% of asset failures are age related. How can you identify the frequency of their preventive maintenance activities? Do you have good data to determine this frequency? If you have, then most asset failures have been correctly documented and coded in the CMMS/EAM [computerized maintenance management system/enterprise asset management]. We find that 98% of companies lack good failure history data.
2. Most reliability studies show that over 80% of asset failures are random. How do you prevent random failure? In many cases, it is possible to detect early signs of random failure by monitoring the right health indicators. In simple terms, how much has the asset degraded and how long before it no longer functions? This approach allows time to take the corrective action, in a scheduled and proactive manner.

Let us take this statement a step further. Preventive maintenance for random failures usually focuses on the health of the asset (through monitoring indicators such as temperature, tolerance, and vibration) to determine where an asset is on the degradation or PF curve (Figure 1.3). Point P is the first point at which we can detect degradation. Point F, the new definition of failure, is the point at which the asset fails to perform at the required functional level.

The amount of time that elapses between the detection of a potential failure (P) and its deterioration to functional failure (F) is known as the PF interval. A maintenance organization needs to know the PF curve on critical equipment to maintain reliability at the level required to meet the plant's needs. An example of a potential (partial) failure is a conveyor that is supposed to operate at 200 meters per minute but, because of a problem, can run at only 160 meters per minute. Full functional failure occurs when the conveyor ceases to run.

However, a few barriers prevent a plant from obtaining a higher level of reliability of its assets :

- Most maintenance and production departments consider failure only when the equipment is broken. A true failure occurs when an asset no longer meets the function required of it at some known rate of standard. For example, if a conveyor is supposed to operate at 200 meters per minute, when the conveyor's speed no longer meets this requirement, it has failed functionally, causing an immediate loss of revenue for the company.

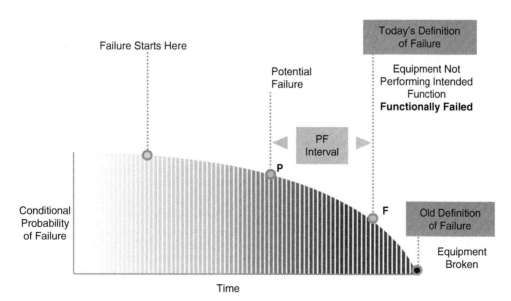

FIGURE 1.3.   PF curve. (Courtesy of *Plant Services Management*.)

• The maintenance department does not get involved when quality or production rate issues arise in the plant. In most cases, when an asset has failed functionally in a plant, the maintenance department is not informed.
• Most maintenance departments do not know the performance targets of the plant equipment and do not understand why it is important that they understand them. This not a failure of the maintenance department but a breakdown in communication of the company's goals.

Overcoming these barriers is essential to rapid performance in reliability. If an understanding and focus on functional failure is applied by all plant personnel, higher asset reliability will rapidly follow. The focus must be on the alignment of the total plant in meeting performance targets of each asset. These performance targets and current performance rates need to be posted, so everyone is aware if a gap occurs in asset performance. The production and maintenance departments must know when an asset no longer meets the performance target. Both departments must accept responsibility for actions to mitigate the performance losses.

1.3.3. A New Way to View Failure

Inside every operation is the "hidden plant," made up of all equipment operating below defined performance targets. These need immediate maintenance. By focusing on the equipment performance targets, asset reliability could be increased rapidly.

A plant must apply RCM methodology to meet the goal of "optimal reliability at optimal cost." That RCM methodology could be RCM II, Streamlined RCM, failure modes and effects analysis (FMEA), or maintenance task analysis (MTA). Be careful which methodology you use if you want rapid performance.

To know what is needed, a company must review the way it manages equipment performance. If equipment continues to fail after preventive maintenance or overhauls, then clearly a change is needed. As a starting point, everyone in a plant should understand what is meant by *reliability* and what it means to the success of the company. *Reliability* should be the plant's collective buzzword.

1.4. MAINTENANCE/RELIABILITY ASSESSMENT

The biggest challenge organizations face is knowing the status of their current maintenance and reli-

ability process and developing a plan to close the gap between current and desired performance. Tables 1.2–1.8 list specific questions to answer in order to define the gap between the current performance of maintenance and reliability processes and those considered "best practices." All answers should be "yes"; if not, this area may need to be explored in more depth. If you are unsure about the question or answer, then the question answer is "no." Remember this assessment is used only to help an organization determine the gaps between its current and desired

TABLE 1.2.   Key Maintenance and Reliability Process Principles

Questions	Yes/No
Does management have roles and responsibilities defined for all the maintenance staff?	
Does management know and manage with leading key performance indicators?	
Are the work flow processes defined for all elements of the maintenance and reliability process, such as planning, scheduling, and work execution?	
Have the critical assets been defined based on consequence and risk to the business, weighting values to determine asset criticality in areas such as	N/A
Safety	
Environmental	
Capacity	
Cost	
Other criteria	
Has the management team defined the gap between the current and desired plant performance and determined the financial opportunity identified?	
Is the current PM/PdM (predictive maintenance) program on critical assets based on RCM methodology (RCM, FMEA, etc.)?	
Does the organization define failure based on the functional failure of the asset?	
Does your organization have dedicated planner/schedulers?	
Does the whole organization accept responsibility for reliability?	
Does your organization use mean time between failure (MTBF) to determine the reliability of your assets?	
Total "yes" answers times 10 = (possible 150 points)	

TABLE 1.3. Reliability

Questions	Yes/No
Are PM/PdM results entered into reliability software to identify P on the P-F interval?	
Does maintenance and production management understand the P-F interval and how it functions to manage asset failure?	
Does maintenance and production management focus on a *formal* critical assets matrix when determining shutdowns and maintenance work?	
Is a work order written for all functional failures (partial and total)?	
Is a formal root cause failure analysis (RCFA) or root cause analysis (RCA) process defined and executed on major and multiple similar asset failures?	
Is a maintenance or reliability engineer on staff?	
Does the maintenance or reliability department focus strictly on reliability of the assets and not "project engineering" work?	
Does your company apply the Six Sigma DMAIC process in its reliability engineering efforts?	
Is an RCM methodology (RCM, RCM II, RCM Turbo, FMEA, etc.) used to determine the maintenance work strategy for your assets?	
Do you know what percentage of your assets your maintenance strategy is RTF (run to failure), PdM, detective maintenance, PM (restorations, replacements, lubrication, etc.), and so forth?	
Total "yes" answers times 30 = (possible 300 points)	

TABLE 1.4. Reliability Engineering

Questions	Yes/No
Does your plant have a reliability engineer?	
Does your plant apply a formal RCFA process on high-value reliability problems?	
Does the reliability engineer perform only reliability engineering work?	
Are all projects reviewed and approved to ensure reliability of the assets?	
Is there validation by engineering, maintenance, and production to ensure a project is completed and operating to design specifications before the project's responsibility is turned over to production?	
Are projects' costs allowed to continue even after a project has overrun its projected cost? (Are project cost paid for by engineering and not maintenance?)	
When a project is completed, has an RCM methodology been applied to the equipment to ensure the correct maintenance strategy has been applied to the assets?	
When a project is completed have all equipment, parts, and the like been entered into the CMMS/EAM?	
Does reliability engineering focus less than 20% of their time on "bad actors"?	
Are reliability engineers *not used* as project engineers?	
Total "yes" answers times 15 = (possible 150 points)	

TABLE 1.5. Planning

Questions	Yes/No
Are preplanned job packages developed for most of the maintenance work scheduled (all specifications, procedures, parts, labor, etc. identified)?	
Does the planner use the maintenance staff to assist in the development of preplanned job packages?	
Is a planner/scheduler (or just a planner) performing the day-to-day job ever called upon to rush in parts for a breakdown?	
Does the planner identify backlog based on categories (e.g., ready to schedule, waiting on parts, waiting on engineering, waiting to be planned)?	
Does the planner validate whether a work request is valid?	

(Continues)

TABLE 1.5. (*Continued*)

Questions	Yes/No
Does the planner provide feedback to the requester when a work request or notification has been entered into the CMMS/EAM system?	
Does the planner visit the job sites of work to be planned on at least 30% of jobs?	
Can the planner check status of planned work parts on the CMMS/EAM within five minutes of any job?	
Does the planner validate work request in three days or less?	
Do you have at least one planner or planner/scheduler for every 7–25 maintenance personnel?	
Total "yes" answers times 10 = (possible 100 points)	

TABLE 1.6.   Scheduling

Questions	Yes/No
Is someone responsible for scheduling, either a full-time maintenance scheduler or full-time planner/scheduler?	
Do planner/schedulers or schedulers work closely with production to schedule maintenance work?	
Is maintenance work scheduled one week ahead at least?	
Is maintenance work scheduled by day?	
Is maintenance work scheduled with a maintenance person's name assigned?	
On large outages do maintenance personnel provide input into the schedule?	
Does the scheduler or planner/scheduler facilitate the maintenance scheduling meeting?	
Does the scheduler or planner/scheduler not report to maintenance supervision? Reporting to a maintenance manager is acceptable.	
Is next week's schedule posted at least the prior Friday for all to view to include maintenance and production?	
Is schedule compliance above 80%?	
Total "yes" answers times 10 = (possible 100 points)	

TABLE 1.7.   Key Performance Indicators (KPIs)

Questions	Yes/No
Are the work flow processes mapped in your maintenance and reliability process with leading and lagging KPIs defined at specific points in these processes?	
Does the maintenance department measure the following?	N/A
Scheduled compliance	
Percent of planned work	
Rework	
Mean time between failures	
Percent of time (by vendor) vendors do not deliver on time	
Percent of time vendors deliver the wrong part	
Stockouts	
Percent of assets, ranked based on criticality	
Percent of assets for which RCM methodology is applied and the maintenance strategy changed based on the data	
Bad actors report	
Number of potential failures identified	
Percent of assets for which functional targets are identified	
Percent of work proactive	
Percent of work reactive	
Maintenance cost as a percent of return on asset value (RAV)	

(*Continues*)

TABLE 1.7. (*Continued*)

Questions	Yes/No
Maintenance material in stores as a percent of RAV	
Maintenance cost per unit produced	
Are at least 50% of the KPIs just listed posted for all to see in the maintenance department?	
Are at least 25% of the KPIs just listed posted for all to see in the maintenance department?	
Are at least 10% of the KPIs just listed posted for all to see in the maintenance department?	
Are at least 10% of the KPIs just listed posted for all in production department to see?	
Are their targets and goals established for over 75% of the KPIs listed above?	
Are your KPIs listed as leading or lagging?	
For each KPI the maintenance department uses, is a standard definition, objective, calculation, example calculation, roles, and responsibility assigned?	
Total "yes" answers times 10 = (possible 250 points)	

TABLE 1.8. **Education and Training**

Questions	Yes/No
Have all management personnel been trained in the basics of reliability?	
If all management personnel have been trained in the basics of reliability, do they demonstrate this knowledge in their job?	
Have all plant personnel been trained in the basics of reliability?	
Have executive management personnel been trained in the basics of reliability?	
Have all maintenance personnel been trained in the basics of root cause failure analysis?	
Does the plant have a skills training program for all maintenance personnel and is it based on a skills assessment?	
Does the company have an apprenticeship or entry-level training program for maintenance and production?	
Are skilled maintenance personnel hired from inside and outside the company administered a written and "hands-on" test?	
Does the plant have a maintenance training budget that equals or exceeds 6% of the maintenance labor budget?	
Do skilled maintenance workers use preplanned job packages with procedures over 90% of the time?	
Total "yes" answers times 10 = (possible 100 points)	

performance. We recommend this assessment to include representatives from all involved departments: plant manager, maintenance and engineering manager, production management, operators, maintenance personnel, maintenance planners, storeroom management, and so forth.

The final score totals are interpreted as follows:

0–500 = Total reactive (reliability principles are not understood or applied). The company needs to educate all management and engineering personnel in reliability and develop a reliability strategy for serious change. It needs to develop a business case to define the opportunity immediately. Read the book.

501–700 = Emerging (long way to go). The company needs to develop a business case and reliability strategy with a timeline, targets, and objectives.

701–850 = Proactive (continue the journey, you are headed in the right direction). The company needs to ensure that a continuous improvement process is built into its asset reliability process. Identify gaps in the assessment and fill the gaps.

851–1000 = World Class. The company should hire an outside reliability consulting firm to assess its current status and make recommendations for any change required. Great job.

This next section was written by Steve Thomas (Changemgt.net), one of the leaders in change management in the world of reliability today. His book, *Improving Maintenance and Reliability through Cultural Change*, is used by companies around the world to solve their change management problems. Below, he provides insight into this challenging issue (see Section 1.5).

1.5. INTRODUCTION TO CHANGE MANAGEMENT*

One of the major areas of focus in industry today is improving equipment reliability. Why? To insure that production is always available to meet the demand of the marketplace. One of the worst nightmares of any company and those who manage it is to have a demand for product but not be able to supply it because of equipment failure. Certainly this scenario will reduce company profitability and could ultimately put a company out of business.

For some firms, poor reliability and its impact on production are far more serious than for others. For those that operate on a continuous basis—they run 24 hours per day seven days per week—there is no room for unplanned shutdowns of the production equipment; any loss of production is often difficult or even impossible to make up. For others that do not operate in a 24/7 mode, recovery can be easier, but nevertheless time consuming and expensive, reducing profits.

Many programs available in the industry are designed to help businesses improve reliability. They are identified in trade literature, promoted at conferences and over the web, and quite often they are in place within the plants in your own company. Most of these programs are "hard skill" programs. They deal with the application of resources and resource skills in the performance of a specific task aimed at reliability improvement. For example, to improve preventive maintenance, you train your workforce in preventive maintenance skills, purchase the necessary equipment, and roll out a PM program accompanied by corporate publicity, presentations of what you expect to accomplish, and other forms of hype to get buy-in from those who need to execute it. Then you congratulate your team for a job well done and move on to the next project. Often at this juncture, something very significant happens. The program you delivered starts strongly, but immediately things begin to go wrong. The work crews assigned to preventive maintenance get diverted to other plant priorities; although promises are made to return them to their original PM assignments, this never seems to happen. Equipment scheduled to be out of service for preventive main-

tenance can't be shut down due to the requirements of the production department; although promises are made to take the equipment off-line at a later time, this never seems to happen. Finally, the various key members of management who were active advocates and supporters at the outset are the very ones who permit the program interruptions, diminish its intent, and reduce the potential value. Often, these people do make attempts to get the program back on track, but these attempts are half-hearted. Although nothing is openly said, the organization recognizes what is important, and often this is not the preventive maintenance program.

[Thomas simplifies the demise of the preventive maintenance program in an example.] Yet this is exactly as it happens, although much more subtle. In the end, the result is the same. Six months after the triumphant rollout of the program, it is gone. The operational status quo has returned and, if you look at the business process, you may not even be able to ascertain that a preventive maintenance program ever existed at all.

For those of us trying to improve reliability or implement any type of change in our business, the question we need to ask ourselves is Why does this happen? The intent of the program was sound. It was developed with a great deal of detail, time, and often money; the work plan was well executed. Yet in the end there is nothing to show for all of the work and effort.

Part of the answer is that change is a difficult process. [Note that he did not say program, because a program is something with a beginning and an end.] A process has a starting point—when you initially conceived the idea—but it has no specific ending and can go on forever.

Yet the difficulty of implementing change isn't the root cause of the problem. You can force change. If you monitor and take proper corrective action, you may even be able over the short term to force the process to appear successful. Here, the operative word is you. What if you implement the previously mentioned preventive maintenance program and then, in order to assure compliance, continually monitor the progress. Further suppose that you are a senior manager and have the ability to rapidly remove from the process change any roadblocks it encounters as it progresses. What then? Most likely the change will stick as long as you are providing care and feeding. But what do you think will happen if after one month into the program you are removed from the equation. If there are no other supporters to continue the oversight and corrective action efforts, the program will most likely lose

*Section 1.5 is taken from Steve Thomas, "Introduction to Change Management," in Improving Maintenance and Reliability through Cultural Change (New York: Industrial Press, 200X). Reprinted by permission for the author and publisher.

energy. Over a relatively short time, everything will likely return to the status quo.

The question we need to answer is Why does this happen to well-intentioned reliability-driven change process throughout industry? The answer is that the process of change is a victim of the organization's culture and failure to address the "soft skills" of the change process.

The culture is the hidden force, which defines how an organization behaves, works behind the scenes to restore the status quo unless specific actions are taken to establish a new status quo for the organization. Without proper attention to organizational culture, long-term successful change is not possible.

Think of an organization's culture as a rubber band. The more you try to stretch it, the harder it tries to return to its previously un-stretched state—the status quo. However if you stretch it in a way that it can't return and leave it stretched for a long time, once released it will sustain the current stretch you gave it and not return to its original dimensions. That is going to be our goal—to figure out how to stretch the organization but in a way that when the driver of the process is out of the equation, the organizational rubber band won't snap back.

Changing the organization's culture in order to promote long-lasting change benefits everyone from the top of the organization to the bottom. It benefits the top by providing a solid foundation on which to build new concepts, behaviors, and ways of thinking about work. To accomplish radical changes in how we think about or execute our work requires that those who are part of the culture support it. Senior management can only take this so far. They can set and communicate the vision and they can visibly support the effort, but the most important thing they can do is empower those in the middle to make it happen.

As middle managers, we all know that we have many more initiatives on which to work than there is time in the day (or night). This spreads our focus. If we don't collectively embrace the new change, then no threat, benefit or any other motivational technique will make the change successful over the long term. Although this book can help educate senior management so that they can empower the rest of us, its real benefit is for middle managers. It will help them understand this very complex concept in a way that will enable them to deliver successful change initiatives.

There is also benefit for those at the bottom tier of the management hierarchy. The term bottom is not meant to demean this roll because this is where the "rubber meets the road." All cultural change and their related initiatives end here. This is where all of the plans, training, and actual work to implement end. If it doesn't work here—failure is the outcome.

As with any effort, getting started is often difficult. However, the value in a successful effort is well worth the time, energy, and commitment required. The reason behind this is that you are not just changing the way a process is executed, or a procedure is followed. By changing the organization's culture, you are in essence changing the very nature of the company. This book addresses this change of culture in a general sense that can be applied across many disciplines. But more specifically, it is targeted toward changing the culture as it pertains to reliability. If we are successful in this arena, the result will be a major shift in how work is performed. In addition, a culture that is focused on equipment reliability reaps other closely associated benefits. These include improved safety and environmental compliance. Both of these are tied closely to reliability in that reliable equipment doesn't fail or expose a plant to potential safety and environmental issues.

Changing a culture to one focused on reliability also has spin-off benefits in many other areas. Reliability or the concept of things not breaking can easily be applied to other processes that are not related to equipment efficiency and effectiveness. However, reliability or any other type of change is very difficult. After all you are trying to alter basic beliefs and values of an organization. These are the behaviors that have been rewarded and praised in the past and may even be the reason that many in the organization were promoted to their current positions. Change may also seriously affect people's jobs because things they did in the past may no longer be relevant in the future.

Improving Maintenance and Reliability through Cultural Change will prove to be a very beneficial book, not just as a place to start or as a test to provide the initial information for your effort, but also as a book which will help you through all the phases of the process and into the future. The following quotation from Niccolo Machiavelli written over 500 years ago clearly describes the issues and hurdles that surround any change initiative which you are undertaking.

> There is nothing more difficult to take in hand, more perilous to conduct, or more uncertain in its success than to take the lead in the introduction of a new order of things, because the innovator has for enemies, all of those who have done well under the old conditions, and luke-warm defenders in those who will do well under the new. (*The Prince*)

Implementing "hard skill" change in an organization in areas such as planning and scheduling, work execution and preventive or predictive maintenance also needs to address "soft skill" change. Making "hard skill" changes without considering the supporting elements of change may not cause your initiative to fail in the short term, but over the long term failure can practically be guaranteed.

The "soft skills," which include leadership, work process, structure, group learning, technology, communication, interrelationships, and rewards are referred to as the Eight Elements of Change®. While each of these is individually important, when collectively addressed, they provide a solid base upon which "hard skill" change can successfully be built.

Below the "soft skills" is the foundational level of organizational culture, already discussed. This level includes organizational values, role models, rites and rituals, and the cultural infrastructure. These are called to Four Elements of Culture®.

When addressed as a composite, the Eight Elements of Change and the Four Elements of Culture help to successfully drive any change initiative. The question is this: With all of these elements interacting within the context of a change initiative, how do you know where you are doing well and where opportunities lie for improvement? The answer is the Cultural Web of Change®, shown in Figure 1.4.

This tool is a survey with 16 statements related to each of the Eight Elements of Change. Score each of the statements from 0 points if you strongly disagree with the statement through 4 points if you strongly agree. The resultant scores then are plotted on the web diagram, as in Figure 1.5. As you can see the Eight Elements of Change are the radial spokes. The Four Elements of Culture are addressed by the survey questions providing cultural information related to each element of change.

It is not possible to provide the entire Cultural Web of Change survey in this section. For that, go to Steve Thomas's Web site Chnagemgt.net, where it is provided along with a detailed explanation. However, Figure 1.4 shows a sample, filled out in Figure 1.5. For this web, there is only one question per element. Score each question and draw the web. The low scoring areas provide some "directional" insight for areas of improvement.

1.6. DEVELOPING A BUSINESS CASE FOR A RELIABILITY INITIATIVE

Developing a business case for a reliability initiative is a must if that initiative is to be success-ful. The objective of a business case is to identify the financial opportunity a reliability improvement initiative provides. A business case must identify the projected financial outcome to accrue from optimizing reliability—increased capacity, reduced maintenance labor and material cost, increased asset utilization, and more—these must be in "hard dollars." Management must be part of the business case development, including a finance person, who could be a comptroller or chief financial officer, for example. The outcome of a business case will contain the current costs, current losses, and future savings associated with a specific time line and project plan. The business plan must be achievable— "undercommit, overdeliver."

Step 1. Educate the plant management on the basics of reliability, which involves how equipment fails, how to monitor equipment health, and how to make the right decision at the right time so as not to see equipment failure.

Step 2. Assemble all the information for the business case. This information could be sketchy dependent on how reactive an organization truly is. The information must include

- Asset reliability data (plant, production line, critical assets' mean time between failure). This information should be available from CMMS/EAM if work orders have been written for most failures.
- Maintenance labor costs: overtime percentage-plus cost (do not include "full burden" cost), yearly labor cost for current-year budget, yearly labor cost for past three to five years, contractor cost used to assist in maintenance projects or maintenance work, and any other labor cost that may be important.
- Maintenance material costs (specify a dollar amount): maintenance storeroom inventory level, maintenance material expenditures for the past three to five years, overnight delivery costs for the past three to five years, and additional costs associated with maintenance parts storage and procurement.
- Current plant capacity and capacity targets and goals for the current year and past three to five years.
- Marketing forecast for next one to three years (if available) in total sales.
- Quality losses for the current year and past three to five years.
  Note: Other costs could be useful in the development of the business case, just be sure any cost the business case claims will be financially affected by the reliability initiative can be validated by the financial person.

Sample Web of Change®

The Web of Change survey below is a simplified example. It is designed to provide you with a graphical representation of your company's strengths and weaknesses in the change arena. The eight elements listed are the critical **eight elements of change**. Read the statement and mark your answer in the right hand column. Then draw your web diagram to see how you have done. The closer to the outer edge of the web the better. Low scores indicate areas for improvement.   **Scores**

0 = **I strongly disagree**
1 = **I disagree**
2 = **I'm neutral**
3 = **I agree**
4 = **I strongly agree**

Category	Statement	Score
Leadership	The leadership of my organization is focused on continuous improvement and supports our change efforts in both word and deed.	
Work Process	The reliability and maintenance work processes are focused on proactive (reliability focused) work as opposed to a "break it-fix it" repair strategy.	
Structure	The organization structure is built with a focus on reliability. We have the right number of maintenance engineers, planners, foremen and skilled mechanics.	
Group Learning	The organization learns from their successes and their failures. This information is fed back into the process so that corrective action can take place and we can get better.	
Technology	We have in place computerized reliability and maintenance systems that support the work effort. The data in the system is accurate and allows for reliability-based decision making.	
Communication	The communication within the reliability and maintenance organization is excellent. People know what needs to be done and why. Additionally communication with other departments is excellent as well.	
Interrelationships	The interrelationships within the organization are very positive and support a continuous improvement reliability-based work strategy.	
Rewards	The reward structure and the related processes clearly provide rewards to those individuals and groups who promote reliability-based practices and continuous improvement.	

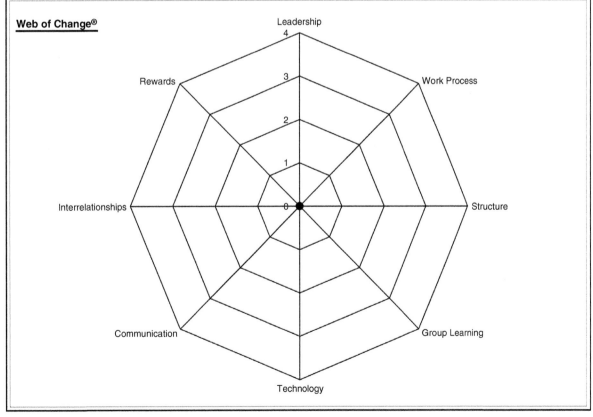

Web of Change®

FIGURE 1.4. Web of change.

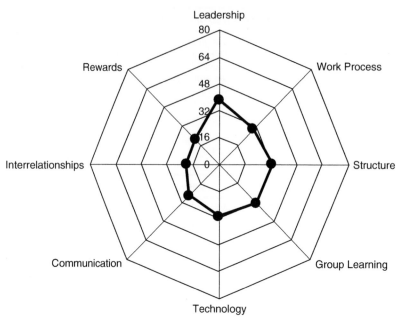

FIGURE 1.5.   Web of change sample.

Step 3. Develop the business plan with plant management. Expect this exercise to take one to two days, depending on the availability of the information. If any of the required members cannot participate fully, then do not move forward with this process. Required team members are the plant manager (do not move ahead without him or her); production manager; chief financial officer, comptroller, or plant accountant; maintenance and engineering manager; safety director (or person responsible); and environmental director (or person responsible).

Step 4. Calculating savings. Once all the data are compiled, begin identifying past losses and potential savings for each of the following areas, based on current reliability of the assets against future state. For example,

- Current capacity: 88% (over the past three years, the number generally has been the same).
- Sales projections: Could sell all products produced—1% increase in capacity increases the bottom line by $150,000 a day; plant operates 365 days a year = $150,000 × 365 days = $54.75 million.
- Potential savings: What is the impact of reliability on production? In most plants, reliability increases capacity by as much as 98%; say 50% to be conservative. New calculations for savings = 50% × $54.75 million = $27.375 million.

- Time until this savings affects the bottom line (educated guess): Aggressive reliability initiative = ("undercommit, overdeliver")
- Year 1 = 10% improvement = $2,737,500.
- Year 2 = 20% (plus last years 10%) = $8,212,500.
- Year 3 = 30% (plus last two years 30%) = $16,425,000.

The savings or just the increase in capacity equals $16,425,000 in a three year period. After adding savings from a reduction in maintenance costs, including maintenance labor and materials, recalculate the savings. In a short time, it should equal or exceed the plan's cost. In addition to the profit from increased capacity, also identify the following:

- Decrease in maintenance labor cost over three years = 30% = $4 million.
- Decrease in maintenance material cost over three years = 20% = $1.2 million.

The business plan would show the savings listed in Table 1.9.

Step 5. Develop an action plan with a time line that identifies the major task defined; defined roles and responsibilities by position, from the floor level to boardroom for each task; defined key performance indicators to measure success of the project (establish targets and goals); established measures of success for specific positions, which align with the roles and

TABLE 1.9.   Projected Savings

	Year 1	Year 2	Year 3	Accumulated Savings
Maintenance labor savings	$800,000	$1,200,000	$2,000,000	$4,000,000
Maintenance materials savings	−$200,000	$300,000	$1,100,000	$1,200,000
Capacity net profit	$2,737,500	$5,475,000	$8,212,500	$16,425,000
Total savings	$3,337,500	$6,975,000	$11,312,500	$21,625,000
Reliability initiative cost	−$3,200,000	−$600,000	−$120,000	−$3,920,000
Net savings	$137,500	$6,375,000	$11,192,500	$17,705,000

responsibilities; and defined milestones with cost and savings identified.

Step 6. Present the business plan to senior management. Spend time to do it right, make it professional and appealing. Rehearse the presentation. Remember you only have one chance at this sales presentation. Use a PowerPoint presentation (no more than 10 slides) to show the benefits (it is all about the money), costs (play up the return on investment), and project plan abbreviated (give an overview); then present a large chart with the project plan attached—make it shine, make a statement with it.

Once this project has been approved, have consultants ready to assist in training and coaching the staff to make this reliability project effective. A consultant will have experienced many failures and many successes. Do not try to reinvent the wheel.

1.7. CALCULATING RETURN ON INVESTMENT

Before any reliability project or initiative can begin, a business case must be established, with the financial benefit or risk determined. In this section, we deal more with financial risk rather than business risk as it pertains to safety or environmental issues.

To determine the return on investment (ROI), it is important to first establish the objective for the reliability initiative. The initiative's objective should have a financial value assigned; for example, "Reliability Objective: Increase reliability of production assets on a specified production line by 8% with a financial value of $8 million the first year."

Developing the initiative's objective and assigning value to it is important in order to receive manage-

ment's approval. The preceding objective could be broken down further into more defined measurements, to include

- Increase number of units produced in year 1, year 2, year 3, and so on.
- Reduce cost per unit in year 1, year 2, and so on.
- Increase in quality yield by X% in year 1, year 2, and so on.
- Reduce maintenance cost by 10% in year 2, 20% in year 3, and 30% in year 3 based on current maintenance material, labor, and contractor costs.

1.7.1. Leadership of the ROI Team

The first step in this process is to develop a team of professionals to jointly develop the business case, which includes the ROI. This team must consist of the plant or corporate financial person, production leaders, maintenance and engineering leaders, and plant or executive management. Many times some of these people will try to not participate in this process; however, if they do not participate, the chance of the initiative being approved is very low. I always ask managers how much their time is valued. In most reliability initiatives, the value is so high they have no option but to participate.

1.7.2. Case Study

The case is from a true situation, where demand for the plant's product was expected to double in the next year or two. The company was adding additional production lines in the plant to keep up with anticipated production demands. However, according to production management, no additional capacity of current assets was available. Equipment reliability was a concern because downtime was seen as high (no real numbers were available). The skill level of maintenance personnel also was seen as a problem because of reliability issues, so a "pay for skills" program was implemented but seemed to have little effect on the reliability issues. A new maintenance manager was appointed from engineering.

A team was developed by the plant manager to determine a solution to the reliability problem with real concern that the new production lines would have similar production problems if equipment reliability were not improved. The team consisted of the plant manager, production and maintenance manager, engineering manager, plant comptroller, and an outside consultant.

After an evaluation of the situation the following information was found to be valid and agreed upon by the management team.

- Asset reliability problems consisted of only 3% of total downtime. The maintenance department developed a PM program based on manufacturer's recommendations and were very vague. The maintenance department had no repair procedures developed.
- Changeover standards were set too high, resulting in a total downtime of 40%. The changeovers were developed based on plant startup six years earlier with no further evaluation for improvement.
- Quality losses exceeded 9% of the production rate due to a lack of good production operating procedures. Operating procedures were developed after the initial startup and were never updated.
- Little if any data tracking system could be found. The computerized maintenance software system was used very little. The production software system was used frequently; however, the data being reported were not valid.

This meant that

- The plant was operating at 57% of true capacity. The changeover standards were set too high (the changeover time went from four hours to 45 minutes) and changeover between shifts was costing an average of 15 minutes each for two shifts a day.
- Industry known product quality benchmark losses were found to be 3%, whereas this plant's quality losses were at 9% resulting in a difference of 6%, where 78% of quality losses were caused by operator error on each production line and the other losses were a result of many factors.

Acting on these findings, plant capacity increased to 94% within six months (the plan was sold to corporate management as a two-year project—"undercommit, overdeliver"), resulting in an increased net profit of $12 million and capital expenditure halted, resulting in a savings of $4 million. Quality losses were brought down to 4% in one year, resulting in an increase in net profit of $800,000. Maintenance costs increased in the first year by 10% but decreased over the next two years by 34% (a $420,000 saving).

The total cost of the project (over two years) was

Production consulting services = $400,000

Maintenance consulting services = $800,000

Skills Training = $150,000

Other cost (training for managers, visits to other plants, etc) = $120,000

Total cost = $1.1 million (in year 1) + $250,000 (in year 2) = $1.35 million

Increase in revenue = $13 million (in year 1) + $4.22 million (in year 2) = $17.22 million

In the final analysis the plant spent $1.35 million for a return of $17.22 million. Note that the cost was front-end loaded. Corporate management was told the expected return would be $3.2 million and not $17.22 million. The cost was stated as an estimate. The plant team "undercommitted but overdelivered."

The following lessons were learned:

1. Develop a leadership team to identify the ROI.
2. Have the primary financial person on the team provide hard validation of the cost and ROI.
3. Determine where all the losses may be occurring and focus on the losses worth going after. Some losses will always occur but must be determined and accepted as part of management's operating strategy. For reliability, allow for scheduled downtime (weekly, monthly, yearly schedules based on time and not asset health), unscheduled downtime (for rework, a possible ineffective PM program, total functional failure), and maintenance, which will reduce as a result of an increase in reliability and reduction in production reactivity (overtime, labor lost through attrition, maintenance materials purchased and their delivery, storeroom value, contractors, and capital for replacing equipment not maintained appropriately). For production—if the equipment is not operating to the functional requirements then it is producing a loss—allow for product startups and changeovers, personnel breaks, product quality losses, operator error, and partial and total functional failures.

4. Determine the objective of the project or initiative and quantify it.
5. Undercommit, overdeliver every time.

1.8. PLANNING AND SCHEDULING

The goal in this section is to present a concise description of the planning and scheduling process for reliability engineers to become familiar with the process and know how to interact with it to leverage their reliability improvement efforts. For brevity, I keep this at a very high level and delve into more detail only where it will be of particular interest to the reliability professional.

A properly designed and executed planning and scheduling (P&S; see Figure 1.6) system can be one of the most effective means of aiding reliability initiatives and inculcating reliability practices into the daily execution of maintenance activities. A reliability engineer is well served in having a good understanding of how maintenance planning and scheduling works ideally, as well as how it has been applied in a plant. Most important, someone who understands the P&S process can use it to leverage reliability improvement efforts.

The purpose of any P&S system should be to eliminate delays in the maintenance process and coordinate the schedules of the maintenance resources and the production schedule.

All jobs are not good candidates for planning. Very simple jobs normally would not go through the planning process, as little value is added. Additionally, jobs that have low predictability are not good candidates, since a planner would be unable to accurately predict the job's resource and part needs. Every organization involved in a P&S effort should identify the jobs that should not be planned. However, all jobs are good candidates for scheduling as long as a reasonable time estimate can be made.

As far as P&S is concerned, work, basically, is either urgent or nonurgent. This sounds simple enough; urgent work must be attended to without delay, whereas nonurgent work allows taking the time required to plan and schedule it if a tremendous savings can be created, and that is the ultimate goal of P&S. Urgent work is fraught with delays, misdirection, and confusion; but by its very nature, we cannot take the time to investigate the needs of the job, estimate the resources, develop a plan, and obtain the necessary parts. Instead, we run headlong into the fray and figure out what we need as we go. Sometimes, we have to back up and try a different route.

This approach adds an unnecessary cost to the job, but when production has been put on hold or threatened, you must demonstrate a bias for action. Putting an end to work such as this is a goal that is common to both reliability and planning and scheduling, and the two should be linked together.

The primary people in a P&S system are the planner, the scheduler, and the maintenance coordinator. In general, the planner identifies everything that will be needed to execute the job, the scheduler arranges and communicates all timing aspects of the job, and the maintenance coordinator enables maintenance to attend to the most important work at the optimum time relative to production scheduling. Usually, both the planner and scheduler report through the maintenance department, whereas the maintenance coordinator should report through the operating department.

Some organizations have one or more persons assigned to each of these roles: at the other end of the spectrum, some may have individuals responsible for more than one of these roles. Either situation can work effectively as long as the roles are looked at as distinct and ample time is allotted to fulfill each one.

As maintenance work requests are made from the operating department, it should be a continual process for them to be "funneled" through the maintenance coordinator. As mentioned, one of the maintenance coordinator's primary tasks is to enable maintenance to work on the most important work. When all work for a given maintenance organization goes through a single operating department person (the maintenance coordinator), priorities can be leveled and unnecessary or duplicate work eliminated. The benefit this provides to the maintenance organization cannot be overstated. Rather than having everyone in the production department who initiates maintenance requests setting a priority based on personal current needs and a limited perspective, a maintenance coordinator, who has a larger perspective by virtue of seeing all maintenance requests and knowing the overall production scheduling needs and constraints, can provide a much clearer focus for response by the maintenance organization. The maintenance coordinator deletes unnecessary or duplicate requests and adjusts the priorities of the remaining work requests to reflect the needs of the production department, while recognizing the limited capacity of the maintenance organization. Then, the requests are sent to the planner.

The planner's most important role is to identify and quantify the resources required to execute a given job.

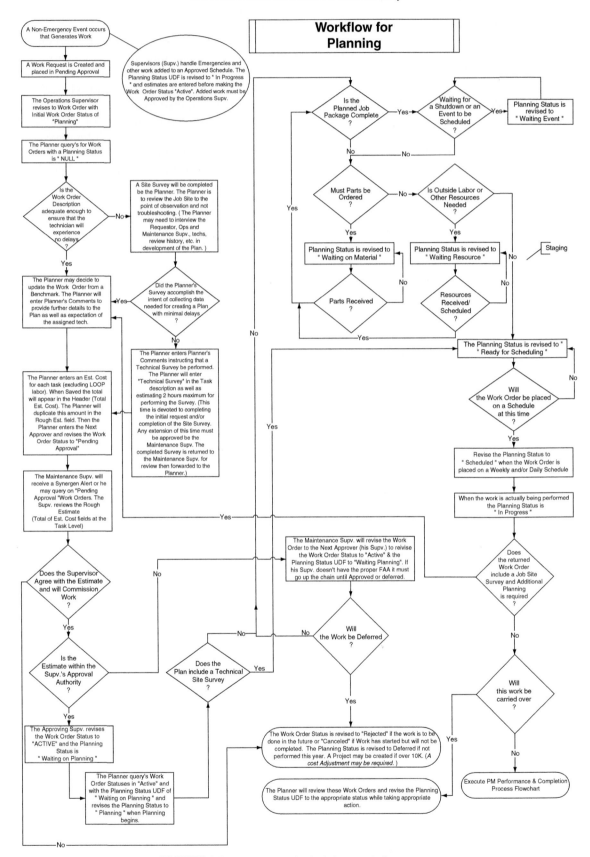

FIGURE 1.6.   Planning and scheduling work flow process.

Categorically, these resources are the (1) number of maintenance personnel and work hours required by skill level and craft, (2) steps required to complete the job, (3) part needs, (4) tool and equipment needs, and (5) information needs, including drawings and specifications. In so doing, planning prevents most delays from materializing and thereby greatly improves maintenance effectiveness. The planner usually is one of the best artisans, who also has the additional skills of excellent written communication, information management, computer skills, managing multiple priorities, and organization, to name a few.

The planner reviews the work requests to see that they are written on the correct item in the computerized maintenance and management system, otherwise corrects them and then determines if a field visit is needed. The planner should conduct a field visit on any work that is not obviously routine and straightforward. During a job site inspection, the planner identifies the resources required by the job and notes ancillary repairs and equipment needs that should be resolved during work order execution. Any other delays likely to be present also are noted. Sources of other delays can be job preparation, permits, lockout or tag out, and physical access problems. The planner determines how best to prevent each delay from materializing. For example, a job may have very difficult access that scaffolding could resolve; the planner identifies the need for scaffolding and an estimate of the time required to erect the scaffolding. All this information is documented in a standardized form, called a *job plan*, for each planned work order. For example, when the job came up to be scheduled, the need for scaffolding would be recognized and the schedule would allow sufficient time for the assembly of the scaffolding. A planner who believes that a particular job is likely to be repeated would likely store the plan, thus cutting the time required to process similar work orders. The reliability engineer should be familiar with all the "stored plans" on equipment involved in a reliability focus. Sometimes, stored plans are for a particular piece of equipment but more often they are for a category of equipment. An example of the second type would be a stored plan to rebuild an ANSI mark III group 2 pump, which could apply to dozens of individual pieces of equipment.

For the reliability engineer, here is a critical point to use the P&S system to leverage professional efforts. These job plans contain instructions on the steps required to complete the job. The level of detail varies, depending on the job. The instruction contained in the job plan may or may not support reliability. Using seal replacement as an example, a reliability engineer who is trying to implement a plan to improve seal life

and has identified best practices can get with the planner, review historically documented seal replacement job plans, and amend them with those best practices. Past plans may have lacked sufficient detail, listing only the major steps such as "replace seal." Or, the reliability engineer may want to specify a new seal type or brand or require laser alignment. The point is that, having identified a change in practice, second only to communicating the change to the field personnel themselves, the reliability engineer must work with the planner to incorporate the new practices into the job plans. These revised job plans will serve as an excellent reminder to field personnel of what you want them to do and how you want it done. These plans also help the planner allow for differences in the time required to do the job or changes in part needs. Usually, job plans need not be very detailed but should simply remind field personnel of what needs to be done, particularly on routine jobs. The less routine the job is, the more that details should be included in the job plan.

Job status codes are used throughout the P&S process to segregate work into its various stages as it progresses through the process. Common status categories are

In planning
Awaiting parts
Waiting for approval
On hold
Ready to be scheduled—planning complete
Scheduled

The planner normally reserves the parts that will be needed for jobs and places orders for any that will come from the plant site. Once the job plan is complete and all part needs have been resolved, meaning that they can be available in less than 24 hours, the job is coded as "ready to be scheduled," signifying that the planning phase is now complete. The on-site parts actually are ordered and all required parts put in a kit a day or two before the date the job is scheduled to start.

The job now goes to the scheduler, who, working with the maintenance coordinator, develops a schedule that optimizes production needs and schedules to the availability and capacity of the maintenance resources. Different organizations use different strategies, some use only weekly schedules, some only daily schedules, and still others use both. The nuances of these various strategies are beyond the scope of this chapter, so we will just assume that a schedule is created and communicated to all parties.

A common problem reported by reliability professionals is that, too often, preventive maintenance work and corrective maintenance work identified by predictive technologies linger in the maintenance backlog unattended until a failure develops. The scheduling process presents another opportunity to aid reliability and prevent this malady.

Schedules are built by assigning dates to the most pressing problems first, then the large majority of the available time remaining in the schedule is filled with jobs selected due to management interest, secondary importance, age, or ease of execution. It should not be difficult to make the case for regular inclusion of reliability-related work orders. Both preventive maintenance and corrective maintenance are very predictable activities, perfect examples of work that can be effectively planned and scheduled. Additionally, this work does not have to be done today or even tomorrow. This being the case, provide a list each month designating the work to be scheduled some time during the month. Often there is a month or more to complete the work before risk to reliability starts to increase. These factors make such tasks ideal to build an effective schedule. Talk with the maintenance scheduler and the maintenance coordinator separately to get their perspectives on the priority given to this type of work for placement into the schedule. If you are not fully satisfied that the appropriate level of priority is being given, then ask to attend the scheduling meeting. Such attendance provides the opportunity to lobby firsthand for inclusion of reliability-related work and hear firsthand the reactive work making scheduling nonreactive work difficult. The reliability engineer may find new candidates for reliability improvements as well as improve the perspective on the issues involved with keeping the plant running.

Once a job is scheduled, as stated previously, the parts will be kitted, placed in bin, or staged in an appropriate area and the job plan package delivered to the individual(s) that will execute the job. When the scheduled time for the job arrives, the maintenance personnel will have every thing they need. The job should already be prepared by operations, meaning that the equipment has been shut down, flushed, or cleaned if necessary; tagged out and at least ready for lockout; and the permits already initiated. There should be no delays when the maintenance personnel arrive at the job site, they should only have to complete the permits and lock the equipment out before starting the job.

Progress of the job should be similar with no need to leave the job site for anything other than breaks or lunch. Prework that should have been completed by other craftspeople, such as insulation removal or the building of scaffolds, should have been completed.

Any required help should arrive when needed or be prepared to arrive upon notification.

When a job can be prepared for and executed without delay, maintenance effectiveness can be multiplied. It is not uncommon for the effectiveness of planned and scheduled jobs to be 25% or more than the same job without P&S. To see just how critical maintenance effectiveness is, consider the following. The term *wrench time* refers to a unit of measure for the amount of time maintenance personnel spend doing the actual work for which they are responsible. Wrench time usually is expressed as a percent, and national studies typically put this number between 25 and 50% for North American industries. As an example, the time a maintenance mechanic takes to replace a mechanical seal and do a laser alignment on a pump would count as wrench time. However, the time spent leaving the job to get the seal from the storeroom and the time spent away from the job site to obtain additional shims for aligning the pump would not count as wrench time. Wrench time is a measurement of effective time and excludes wasted or unnecessary time. The goal is to eliminate all delays and nonproductive work, so that maintenance personnel can work effectively nonstop, never leaving the job once started until it had been completed, except for breaks, lunch, or the end of the day. In that ideal situation, the only time that would not be counted as wrench time would be things like safety and other meetings, break and lunch times, and travel time to and from the job. While this measure is somewhat idealistic, it does provide a clear way to assess overall effectiveness. P&S is the most effective way to improve an organization's wrench time. The power in improving wrench time is considerable. For example, if you have a crew of 10 people that has a wrench time of 30% and it is improved to 40%, one mechanic effectively will have been added, and this mechanic has a wrench time of 100%.

When P&S and reliability are implemented together, a synergistic relationship develops. A properly designed and operated P&S system increases a maintenance organization's effectiveness, enabling it to complete a given amount of work in less time. This shifts the balance of the work system, creating a void to fill in the remaining time. If well-thought-out reliability initiatives are used to fill that void, reactive work is reduced, which in turn enables more planned and scheduled work to be completed, a virtuous cycle that continues to improve until a new balance point is achieved. Planning enables scheduling, and scheduling enables effectiveness. Effectiveness enables reliability, and reliability enables planning. Without planning, scheduling loses its ability maximize the use of time. Without scheduling, planning lacks a means to orchestrate the activities required by the job. Without P&S, reliability-related work is neglected for sake of the urgent work. Without reliability-related work, the decreased flexibility in scheduling the job makes it more difficult to schedule 100% of the available personnel resources.

Ideally, this overview of the maintenance planning and scheduling process has provided some insight into the value of P&S and how to use the process to facilitate reliability efforts.

2

The Functional Maintenance Organization and Its People

2.1. FUNCTIONAL MAINTENANCE ORGANIZATIONAL STRUCTURE

Figure 2.1 represents the organizational structure from the maintenance department's point of view. Each element plays a role in achieving the vision and mission of maintenance and the success of each element ensures the success of the maintenance effort:

Executive staff—Organizational vision and mission statements are fashioned by the executive staff to set the direction for the organization as a whole. By monitoring goals and objectives, assurance is given that the vision and mission of the organization will succeed.

Vice president of operations—Playing a key role, the VP of operations must be familiar with not only the goals and objectives of the executive staff but those of maintenance to ensure continuity. Conflicts between the organization vision and mission and the maintenance effort are best solved by working closely with the plant manager.

Plant manager—Providing direction, support, and commitment to the maintenance effort, the plant manager works closely with the maintenance staff to have knowledge of issues facing the maintenance effort. Solving these issues are best handled by the plant manager working closely with the VP of operations.

Accounting—Accounting produces the portion of the overall budget earmarked for maintenance. This is accomplished by facilitating communications with general manager, vice president, and maintenance to recognize the needs of the maintenance effort and ensure

that these requirements are proposed during budget approval. Once the budget is in place and activated, accounting monitors expenditures and works closely with the plant manager to ensure budget management processes are administered.

Business development—Maintenance looks for business development to supply new business and maintain current customer relations. By ensuring availability, reliability, capacity, and quality and by stabilizing fee per unit through maintenance cost control, the maintenance effort lends leverage to business development in obtaining new customers.

Scheduling—Scheduling is a customer of maintenance. Maintenance ensures that any equipment needed in production coordination is available, reliable, and at required operational capacity when needed.

Operations—Operations approves the weekly and daily maintenance schedules and sets priorities to ensure scheduling production requirements are met.

Maintenance manager—Ensuring the success of the maintenance effort, the superintendent focuses on those requirements needed to meet the maintenance vision and mission.

Supervision—By managing the maintenance schedule approved by operations, supervision ensures the expectations of operations are achieved. Each day, supervision executes the maintenance mission and ensures success of the maintenance effort.

Planning—The planners plan approved jobs to increase the efficiency and effectiveness of the maintenance technician. Each plan is completed to meet or exceed operational expectations by the required date on the work order.

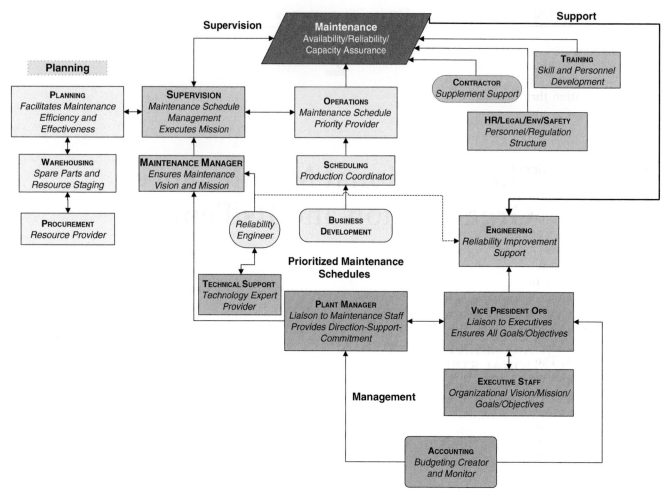

FIGURE 2.1.   Functional maintenance organizational structure.

Warehousing and procurement—Working with the plans derived from the planners, all required resources are readied for the efficient execution of work orders. Warehousing also stages those items as conveyed by the planners.

Human resources/legal/environmental/safety—Human resources provides the personnel with the skills necessary to achieve the maintenance vision and mission. The legal, environmental, and safety departments monitor and ensure that regulatory requirements are achieved.

Reliability engineering—Thought of as a function and not as a department, engineering provides support in ensuring the reliability of field equipment. As experts in engineering and reliability-centered maintenance practices, the maintenance staff looks to these engineers for support in rectifying identified reliability problems. The engineers work closely with field personnel in identifying areas for improvement.

Technical support—As with engineering, the technical support function fills those voids in the maintenance effort that require addition expertise. Supporting reliability engineering by keeping the backbone of equipment functionality current with the technology available, the technical support function plays a vital role in ensuring the efficiency and effectiveness of the maintenance effort.

Contractor—Contractors play multiple roles in the maintenance effort. From expert consulting to general laborers, each acts as a resource to supplement maintenance needs.

Training—The key to success is knowledge. Through personnel development and skills training, not only does the maintenance effort succeed but so too does the organization.

Several of these functions are examined in the remainder of the chapter.

2.2. MAINTENANCE SUPERVISOR

The maintenance supervisor manages the daily activities of the maintenance department in assigned area(s) to ensure the safe, effective, efficient, continuous plant operation through maintenance and repair of assigned facilities and equipment.

2.2.1. Responsibilities

The position functions under the direction of the maintenance manager, subject to established policies, precedents, and practices. The position involves working with capital projects, monitoring cost, and ensuring that contractors comply with safety rules and practices. The maintenance supervisor is allowed considerable freedom of action to plan, organize, and direct others to achieve department objectives and goals and must have the ability to plan, organize, and direct maintenance and repair operations. Extensive problem solving is associated with maintenance and repair of large and sophisticated equipment. The maintenance supervisor also must have highly developed human relation skills to direct and aid subordinate personnel for self-motivation, addressing competing and conflicting situations, and communicating with contractors and vendors. He or she manages the total maintenance efforts for all assigned area(s).

The position involves working closely with the planning, engineering, production, environmental, and other departments to build and sustain effective working relationships that reinforce a managerial role versus a foreman role. Therefore, the maintenance supervisor must maintain effective communication, meet emergencies, assess and provide for the training needs of crew members, and perform employee evaluations.

The position requires a high school diploma or equivalent and five years' experience in industrial maintenance. Supervisory or managing experience is preferred. Proficiency in the use of a personal computer is required, as the maintenance supervisor must be able to store, retrieve, and analyze data from the CMMS and network applications.

The maintenance supervisor's responsibilities include

- Ensuring that individual safety is a top priority by adhering to the plant's safety and environmental objectives, attending weekly safety meetings, daily toolbox meetings, monthly review of the "Safe Comment Report," and insisting that all safety rules are followed. Evidence of this is achieved by no lost-time accidents.
- Using CMMS data, reports, and graphics to pinpoint potential problem areas and enhance productivity of assigned crew(s).
- Utilizing a program of equipment evaluation, modification, and improvement and using employees to make it successful.
- Assessing and providing for the learning and training needs of crew members.
- Monitoring apprentice progress and logbooks, evaluating and counseling as required.
- Supporting company policies, departmental instructions, and objectives in a positive manner with employees.
- Spending time with each crew member to develop personal and working relationships and listen to new ideas, problems, and suggestions.
- Setting a leadership example supported by crew participation in decisions and activities.
- Demonstrating a long-term view and approach to work.
- Working with planning and a computerized maintenance management software on a systematic basis to improve and update work procedures, time estimates, and preventive and predictive maintenance frequencies.
- Responding appropriately to off-shift and off-day problems and maintenance shift problems.
- Setting, tracking, and achieving crew goals and performance. Averaged planned maintenance should meet or better the goal.
- Where minimized breakdowns is a goal, measuring and documenting trends through mean time between failures.
- Managing overtime at the company's goal or lower.
- Responsibly conducting annual performance reviews with regards to performance objectives for crew members.
- Utilizing the resources of the maintenance technical advisors, reliability engineers, and others in solving problems and improving productivity. Soliciting their help to increase predictive maintenance and reduce unnecessary asset activities.
- Consistently completing tasks and providing information in requested time.
- Assuring the credibility of maintenance functions by meeting committed downtimes for equipment.
- Keeping customers knowledgeable of the proactive maintenance process and aware of maintenance activities through the weekly planning meetings.
- Ensuring appropriate and consistent discipline is administered as needed.
- Assisting in staffing the department with the best skilled craftspeople and apprentices possible.
- Promoting integrated work between mechanical and electrical craftspeople.
- Working with operating departments to maintain optimum operating conditions and compatible working relations.

- Ensuring mandated regulatory training is 100% crew compliant.
- Preparing period reports as required.
- Filling in for department management as required.
- Responsibly preparing the crew budget(s) and controlling costs to within 2% and being prepared to explain variances greater than $2,000.
- Working directly with inventory control, changing minimum/maximum quantities, controlling rebuild units, surplus and obsolete items, quality control, and reduction of inventory and capital spares to reduce equipment costs.
- Working with all personnel in the promotion of a good labor relations. Promoting the use of the open door policy, communicating with department personnel in keeping them updated on all company policies and procedures, and providing timely answers to all questions and concerns.

2.2.2. Environmental, Health, and Safety Aspects

This position requires awareness of the plant's environmental policy, waste minimization/pollution prevention policies, and environment, health, and safety policies and principles. The maintenance supervisor must be knowledgeable of the environmental aspects of processes, activities, and services in the job area and appropriate measures to control their impact on the environment. He or she must support plant policies for waste minimization and pollution prevention by conformance to plant policies, taking standard environmental control measures when performing work, and promote department environmental objectives and plant environmental goals, such as zero reportable spills and minimal landfill waste.

2.3. MAINTENANCE PLANNER/ SCHEDULER

The maintenance planner/scheduler works in a team environment assisted by a maintenance supervisor to

- Inspect jobs that are candidates for planning or scheduling.
- Determine the various craftspeople that may be required to complete the job.
- Estimate the work hours required to complete the job by each craft.
- Anticipate the permits necessary to execute the job.
- Determine the parts, tools, and equipment necessary to complete the job.

- Order parts.
- Write job plans.
- Anticipate and plan for the resolution of the various coordination requirements among the various craftpeoples, contractors, operations, and stages of jobs.
- Update SAP bills of materials.
- Publish a weekly schedule for each area of the plant.
- Utilize Microsoft Outlook, Microsoft Projects, and CMMS/EAM systems to enter and retrieve data as necessary.

2.3.1. Responsibilities

Maintenance planner/schedulers must be able to perform all of the essential elements of the job, with or without reasonable accommodations; have successfully completed either a mechanical or electrical maintenance apprenticeship program or an equivalently recognized program; be capable of clear, concise verbal and written communication; possess and practice good organizational skills; be capable of interacting with various levels of company management, employees, and external contacts; be a team player; have, or have the ability to learn, basic computer skills; and be willing to work overtime as directed by the team manager.

The following job duties are not inclusive; however, they should be used as a guideline for specific job duties for a planner/scheduler:

- Inspect and estimate jobs that are candidates for planning or scheduling.
- Utilize CMMS/EAM, Microsoft Outlook, and Microsoft Projects to enter, retrieve, and communicate the various planning and scheduling information and parts ordering.
- Determine the scope, craft, personnel, tools, parts, and equipment needs of each job and document this job plan as a means of reducing delays to the maintenance activities.
- Develop stored plans for equipment, based on criticality, frequency, and historical costs.
- Assemble job boxes that contain the documentation, parts, and tools required by the maintenance personnel to complete the job without delays.
- Communicate effectively with operations and maintenance personnel orally, written, and electronically to determine priorities, available personnel, and schedules.
- Schedule and lead weekly scheduling meetings for each area of the plant.
- Work from blueprints, factory manuals, written specifications, sketches, oral instructions, and personal

knowledge of the work being performed to plan and schedule effectively.

- Attend special schools or training sessions, on and off site, concerning maintenance planning and scheduling techniques, skills, or methods and train others in the techniques thus acquired.
- Work with other maintenance planner(s) to distribute the planning and scheduling work load to best meet the needs of operations and maintenance.
- Coordinate with other planner(s) to provide coverage during sickness and vacation to avoid interruption of planning and scheduling activities.
- Analyze significant differences between estimated and actual work hours, materials, and equipment and determine if standards for time, materials, or equipment should be revised to reflect these differences.
- Be able to work near rotating and moving machinery.
- Work overtime and respond to emergencies after hours as needed.
- Be able to operate motor vehicles.
- Follow all applicable safety rules and use all safety equipment as designated.
- Perform other duties as assigned by supervision.

2.4. MAINTENANCE AND ENGINEERING MANAGER

Reporting to the plant manager, the maintenance and engineering manager ensures continuous, effective, efficient, and safe plant capacity and operations through the application of total proactive maintenance management systems and the principles of maintenance reliability, equipment modification, and supply of operating and maintenance materials to the plant. Also, this manager is expected to improve plant operations and efficiencies through the administration of capital projects encompassing purchase, design, and replacement of existing equipment and revisions to plant layout.

The position requires an in-depth knowledge of mechanical and electrical engineering principles as well as maintenance and repair. It also requires specific detailed knowledge of plant operations to integrate engineering and maintenance with the plant's repair and maintenance requirements. The position requires highly developed organizational and management skills to effectively and efficiently direct and coordinate department functions. Highly developed human relations and oral and written communications skills are required to direct and motivate subordinate personnel to meet the stringent goals and objectives of the department.

Maintenance and engineering managers typically have a bachelors degree in engineering, most commonly mechanical or electrical engineering, and 12 to 15 years of progressively responsible experience in process plant maintenance and engineering, including project leadership.

2.4.1. Responsibilities

This position functions under broad management guidance and general direction from the plant manager. The maintenance and engineering manager operates independently to meet the goal of providing production capacity equal to or exceeding plant design specifications through effective repair and maintenance revisions to plant equipment.

The plant operates 365 days per year, which requires extensive planning, organization, and effective delegation and communication to subordinates to meet established production goals. The repair and maintenance of large and sophisticated mechanical and electrical equipment calls for extensive problem solving. For example, the plant has its own electrical power substation, which is maintained, repaired, and modified by in-house expertise. Those in this position must also manage the department to reach the goal of the capacity requirements of the three major production units while balancing preventive maintenance, quality of maintenance, and maintenance costs.

The maintenance and engineering manager must have highly developed human relations skills to direct and motivate subordinates and deal with personnel at all levels. Major interaction takes place at the superintendent and manager level.

This manager also is expected to

- Maintain and repair plant and equipment at the least possible cost while ensuring continuous plant operations 365 days per year.
- Prepare and monitor the maintenance and repair annual budget for the department to ensure costs are in line with approved budgets and determine variances and the cause of variances.
- Participate in the capital improvement and replacement budget for the entire plant to meet goals and objectives for asset management.
- Manage, direct, and motivate subordinate personnel to achieve annual and standing department goals and objectives in a timely and safe manner, while continuously improving subordinate skills.
- Manage, direct, coordinate, and organize the engineering and maintenance departments to meet the goal of providing production capacity equal to or exceeding plant design specifications at the least cost in the most efficient, effective, and safe manner.

- Act as the natural disaster coordinator for the plant during a natural disaster with responsibilities.
- Review and assist in the engineering of major projects.
- Manage and direct the maintenance supervisors, who are responsible for repairing and maintaining electrical and mechanical plant and equipment and site facilities.
- Be responsible for all equipment at the plant, including operating and maintaining the substation, distribution system, instrument controls, production computers, buildings, and support facilities.
- Manage and direct the planning and warehouse supervisor, who is responsible for maintenance planning and warehouse and inventory control.
- Be responsible for all planning and scheduling of work; daily, weekly, and annual overhauls; and provide the materials to meet the schedules at maximum inventory turnover.
- Carry out warehouse and inventory control operations for the plant.
- Manage and direct the technical group personnel.
- Maintain and foster positive labor relations and communication, both within the maintenance and engineering departments and among the plant's various departments.
- Monitor and improve processes, procedures, and equipment to reduce or eliminate safety hazards and risks in maintenance and engineering with a goal of achieving a zero lost time accident record.
- Chair apprenticeship and craft training programs.
- Serve as coordinator of insurance facility requirements for the plant and port.
- Provide port support for engineering and project activities.
- Plan for and coordinate energy conservation within the plant.
- Support plant environmental programs and requirements with engineering solutions and day-to-day compliance.
- Work with outside companies to promote advanced maintenance management procedures, philosophies, and work culture considerations.
- Support all aspects of plant and departmental quality. Perform department management reviews.
- Observe and abide by company policies, practices, and rules and be alert to safety when in operating areas.

The maintenance and engineering manager supervises

Planning and warehouse supervisor
Supervisors
Administrative assistant

Project engineers
Contractor safety coordinator
Drafters
Reliability engineers
Technical advisors

2.4.2. Environmental, Health, and Safety Aspects

This position requires awareness of the plant's environmental policy, waste minimization/pollution prevention policies and the corporate environment, health, and safety policies and principles. The maintenance and engineering manager must be knowledgeable of the environmental aspects of processes, activities, and services in the job area and the appropriate measures to control the impact on the environment.

He or she must support plant policies for waste minimization and pollution prevention by conformance to plant policies on waste minimization, taking standard environmental control measures when performing work, and promoting department environmental objectives and plant environmental goals.

2.5. AREA MANAGER OF WAREHOUSE AND INVENTORY CONTROL

Reporting to the maintenance and engineering manager, the area manager of warehouse and inventory control provides supervision and direction for the operation of the total warehouse and inventory control function.

2.5.1. Responsibilities

The area manager of warehouse and inventory control is expected to

- Supervise inventory control and warehouse personnel in the performance of their duties, make the necessary job assignments, and schedule workloads as appropriate to meet the plant demands.
- Coordinate and work closely with all supervision throughout the plant to ensure a totally effective material control system.
- Provide or arrange for the training of an efficient warehouse and inventory control staff and develop leadership capabilities for potential advancement.
- Inform and counsel subordinates on policy or procedural matters, handle suggestions and work-related problems in accordance with established policies and procedures, appraise individual job performance, and maintain the required records.

- Continuously evaluate the operation of the warehouse and inventory control function and make recommendations to the engineering and maintenance manager to ensure maximum efficiency.
- Maintain an active safety program, ensuring the observation of safety precautions and safe work methods by all personnel. Ensure that housekeeping standards are maintained at the highest level possible.
- Ensure that all inventory control records, receiving, dispersal, and existing inventory are correctly maintained through inventory control module system.
- Notify all personnel, including the engineering and maintenance manager, planning, accounts payable, and purchasing personnel of all overages, shortages, and damaged shipments.
- Assist in determining and ensuring that good established order point and order quantity inventory levels are established through the maintenance control system.
- Work directly with the engineering and maintenance manager and the M.I.S. department on all new developments to improve the warehouse inventory control system.
- Develop, supervise, and maintain an effective inventory control system that utilizes total computerized techniques.
- Work with the warehouse supervisor in the development and maintenance of a procedures manual (to include flow diagrams) for the total warehouse function.
- Supply all the necessary inventory historical information requested by maintenance and management.
- Coordinate and work closely with the warehouse supervisor and engineering and maintenance manager to ensure a totally effective material control function.
- Report all improvements, problems, and recommendations in a written period-end report to the engineering and maintenance manager.
- Establish, set schedule, and direct a periodic-cycle physical inventory of all warehouse items, including rebuilt repaired items. Reconcile all adjustments to maintain minimal errors.
- Respond to requests for information, assistance, or administrative direction outside of normal working hours; assure availability for such purposes on certain weekends as assigned.
- Maintain and update all graphs as assigned, through use of the network graphics system.

The area manager of warehouse and inventory control supervises all inventory control and warehouse personnel, including

Inventory quality specialist
Inventory control specialists
Warehouse supervisor

In addition, the area manager of warehouse and inventory control maintains contacts and relationships with

- M.I.S. personnel to develop and improve the computerized inventory control and warehouse programs.
- Maintenance personnel to coordinate inventory levels and material control.
- Accounting personnel concerning inventory cost reporting methods.
- Engineering personnel concerning blue print involvement for manufactured and fabricated spare parts.
- Visitor representatives from other plants to discuss warehousing and inventory control functions (on the approval of the engineering and maintenance manager).
- The supervision of the production service and maintenance departments to coordinate warehousing functions.

2.6. RELIABILITY ENGINEER

First, a word of advice: If the company is struggling with hiring a reliability engineer and might never hire one, then identify the smartest maintenance technician and provide all the training recommended in the Introduction to this book. Reliability engineering technicians, if trained properly, have a dramatic impact on the reliability in an organization.

The reliability engineer reports to the maintenance and engineering manager and is responsible for implementing advanced reliability tools and work processes and supporting reliability efforts in manufacturing plants. Proven expertise in basic engineering skills is required: generally a bachelor's degree in mechanical or electrical engineering; a master's degree engineering (take out business) is a plus. Professional certification is available and should be pursued. Hands-on experience in a craft skill area would be beneficial but is not a requirement. This engineer works to improve equipment and process performance by applying reliability engineering principles, statistical data analysis, and supporting work process.

For example, the reliability engineer is expected to

- Provide engineered solutions to complex equipment problems.
- Identify and set priorities on opportunities for reliability improvement by mining information using the reliability databases.
- Statistically analyze equipment and process performance to assess reliability, maintainability, and availability.

- Lead maintenance planning and process procedure improvements via reliability tools.
- Translate reliability opportunities into bottom-line cost savings and top-line growth (e.g., implement Six Sigma work processes in support of reliability projects and high-impact projects to deliver bottom-line cost reductions).
- Provide training and facilitation on root cause failure analysis with maintenance and production personnel.
- Provide training and facilitating failure modes and effects analysis (FMEA), a process by which maintenance strategies are developing identifying known and likely failure modes and what there effects on a system, with maintenance and production personnel.
- Provide "design for reliability" input to capital projects.
- Facilitate and manage ad hoc work teams.
- Provide technical leadership at all levels of the plant organization with special emphasis on interacting with maintenance crews.

Serving in a staff capacity, the reliability engineer relieves maintenance supervisors and planner/schedulers of those responsibilities that are technical and engineering in nature. Maintenance and reliability engineering is different and distinct from plant, project, and process engineering. Plant and project engineering support the longer-range capital program of new installations and alterations. Process engineering typically reports to and focuses on problems of production. Reliability engineering is dedicated to the preservation of day-to-day asset reliability.

2.6.1. Responsibilities

- Perform detailed engineering calculations and correlate data for analysis.
- Support operations and maintenance to improve equipment reliability.
- Perform analysis of failure data using statistical analysis tools.
- Lead problem solving teams to identify, design, and implement improvements.
- Work closely with maintenance to target maintainability issues, lower cost, and improve productivity.
- Identify value creation opportunities using reliability data collection and reliability business case tools.
- Work closely with operations and maintenance to communicate the opportunities of the most value-adding jobs.

- Promote the use of reliability centered maintenance and reliability health monitoring software and other support tools.
- Manage reliability projects, interact with vendors, and work closely with project engineers.
- Provide equipment, safety, and environment risk mitigation leadership.

2.6.2. Job Skills

- Strong basic engineering skills (mechanical, electrical, chemical, industrial, and materials).
- Solid verbal and written communication skills.
- Robust computer skills (at a minimum, Word, Access, Advanced Excel, PowerPoint, and SAP).
- Effective interpersonal skills and the ability to work with and lead loosely knit, cross-functional working teams.
- Basic skill in root cause analysis work process methods and tools.
- Proven facilitation and group leadership experience and skills.
- Skills in basic reliability engineering principles, such as mechanical components and systems; properly categorizing and analyzing failure data and modes; translating failure data to cost; FMEA; fault tree analysis; data mining and management; reliability growth methods (Crow-AMSAA); failure reporting and analysis (FRACAS); basic reliability models, block diagrams, impact on reliability; and availability, reliability and maintainability.
- Practical statistical analysis skills, including failure statistics (normal, exponential, lognormal, binomial, Poisson); failure rates, hazard rates, expected life, MTTF, MTBF, MTTR, MTBM; and Weibull analysis.
- Basic understanding of common predictive technologies: vibration analysis, lubrication, thermography, motor analysis, ultrasonics, machinery analysis, alignment and balancing procedures, and fixed equipment diagnostic technologies.
- Project value identification, including the ability to convert the cost of unreliability into bottom-line profit, and apply reliability business case tools.
- Lean and Six Sigma skills are a plus. (This is specialized and requires additional training).

Lean is a process that looks for ways to eliminate waste in any process. The goal is to eliminate steps in a process that add no value and develop new steps that improve efficiency in a process.

Six Sigma is a system of practices originally developed by Motorola to systematically improve processes by eliminating defects. Defects are defined as units that are not members of the intended population. Since it was originally developed, Six Sigma has become an element of many quality management initiatives.

The process, pioneered by Bill Smith at Motorola in 1986, originally was defined as a metric for measuring defects and improving quality and a methodology to reduce defect levels below 3.4 defects per million opportunities

DMAIC is a five-step process using Six Sigma methodology:

Define—This is the first phase of the DMAIC process. It involves clearly defining the opportunity or problem and validating it in measurement terms.

Measure—The measure phase of DMAIC is dedicated to assembling a data collection plan, executing that plan, and verifying the data collection is performed properly.

Analyze—In the analyze phase, first, develop hypotheses about the causes of the defects. Next, analyze and process the data using statistical and nonstatistical methods. Then, prove or disprove the hypothesis.

Improve—In the improve phase, confirm the key process inputs that affect the process outputs, causing defects. Identify the acceptable range of each input, so that the output stays within the specified limits. Implement changes and install and validate the measurement systems used to improve the process in order to verify the new process is working.

Control—In the final phase of the DMAIC process, implement the control phase, in which controls are used to continuously improved a process. These tools are statistical in nature and are used to control variation in a process.

2.6.3. Reliability Engineering Dashboard—Key Performance Indicators

A reliability engineer must be able to measure reliability and identify areas of opportunity where his or her technical ability can have the most impact on an operation. The most widely used measurements or KPIs (key performance indicators) used by reliability engineers are the following:

- Bad actors report. Top five critical assets with the highest total maintenance cost and worst reliability (MTBFF, mean time between functional failure).
- MTBFF by production line and area.
- Percentage of critical assets with a maintenance strategy developed using an RCM methodology.
- Percentage of new assets with a maintenance strategy developed using an RCM methodology and ranked based on risk to the business.

These four KPIs can be identified as the "KPI dashboard" for a reliability engineer. A KPI dashboard is a tool that provides data to a reliability engineer in order to measure their impact on an operation and provide direction where problems may exist.

3

Preventive Maintenance Program

The preventive maintenance program is developed using a guided logic approach and is task oriented rather than maintenance process oriented. This eliminates the confusion associated with the various interpretations across different industries of terms such as *condition monitoring, on-condition,* and *hard time.* By using a task-oriented concept, it is possible to see the whole maintenance program reflected for a given item. A decision logic tree is used to identify applicable maintenance tasks. Servicing and lubrication are included as part of the logic diagram, as this ensures that an important task category is considered each time an item is analyzed.

The content of the maintenance program itself consists of two groups of tasks: preventive maintenance tasks and nonscheduled maintenance tasks.

The preventive maintenance tasks, which include failure-finding tasks, are scheduled to be accomplished at specified intervals or based on condition. The objective of these tasks is to identify and prevent deterioration below inherent safety and reliability levels by one or more of the following means:

- Lubrication and servicing
- Operational, visual, or automated checking
- Inspection, functional test, or condition monitoring
- Restoration
- Discard

This group of tasks is determined by RCM analysis, that is, it constitutes the RCM-based preventive maintenance program.

The nonscheduled maintenance tasks result from

- Findings from the scheduled tasks accomplished at specified intervals of time or usage.
- Reports of malfunctions or indications of impending failure (including automated detection).

The objective of the second group of tasks is to maintain or restore the equipment to an acceptable condition in which it can perform its required function.

An effective program schedules only those tasks necessary to meet the stated objectives. It does not schedule additional tasks that increase maintenance costs without a corresponding increase in protection of the inherent level of reliability. Experience has clearly demonstrated that reliability decreases when inappropriate or unnecessary maintenance tasks are performed, due to an increased incidence of maintainer-induced faults.

3.1. RELIABILITY-BASED PREVENTIVE MAINTENANCE

This section describes the tasks in the development of a reliability-based preventive maintenance program for both new and in-service equipment. In the development of such a program, the progressive logic diagram and the task selection criteria are the principal tools. This progressive logic is the basis of an evaluation technique applied to each functionally significant item (FSI) using the technical data available. Principally, the evaluations are based on the items' functional failures and failure causes. The development of a reliability-based preventive maintenance program is based on the following:

- Identification of functionally significant items.
- Identification of applicable and effective preventive maintenance tasks using the decision tree logic.

A functionally significant item is one whose failure would affect safety or could have a significant operational or economic impact in a particular operating or maintenance context. The identification of FSIs is based on the anticipated consequences of failures using an analytical approach and good engineering judgment. Identification of FSIs also uses a top-down approach, conducted first at the system level, then at the subsystem level, and where appropriate, down to the component level. An iterative process should be followed in identifying FSIs. Systems and subsystem boundaries and functions are identified first. This permits selection of critical systems for further analysis, which involves a more comprehensive and detailed specification of system, system functions, and system functional failures.

The procedures in Figure 3.1 outline a comprehensive set of tasks in the FSI identification process. All the tasks should be applied in the case of complex or new equipment. However, in the case of well-established or simple equipment, where functions and functional degradations and failures are well recognized, the tasks listed under the heading of "System Analysis" can be covered very quickly. They should be documented, however, to confirm that they were considered. The depth and rigor used in the application of these tasks also varies with the complexity and newness of the equipment.

3.1.1. Information Collection

Equipment information provides the basis for the evaluation and should be assembled prior to the start of the analysis and supplemented as the need arises. The following should be included:

- Requirements for equipment and its associated systems, including regulatory requirements.
- Design and maintenance documentation.
- Performance feedback, including maintenance and failure data.

Also, to guarantee completeness and avoid duplication, the evaluation should be based on an appropriate and logical breakdown of the equipment.

3.1.2. System Analysis

The tasks just described specify the procedure for the identification of the functionally significant items and the subsequent maintenance task selection and implementation. Note that the tasks can be tailored to meet the requirements of particular industries, and the emphasis placed on each task depends on the nature of that industry.

3.1.3. Identification of Systems

The objective of this task is to partition the equipment into systems, grouping the components contributing to

achievement of well-identified functions and identifying the system boundaries. Sometimes, it is necessary to perform further partitioning into the subsystems that perform functions critical to system performance. The system boundaries may not be limited by the physical boundaries of the systems, which may overlap.

Frequently, the equipment is already partitioned into systems through industry-specific partitioning schemes. This partitioning should be reviewed and adjusted where necessary to ensure that it is functionally oriented. The results of equipment partitioning should be documented in a master system index, which identifies systems, components, and boundaries.

3.1.4. Identification of System Functions

The objective of this task is to determine the main and auxiliary functions performed by the systems and subsystems. The use of functional block diagrams assist in the identification of system functions. The function specification describes the actions or requirements the system or subsystem should accomplish, sometimes in terms of performance capabilities within the specified limits. The functions should be identified for all modes of equipment operation.

Reviewing design specifications, design descriptions, and operating procedures, including safety, abnormal operations, and emergency instructions, may determine the main and auxiliary functions. Functions such as testing or preparation for maintenance, if not considered important, may be omitted. The reason for omissions must be given. The product of this task is a listing of system functions.

3.1.5. Selection of Systems

The objective of this task is to select and rank systems to be included in the RCM program because of their significance to equipment safety, availability, or economics. The methods used to select and rank the systems can be divided into

- Qualitative methods based on past history and collective engineering judgment.
- Quantitative methods, based on quantitative criteria, such as criticality rating, safety factors, probability of failure, failure rate, or life cycle cost, used to evaluate the importance of system degradation or failure on equipment safety, performance, and costs. Implementation of this approach is facilitated when appropriate models and data banks exist.
- A combination of qualitative and quantitative methods.

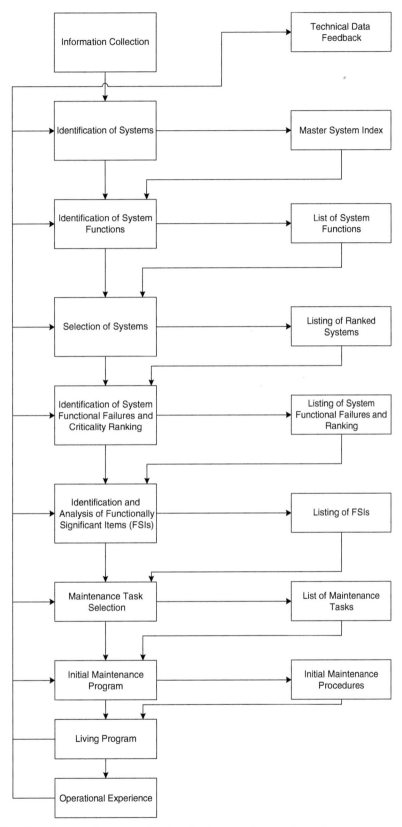

FIGURE 3.1.   Development tasks of a reliability-based preventive maintenance program.

The product of this task is a listing of systems ranked by criticality. The systems, together with the methods, the criteria used, and the results should be documented.

3.1.6. System Functional Failure and Criticality Ranking

The objective of this task is to identify system functional degradations and failures and rank them as to priority. The functional degradation or failure of a system for each function should be identified, ranked by criticality, and documented.

Since each system functional failure may have a different impact on safety, availability, and maintenance cost, it is necessary to rank and assign priorities to them. The ranking takes into account probability of occurrence and consequences of failure. Qualitative methods based on collective engineering judgment and the analysis of operating experience can be used. Quantitative methods of simplified failure modes and effects analysis (SFMEA) or risk analysis also can be used.

The ranking represents one of the most important tasks in RCM analysis. Too conservative a ranking may lead to an excessive preventive maintenance program, and conversely, a lower ranking may result in excessive failures and a potential safety impact. In both cases, a nonoptimized maintenance program results. The outputs of this task are the following:

- List of system functional degradations and failures and their characteristics.
- Ranked list of system functional degradations and failures.

3.2. IDENTIFICATION OF FUNCTIONALLY SIGNIFICANT ITEMS

Based on the identification of system functions, functional degradations and failures and their effects, and collective engineering judgment, it is possible to identify and develop a list of candidate FSIs. As said before, failure of these items could affect safety, be undetectable during normal operation, and have significant operational or economic impact. The output of this task is a list of candidate FSIs.

Once an FSI list has been developed, a method such as failure modes and effects analysis (FMEA) should be used to identify the following information, which is necessary for the logic tree evaluation of each FSI. The following examples refer to the failure of a pump providing cooling water flow:

Function. The normal characteristic actions of the item (e.g., to provide cooling water flow at 100 I/s to 240 I/s to the heat exchanger).

Functional failure. How the item fails to perform its function (e.g., pump fails to provide required flow).

Failure cause. Why the functional failure occurs (e.g., bearing failure).

Failure effect. What is the immediate effect and the wider consequence of each functional failure (e.g., inadequate cooling, leading to overheating and failure of the system).

The FSI failure analysis is intended to identify functional failures and failure causes. Failures not considered credible, such as those resulting solely from undetected manufacturing faults, unlikely failure mechanisms, or unlikely external occurrences, should be recorded as having been considered and the factors that caused them to be assessed as not credibly stated.

Prior to applying the decision logic tree analysis to each FSI, preliminary worksheets need to be completed that clearly define the FSI, its functions, functional failures, failure causes, failure effects, and any additional data pertinent to the item (e.g., manufacturer's part number, a brief description of the item, predicted or measured failure rate, hidden functions, redundancy). These worksheets should be designed to meet the user's requirements.

From this analysis, the critical FSIs can be identified (i.e., those that have both significant functional effects and a high probability of failure or have a medium probability of failure but are judged critical or have a significantly poor maintenance record).

3.3. MAINTENANCE TASK SELECTION (DECISION LOGIC TREE ANALYSIS)

The approach used for identifying applicable and effective preventive maintenance tasks provides a logic path for addressing each FSI functional failure. The decision logic tree (Figures 3.2 and 3.3) uses a group of sequential "yes or no" questions to classify or characterize each functional failure. The answers to the "yes or no" questions determine the direction of the analysis flow and help determine the consequences of the FSI functional failure, which may be different for each failure cause. Further progression of the analysis ascertains if there is an applicable and effective maintenance task that prevents or mitigates it. The resultant tasks and related intervals form the initial scheduled maintenance program.

Note: Proceeding with the logic tree analysis with inadequate or incomplete FSI failure information

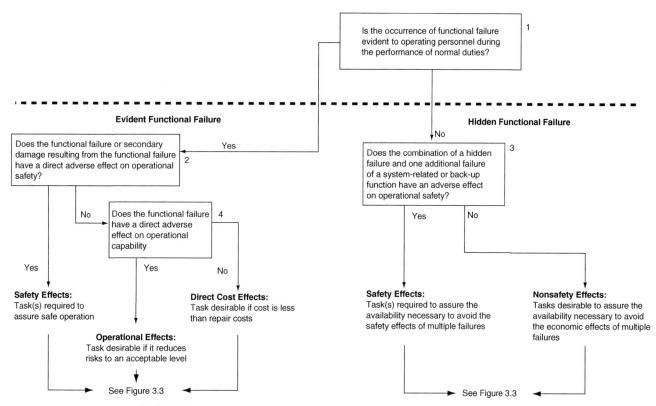

FIGURE 3.2. Reliability decision logic tree, level 1, effects of functional failures.

could lead to the occurrence of safety critical failures, due to inappropriate, omitted, or unnecessary maintenance; to increased costs due to unnecessary scheduled maintenance activity; or both.

3.3.1. Levels of Analysis

Two levels are apparent in the decision logic:

1. The first level (Figure 3.2, questions 1–4) requires an evaluation of each functional degradation or failure to determine the ultimate effect category, that is, evident safety, evident operational, evident direct cost, hidden safety, hidden nonsafety, or none.
2. The second level (Figure 3.3) takes the failure causes for each functional degradation or failure into account to select the specific type of tasks.

First Level Analysis (Determination of Effects)

The consequence of failure (which could include degradation) is evaluated at the first level using four basic questions (Figure 3.2). Note: The analysis should not proceed through the first level unless there is a full and complete understanding of the particular functional failure.

Question 1. *Evident or hidden functional failure?* The purpose of this question is to segregate the evident and hidden functional failures and should be asked for each functional failure.

Question 2. *Direct adverse effects on operating safety?* To be direct, the functional failure or resulting secondary damage should achieve its effect by itself, not in combination with other functional failures. An adverse effect on operating safety implies that damage or loss of equipment, human injury or death, or some combination of these events is a likely consequence of the failure or resulting secondary damage.

Question 3. *Hidden functional failure safety effect?* This question takes into account failures in which the loss of a hidden function (whose failure is unknown to the operating personnel) does not of itself affect safety but, in combination with an additional functional failure, has an adverse effect on operating safety. Note: The operating personnel consist of all qualified staff members on duty and directly involved in the use of the equipment.

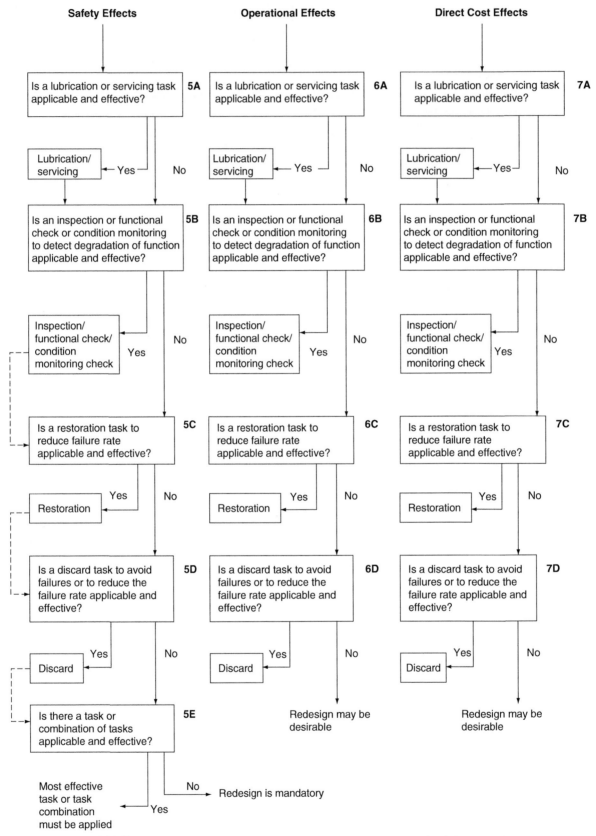

FIGURE 3.3. Reliability decision logic tree, level 2, effects categories and task determination.

Question 4. *Direct adverse effect on operating capability?* This question asks if the functional failure could have an adverse effect on operating capability, requiring either the imposition of operating restrictions or correction prior to further operation or the operating personnel to use abnormal or emergency procedures.

Second Level Analysis (Effects Categories)

Applying the decision logic of the first level questions to each functional failure leads to one of five effect categories, as follows:

Evident safety effects, questions 5A to 5E. This category should be approached with the understanding that a task (or tasks) is required to ensure safe operation. All questions in this category need to be asked. If no applicable and effective task results from this category analysis, then redesign is mandatory.

Evident operational effects, questions 6A to 6D. A task is desirable if it reduces the risk of failure to an acceptable level. If all answers are "no" in the logic process, no preventive maintenance task is generated. If operational penalties are severe, a redesign is desirable.

Evident direct cost effects, questions 7A to 7D. A task is desirable if the cost of the task is less than the cost of repair. If all answers are "no" in the logic process, no preventive maintenance task is generated. If the cost penalties are severe, a redesign may be desirable.

Hidden function safety effects. The hidden function safety effect requires a task to ensure the availability necessary to avoid the safety effect of multiple failures. All questions should be asked. If no applicable and effective tasks are found, then redesign is mandatory.

Hidden function non-safety effects. This category indicates that a task may be desirable to assure the availability necessary to avoid the direct cost effects of multiple failures. If all answers are "no" in the logic process, no preventive maintenance task is generated. If economic penalties are severe, a redesign may be desirable.

Task Determination

Task determination is handled in a similar manner for each of the effect categories. For task determination, it is necessary to apply the failure causes for the functional failure to the second level of the logic diagram. Seven possible task result questions in the effect categories have been identified, although additional tasks, modified tasks, or a modified task definition may be warranted, depending on the needs of particular industries.

3.3.2. Paralleling and Default Logic

Paralleling and default logic play an essential role at level 2 (see Figure 3.3). Regardless of the answer to the first question regarding "lubrication/servicing," the next task selection question should be asked in all cases. When following the hidden or evident safety effects path, all subsequent questions should be asked. In the remaining categories, subsequent to the first question, a "yes" answer will allow exiting the logic. (At the user's option, advancement is allowable to subsequent questions after a "yes" answer is derived but only if the cost of the task is equal to the cost of the failure prevented.)

In the tree, default logic is reflected in paths outside the safety effects areas by the arrangement of the task selection logic. In the absence of adequate information to answer "yes" or "no" to questions in the second level, default logic dictates that a "no" answer be given and the subsequent questions be asked. As "no" answers are generated, the only choice available is the next question, which in most cases provides a more conservative, stringent, or costly route. Re design is mandatory for failures that fall into the safety effects category (evident or hidden) and for which there are no applicable and effective tasks.

3.4. MAINTENANCE TASKS

Explanations of the terms used in the possible tasks are as follows:

Lubrication/servicing (all categories). This involves any act of lubricating or servicing for maintaining inherent design capabilities.

Operational/visual/automated check (hidden functional failure categories only). An operational check is to determine that an item is fulfilling its intended purpose. It does not require quantitative checks and is a failure-finding task. A visual check is an observation to determine that an item is fulfilling its intended purpose and does not require quantitative tolerances. This, again, is a failure-finding task. The visual check could also involve interrogating electronic units that store failure data.

Inspection/functional check/condition monitoring (all categories). An inspection is an examination of an item against a specific standard. A functional check is a quantitative check to determine if one or more functions of an item performs within specified limits. Condition monitoring is a task, which may be continuous or periodic, to monitor the condition of an item in operation against preset parameters.

Restoration (all categories). Restoration is the work necessary to return the item to a specific standard. Since restoration may vary from cleaning or replacement of single parts up to a complete overhaul, the scope of each assigned restoration task has to be specified.

Discard (all categories). Discard is the removal from service of an item at a specified life limit. Discard tasks are normally applied to so-called single-cell parts such as cartridges, canisters, cylinders, turbine disks, safe-life structural members, and the like.

Combination (safety categories). Since this is a safety category question and a task is required, all possible avenues should be analyzed. To do this, a review of the applicable tasks is necessary. From this review, the most effective tasks should be selected.

No task (all categories). It may be decided that no task is required in some situations, depending on the effect. Each of the possible tasks just defined is based on its own applicability and effectiveness criteria. Table 3.1 summarizes these task selection criteria.

3.5. TASK FREQUENCIES/INTERVALS

To set a task frequency or interval, it is necessary to determine the existence of applicable operational experience data that suggest an effective interval for task accomplishment. Appropriate information may be obtained from one or more of the following:

- Prior knowledge from other similar equipment shows that a scheduled maintenance task has offered substantial evidence of being applicable, effective, and economically worthwhile.
- Manufacturer or supplier test data indicate that a scheduled maintenance task is applicable and effective for the item being evaluated.
- Reliability data and predictions.

TABLE 3.1.   Task Selection Criteria

Task	Application Criteria	Effectiveness Criteria		
		Safety	Operational	Direct Cost
Lubricating/servicing	Replenishment of the consumable reduces the rate of functional deterioration	The task reduces the risk of failure	The task reduces the risk of failure to an acceptable level	The task is cost effective; that is, the cost of the task is less than that of of the task the failure prevented
Operational/visual/ automated check	Identification of the failure is possible	The task ensures adequate availability of the hidden function to reduce the risk of multiple failures	Not applicable	The task ensures adequate availability of the hidden function to avoid the economic effects of multiple failures and is cost effective
Inspection/functional check/condition monitoring	Reduced resistance to failure is detectable and rate of reduction in failure resistance is predictable	The task reduces the risk of failure to assure safe operation	The task reduces the risk of failure to an acceptable level	The task is cost effective
Restoration	The item shows functional degradation characteristics at an identifiable age and a large proportion of units survive to that age; the item can be restored to a specific standard of failure resistance	The task reduces the risk of failure to assure safe operation	The task reduces the risk of failure to an acceptable level	The task is cost effective
Discard	The item shows functional degradation characteristics at an identifiable age and a large proportion of units survive to that age	A safe-life limit reduces the risk of failure to assure safe operation	The task reduces the risk of failure to an acceptable level	An economic life limit is cost effective

Safety and cost considerations need to be addressed in establishing the maintenance intervals. Scheduled inspections and replacement intervals should coincide whenever possible, and tasks should be grouped to reduce the operational impact.

The safety replacement interval can be established from the cumulative failure distribution for the item by choosing a replacement interval that results in an extremely low probability of failure prior to replacement. Where a failure does not cause a safety hazard but causes loss of availability, the replacement interval is established in a trade-off process involving the cost of replacement components, the cost of failure, and the availability requirement of the equipment.

Mathematical models exist for determining task frequencies and intervals, but these models depend on the availability of the appropriate data. These data are specific to particular industries, and those industry standards and data sheets should be consulted as appropriate.

If there is insufficient reliability data or no prior knowledge from other similar equipment, or if there is insufficient similarity between the previous and current systems, the task interval frequency can be established initially only by experienced personnel using good judgment and operating experience in concert with the best available operating data and relevant cost data.

4

Predictive Maintenance Program

Like preventive maintenance, *predictive maintenance* has many definitions. To some, predictive maintenance is monitoring the vibration of rotating machinery in an attempt to detect incipient problems and prevent catastrophic failure. To others, it is monitoring the infrared image of electrical switchgear, motors, and other electrical equipment to detect developing problems. The common premise of predictive maintenance is that regular monitoring of the actual crafts condition, operating efficiency, and other indicators of operating condition of machine trains and process systems provides the data required to ensure the maximum interval between repairs and minimize the number and cost of unscheduled outages created by machine train failures.

Predictive maintenance is much more. It is the means of improving productivity, product quality, and overall effectiveness of manufacturing and production plants. Predictive maintenance is not vibration monitoring, thermal imaging, lubricating oil analysis, or any of the other nondestructive testing techniques being marketed as predictive maintenance tools. Predictive maintenance, simply stated, is a philosophy or attitude of using the actual operating condition of plant equipment and systems to optimize total plant operation. A comprehensive predictive maintenance management program utilizes a combination of the most cost effective tools, that is, vibration monitoring, thermography, tribology, and the like, to obtain the actual operating condition of critical plant systems and, based on this actual data, schedules all maintenance activities on an "as needed" basis.

Including predictive maintenance in a comprehensive maintenance management program allows optimizing the availability of process machinery and greatly reduces the cost of maintenance. It also provides the means to improve product quality, productivity, and profitability in manufacturing and production plants.

Predictive maintenance is a condition-driven preventive maintenance program. Instead of relying on industrial or in-plant average life statistics (i.e., mean time to failure) to schedule maintenance activities, predictive maintenance uses direct monitoring of the crafts condition, system efficiency, and other indicators to determine the actual mean time to failure or loss of efficiency for each machine train and system in the plant. At best, traditional time-driven methods provide a guideline to "normal" machine train life spans.

In preventive or run-to-failure programs, the final decision on repair or rebuild schedules must be made on the basis of intuition and the personal experience of the maintenance manager. The addition of a comprehensive predictive maintenance program provides factual data on the actual crafts condition of each machine train and process system that the maintenance manager can use for scheduling maintenance activities.

A predictive maintenance program minimizes unscheduled breakdowns of all crafts equipment in the plant and ensures that repaired equipment is in acceptable condition. The program also identifies machine train problems before they become serious. Most equipment problems can be minimized if detected and repaired early.

Predictive maintenance utilizing vibration signature analysis is predicated on two basic facts: (1) All common failure modes have distinct vibration frequency components that can be isolated and identified and (2) the amplitude of each distinct vibration component remains constant unless there is a change in the operating dynamics of the machine train. These facts, their impact on machinery, and the methods that identify and quantify the root cause of failure modes are developed in more detail in later chapters.

Predictive maintenance utilizing process efficiency, heat loss, or other nondestructive techniques can quantify the operating efficiency of plant equipment or systems. These techniques used in conjunction with vibration analysis can provide the maintenance manager or plant engineer information for achieving optimum reliability and availability from the plant.

A wide variety of predictive techniques and technologies may provide benefit to a facility or plant. In most cases, more than one is needed for complete coverage of all critical assets and to gain maximum benefits from their use. Several nondestructive techniques normally are used for predictive maintenance management: vibration monitoring, process parameter monitoring, thermography, tribology, and visual inspection. Each technique has a unique data set to assist in determining the actual need for maintenance. Some typical technologies successfully utilized by facilities and plants are included in Table 4.1.

Most comprehensive predictive maintenance programs use vibration analysis as the primary tool. Since the majority of normal plant equipment is rotating, vibration monitoring provides the best tool for routine monitoring and identification of incipient problems. However, vibration analysis does not provide the data required on electrical equipment, areas of heat loss, condition of lubricating oil, and other parameters that should be included in the program.

TABLE 4.1   Monitoring Techniques

Technique	Where Used	Problems Detected
Vibration	Rotating machinery (e.g., pumps, turbines, compressors, internal combustion engines, and gearboxes)	Misalignment, imbalance, defective bearings, mechanical looseness, defective rotor blades, oil whirl, and broken gear teeth
Shock pulse	Rotating machinery	Trends of bearing condition
Fluid analysis	Lubrication, cooling, hydraulic power systems	Excessive wear of bearing surfaces and fluid contamination
Infrared thermography	Boilers, steam system components, electrical switchboards and distribution equipment, motor controllers, diesel engines, and power electronics	Leaky steam traps, boiler refractory cracks, deteriorated insulation, loose electrical connections, and hot- or cold-firing cylinders
Performance trending	Heat exchangers, internal combustion engines, pumps, and refrigeration units and compressors	Loss in efficiency and deteriorating performance trends due to faulty components
Electrical insulation tests (e.g., surge tests, comparison testing, rotor impedance testing, dc high-potential testing)	Motor and generator windings and electrical distribution equipment	Trends of electrical insulation condition, turn-to-turn and phase-to-phase short, grounds, and reversed coils or turns
Ultrasonic leak detectors	Steam hydraulic and pneumatic system piping	Valve and system leaks
Fault gas analysis and insulating liquid analysis	Circuit breakers, transformers, and other protective devices	Overheating, accelerated deterioration trends, and hostile dielectric
Protection relay testing and time travel analysis	Circuit breakers, transformers, and other protective devices	Deteriorating or unsafe performance
Stereoscopic photography, hull potential measurements, and diving inspection	Underwater hull	Corrosion, fatigue cracking trends, and hull fouling trends
Material (nondestructive) testing (e.g., ultrasonics, eddy current, borescopic inspection)	Hull structure and shipboard machinery and associated piping systems and mechanical components	Corrosion, erosion, fatigue cracking, delaminations, and wall thickness reduction
Signature analysis, time domain, and frequency domain	Rectifiers, power supplies, inverters, ac and dc regulators, and generators	Degraded solid-state circuits and other electrical components
Wear and dimensional measurement	Sliding, rotating, and reciprocating elements	Excessive wear and proximity to minimum acceptable dimensions that affect performance

4.1. SETTING UP A PREVENTIVE/ PREDICTIVE MAINTENANCE PROGRAM

Setting up an effective preventive/predictive maintenance program is not a trivial exercise. A logical, methodical approach must be used to ensure that the tasks, intervals, and methods selected provide a level of maintenance that supports reliability and an optimum life cycle of the facility's assets.

The process should be predicated on a criticality analysis that defines the relative importance of each asset. Obviously, the more critical assets receive more attention and effort than less critical ones. The first step in the process is to determine the maintenance requirements for those assets to be included in the preventive maintenance program. Based on the criticality analysis, you may elect to omit some of the less critical assets. The assessment of need should include three tasks:

1. *Simplified failure modes and effects analysis.* A SFMEA identifies the more common failure modes for classes of assets, such as compressors or fans. This information should be used to develop specific preventive maintenance tasks (inspections, testing, etc.) that will detect these failure modes before they result in an emergency or breakdown.
2. *Duty-task analysis.* Duty-task analysis is used to determine the specific maintenance activities, interval, skills, and materials needed to properly maintain assets. The process uses a combination of engineering design analysis and information derived from the vendor's operating and maintenance manual or similar sources.
3. *Asset history review.* For existing assets, this step should include a thorough evaluation of the maintenance and reliability history of each. As a minimum, this part of the assessment should include an evaluation of equipment history from the CMMS. Other available records, such as downtime reports and safety investigation documents, should be reviewed as well.

Without infinite resources, all assets cannot be included in the preventive/predictive maintenance program. Therefore, the next step is to determine the specific assets to include. Obviously, the criticality analysis should be the dominant tool for this determination, but the following steps should confirm it:

Inventory assets and equipment. The primary intent of this task is to verify that all assets and equipment have been identified and included in the criticality analysis. There is a tendency to overlook critical assets, such as the infrastructure (e.g., electric power distribution and compressed air), which can have a major impact on plant performance. A subset of this task is to establish a plantwide asset identification method that will be used as part of the preventive maintenance program and is essential for CMMS implementation.

Appraise asset condition. When starting an effective preventive maintenance program, it is essential to know the current condition of all assets in the plant or facility. This is normally accomplished by performing condition assessments. In addition to the information derived from the condition assessment, this step should also answer the following questions:
- Can it be easily maintained in its present condition?
- If not, can it be upgraded to maintainable condition?
- If it can, how can it be taken out of service and how much will it cost?
- If it cannot be, what alternatives are there to choose from?

Choose assets to be included in the initial program. The change process generally is more effective when the program starts small and builds over time. Initially, the preventive maintenance program should concentrate on the most critical assets. As a rule, this initial effort will address the top quartile (25%) of critical assets, as determined by the criticality analysis. Use early successes to justify and sell program expansion, but always remember the preventive maintenance program cannot be considered world class until all critical assets are included.

Using the knowledge gained in the preceding steps, determine the organizational requirements for the upgraded preventive/predictive maintenance program. This step should include an accurate determination of

Effort hours required to implement and maintain the program. This task should define the total craft effort hours, by craft line and skill level, required to execute the activities identified by the duty-task analysis.

Supervision hours required to effectively manage the craft workforce. Adequate supervision is an absolute requirement for effective preventive/predictive maintenance.

Planning hours required to effectively support the preventive maintenance activity.

Support-hours required to effectively support the program. It is important to provide clerical, material handling, and other support for the maintenance organization to prevent overloading the planners, supervisors, and other line managers.

Effort hours required to update the program. These are the effort hours required to write or modify preventive maintenance task lists, work orders, maintenance procedures, and other documents needed to support the upgraded program.

Materials required to support the program. Identification of these should include a determination of items to stock, minimum/maximum stocking levels, and other information required for materials management.

Facilities required supporting the program. This task should include a determination of where to locate and configure shops, tool rooms, maintenance repair and operating inventories, training, and other facilities required to support an effective preventive maintenance program. Predictive tools also are needed to support the program. Data collectors, analysis software, infrared cameras, and the like have to be acquired, as well as training for technicians. Consideration should be given to requiring technicians to gain and hold a certification in the various predictive disciplines.

A fundamental requirement of reliability excellence is changing the reactive environment that permeates most organizations. A key part of the change is clear, definitive policies and procedures that define how the maintenance organization is to perform its day-to-day activities. This step should establish clear, concise definitions, principles, and concepts that govern the operation of the maintenance function, including

- Roles and responsibilities for all functional groups, such as maintenance manager, supervisors, planners, schedulers, and technicians.
- Job descriptions for all employees that clearly define their contribution to a world-class organization. Job descriptions should minimize the use of generic descriptions, such as "and other responsibilities as defined."
- Expectations that clearly establish a vision of what the final objectives of the change process will be.

Using the results from the preceding steps, the plan is presented to plant or facility management. The presentation should be in a format that addresses the financial benefits the enhanced preventive/predictive maintenance program will generate for the facility. If the business plan is presented in terms that facility management can accept, approval and at least general support and commitment should be automatic. This commitment is essential. Without at least general commitment, the program will not have the budget and support needed for effective implementation.

After the initial startup of the program, a long-range plan to upgrade critical assets to improve the ability to effectively maintain their operating condition should be developed. This may require modifications, rebuilds, and other nonpreventive maintenance activities designed to restore the asset condition to a maintainable level or simplify the ability to perform effective maintenance.

The modified organizational structure, preventive maintenance requirements, and maintenance work culture established to support the preventive maintenance program require retraining the workforce. This training is needed at all levels, from maintenance manager to the junior craftsperson in the organization.

4.2. VISUAL INSPECTION

Visual inspection was the first method used for predictive maintenance. Almost from the inception of the industrial revolution, maintenance technicians performed daily "walk-downs" of critical production and manufacturing systems in an attempt to identify potential failures or maintenance-related problems that could affect reliability, product quality, or production costs. A visual inspection still is a viable predictive maintenance tool and should be included in all total plant maintenance management programs.

4.3. VIBRATION ANALYSIS

Since most plants are composed of electromechanical systems, vibration monitoring is the primary predictive maintenance tool. Over the past 10 years, the majority of these programs have adopted the use of microprocessor-based, single-channel data collectors and Windows-based software to acquire, manage, trend, and evaluate the vibration energy created by these electromechanical systems. While this approach is a valuable predictive maintenance method, limitations within these systems may limit potential benefits.

There are several limitations of the computer-based systems, and some system characteristics, particularly simplified data acquisition and analysis, provide both advantages and disadvantages. While providing many advantages, simplified data acquisition and analysis also can be a liability. If the database is improperly configured, the automated capabilities of these analyzers will yield faulty diagnostics that can allow catastrophic failure of critical plant machinery.

Because technician involvement is reduced to a minimum level, the normal tendency is to use untrained or partially trained personnel for this repetitive function. Unfortunately, the lack of training results in less awareness and knowledge of visual and audible clues that can and should be an integral part of the monitoring program.

Most of the microprocessor-based vibrations monitoring systems collect single-channel, steady-state data that cannot be used for all applications. Single-channel data are limited to the analysis of simple machinery that operates at relatively constant speed.

While most microprocessor-based instruments are limited to a single input channel, in some cases, a second channel is incorporated in the analyzer. However, this second channel generally is limited to input from a tachometer or a once-per-revolution input signal. This second channel cannot be used to capture vibration data.

This limitation prohibits the use of most microprocessor-based vibration analyzers for complex machinery or machines with variable speeds. Single-channel data-acquisition technology assumes the vibration profile generated by a machine train remains constant throughout the data-acquisition process. This generally is true in applications where machine speed remains relatively constant (i.e., within 5 to 10 rpm). In this case, its use does not severely limit diagnostic accuracy and can be effective in a predictive maintenance program.

Most microprocessor-based instruments are designed to handle steady-state vibration data. Few have the ability to reliably capture transient events, such as rapid speed or load changes. As a result, their use is limited in situations where these occur.

In addition, vibration data collected with a microprocessor-based analyzer is filtered and conditioned to eliminate nonrecurring events and their associated vibration profiles. Antialiasing filters are incorporated into the analyzers specifically to remove spurious signals, such as impacts or transients. While the intent behind the use of antialiasing filters is valid, their use can distort a machine's vibration profile.

Because vibration data are dynamic and the amplitudes constantly change, as shown in Figure 4.1, most predictive maintenance system vendors strongly recommend averaging the data. They typically recommend acquiring 3–12 samples of the vibration profile and averaging the individual profiles into a composite signature. This approach eliminates the variation in vibration amplitude of the individual frequency components that make up the machine's signature. However, these variations, referred to as *beats*, can be a valuable diagnostic tool. Unfortunately, they are not available from microprocessor-based instruments because of averaging and other system limitations.

The most serious limitation created by averaging and the antialiasing filters are the inability to detect and record impacts that often occur within machinery. These impacts generally are indications of abnormal behavior and often are the key to detecting and identifying incipient problems.

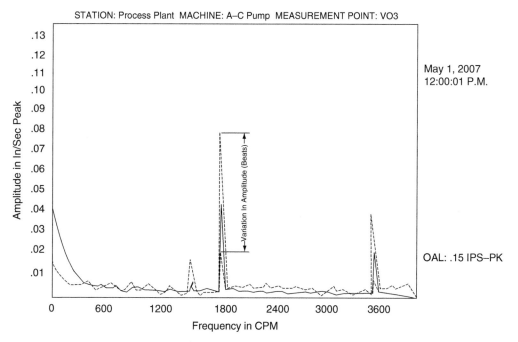

FIGURE 4.1. Frequency chart.

Most predictive maintenance programs rely almost exclusively on frequency-domain vibration data. The microprocessor-based analyzers gather time-domain data and automatically convert it using fast Fourier transform (FFT) to frequency-domain data. A frequency-domain signature shows the machine's individual frequency components, or peaks.

While frequency-domain data analysis is much easier to learn than time-domain data analysis, it does not allow isolating and identifying all incipient problems within the machine or its installed system. Because of this, additional techniques (e.g., time-domain, multi-channel, and real-time analysis) must be used in conjunction with frequency-domain data analysis to obtain a complete diagnostic picture.

Many of the microprocessor-based vibration monitoring analyzers cannot capture accurate data from low-speed machinery or machinery that generates low-frequency vibration. Specifically, some of the commercially available analyzers cannot be used where frequency components are below 600 cycles per minute or 10 Hertz. Two major problems restricting the ability to acquire accurate vibration data at low frequencies are electronic noise and the response characteristics of the transducer. The electronic noise of the monitored machine and the "noise floor" of the electronics within the vibration analyzer tend to override the actual vibration components found in low-speed machinery.

Analyzers especially equipped to handle noise are required for most industrial applications. At least three microprocessor-based analyzers capable of acquiring data below 600 cpm are commercially available. These systems use special filters and data-acquisition techniques to separate real vibration frequencies from electronic noise. In addition, transducers with the required low-frequency response must be used.

All machine trains are subject to random, nonrecurring vibration as well as periodic vibration. Therefore, it is advisable to acquire several sets of data and average them to eliminate the spurious signals. Averaging also improves the repeatability of the data, since only the continuous signals are retained.

Typically, a minimum of three samples should be collected for an average. However, the factor that determines the actual number is time. One sample takes 3–5 seconds, a four-sample average takes 12–20 seconds, and a 1000-sample average takes 50–80 minutes to acquire. Therefore, the final determination is the amount of time that can be spent at each measurement point.

In general, three or four samples are acceptable for good statistical averaging and keeping the time required per measurement point within reason. Exceptions to this include low-speed machinery, transient-event capture, and synchronous averaging.

Many of the microprocessor-based vibration monitoring systems offer the ability to increase their data-acquisition speed. This option is referred to as *overlap averaging*. While this approach increases speed, it generally is not recommended for vibration analysis. Overlap averaging reduces the accuracy of the data and must be used with caution. Its use should be avoided except where fast transients or other unique machine-train characteristics require an artificial means of reducing the data-acquisition and processing time.

When sampling time is limited, a better approach is to reduce or eliminate averaging altogether in favor of acquiring a single data block or sample. This reduces the acquisition time to its absolute minimum. In most cases, the single-sample time interval is less than the minimum time required obtaining two or more data blocks using the maximum overlap-averaging sampling technique. In addition, single-sample data are more accurate.

Perhaps the most serious diagnostic error made by typical vibration monitoring programs is the exclusive use of vibration-based failure modes as the diagnostic logic. For example, most of the logic trees state that, when the dominant energy contained in a vibration signature is at the fundamental running speed, a state of unbalance exists. While some forms of unbalance create this profile, the rules of machine dynamics clearly indicate that all failure modes on a rotating machine increase the amplitude of the fundamental or actual running speed.

Without a thorough understanding of machine dynamics, it is virtually impossible to accurately diagnose the operating condition of critical plant production systems. For example, gear manufacturers do not finish the backside (i.e., nondrive side) of gear teeth. Therefore, any vibration acquired from a gear set when it is braking is an order of magnitude higher than when it is operating on the power side of the gear. Another example is even more common. Most analysts ignore the effect of load on a rotating machine. However, when it is operating at full load, a centrifugal compressor may generate a vibration reading with an overall level of 0.1 ips-peak. The same measurement point would generate a reading in excess of 0.4 ips-peak when the compressor operates at 50% load. The difference is the spring constant being applied to the rotating element. The spring constant or stiffness at 100% load is twice that when operating at 50%. However, spring constant is a quadratic function. A reduction of 50% in the spring constant increases the vibration level by a factor of 4.

To achieve maximum benefits from vibration monitoring, it is imperative that the analyst understand the limitations of the instrumentation as well as the basic operating dynamics of machinery. Without this knowledge, the benefits are dramatically reduced. The greatest mistake made by traditional application of vibration monitoring is its application. Most programs limit the use of this predictive maintenance technology to simple rotating machinery and not on the critical production systems that produce the plant's capacity. As a result, the auxiliary equipment is kept in good operating condition, but the plant's throughput is unaffected.

Vibration monitoring is not limited to simple rotating equipment. The microprocessor-based systems used for vibration analysis can be used effectively on all electromechanical equipment, no matter how complex or what form the mechanical motion may take. For example, it can be used to analyze hydraulic and pneumatic cylinders that have purely linear motion. To accomplish this type of analysis, the analyst must use the time-domain function built in to these instruments. Proper operation of cylinders is determined by the time it takes for the cylinder to complete one complete motion. The time required for the cylinder to extend is shorter than its return stroke. This is a function of the piston area and inlet pressure. By timing the transient from fully retracted or extended to the opposite position, the analyst can detect packing leakage, scored cylinder walls, and other failure modes.

Vibration monitoring must be focused on the critical production systems. Each of these systems must be evaluated as a single machine and not as individual components. For example, a paper machine, annealing line, or any other production system must be analyzed as a complete machine, not as individual gearboxes, rolls, or other components. This methodology permits the analyst to detect abnormal operation within the complex system. Problems such as tracking, tension, and product quality deviations can be easily detected and corrected using this method of analysis.

When properly used, vibration monitoring and analysis is the most powerful predictive maintenance tool available. It must be focused on critical production systems, not simple rotating machinery. Diagnostic logic must be driven by the operating dynamics of machinery, not simplified vibration failure modes.

The proof is in the results. The survey conducted by *Plant Services* in July 1999 indicated that less than 50% of the vibration monitoring programs generated enough quantifiable benefits to offset the recurring cost of the program. Only 3% generated a return on investment of 5%. When properly used, vibration-based predictive maintenance can generate a return on investment of 100–1 or better.

4.4. THERMOGRAPHY

Thermography is a predictive maintenance technique that can be used to monitor the condition of plant machinery, structures, and systems, not just electrical equipment. It uses instruments designed to monitor the emission of infrared energy (i.e., surface temperature) to determine their operating condition. By detecting thermal anomalies (i.e., areas hotter or colder than they should be), an experienced technician can locate and determine a multitude of incipient problems within the plant.

Infrared technology is predicated on the fact that all objects having a temperature above absolute zero emit energy or radiation. Infrared radiation is one form of this emitted energy. Infrared, or below red, emissions are the shortest wavelengths of all radiated energy and are invisible without special instrumentation. The intensity of infrared radiation from an object is a function of its surface temperature. However, temperature measurement using infrared methods is complicated because three sources of thermal energy can be detected from any object: energy emitted from the object itself, energy reflected from the object, and energy transmitted by the object. Only the emitted energy is important in a predictive maintenance program. Reflected and transmitted energies distort raw infrared data. Therefore, the reflected and transmitted energies must be filtered out of acquired data before a meaningful analysis can be completed.

Variations in surface condition, paint, or other protective coatings and many other variables can affect the actual emissivity factor for plant equipment. In addition to reflected and transmitted energy, the user of thermographic techniques also must consider the atmosphere between the object and the measurement instrument. Water vapor and other gases absorb infrared radiation. Airborne dust, some lighting, and other variables in the surrounding atmosphere can distort measured infrared radiation. Since the atmospheric environment constantly changes, using thermographic techniques requires extreme care each time infrared data are acquired.

Most infrared monitoring systems or instruments provide filters that can be used to avoid the negative effects of atmospheric attenuation of infrared data. However, the plant user must recognize the specific factors that affect the accuracy of the infrared data and apply the correct filters or other signal conditioning required to negate that specific attenuating factor or factors.

Collecting optics, radiation detectors, and some form of indicator are the basic elements of an industrial infrared instrument. The optical system collects radiant energy and focuses it on a detector, which converts it into an electrical signal. The instrument's electronics amplifies the output signal and processes it into a form that can be displayed. Three types of instruments generally are used as part of an effective predictive maintenance program: infrared thermometers, line scanners, and infrared imaging systems:

- Infrared thermometers or spot radiometers are designed to provide the actual surface temperature at a single, relatively small point on a machine or surface. Within a predictive maintenance program, the point-of-use infrared thermometer can be used in conjunction with many of the microprocessor-based vibration instruments to monitor the temperature at critical points on plant machinery or equipment. This technique typically is used to monitor bearing cap temperatures, motor winding temperatures, spot checks of process piping temperatures, and similar applications. It is limited in that the temperature represents a single point on the machine or structure. However, when used in conjunction with vibration data, point-of-use infrared data can be a valuable tool.
- Line scanners provide a single dimensional scan or line of comparative radiation. While this type of instrument provides a somewhat larger field of view, that is, the area of machine surface, it is limited in predictive maintenance applications.
- Infrared imaging provides the means to scan the infrared emissions of complete machines, processes, or equipment in a very short time. Most of the imaging systems function much like a video camera. The user can view the thermal emission profile of a wide area by simply looking through the instrument's optics.

A variety of thermal imaging instruments are on the market, ranging from relatively inexpensive black and white scanners to full color, microprocessor-based systems. Many of the less expensive units are designed strictly as scanners and do not provide the capability to store and recall thermal images. This inability to store and recall previous thermal data limits a long-term predictive maintenance program.

Training is critical with any of the imaging systems. The variables that can destroy the accuracy and repeatability of thermal data must be compensated for each time infrared data are acquired. In addition, interpretation of infrared data requires extensive training and experience.

Inclusion of thermography into a predictive maintenance program enables monitoring the thermal efficiency of critical process systems that rely on heat transfer or retention, electrical equipment, and other parameters that improve both the reliability and efficiency of plant systems. Infrared techniques can be used to detect problems in a variety of plant systems and equipment, including electrical switchgear, gearboxes, electrical substations, transmissions, circuit breaker panels, motors, building envelopes, bearings, steam lines, and process systems that rely on heat retention or transfer.

Equipment included in an infrared thermography inspection usually is energized. For this reason, a lot of attention must be given to safety. The following are basic rules for safety while performing an infrared inspection:

- Plant safety rules must be followed at all time. A safety person must be used at all times. Because proper use of infrared imaging systems requires the technician to use a viewfinder, similar to a video camera, to view the machinery to be scanned, he or she is blind to the surrounding environment. Therefore, a safety person is required to assure safe completion. Notify area personnel before entering the area for scanning. Qualified electricians from the area should be assigned to open and close all electrical panels.
- Where safe and possible, all equipment to be scanned will be on-line and under normal load with a clear line of sight to the item. Assets with covers that are interlocked without an interlock defeat mechanism should be shut down when allowable. If safe, their control covers should be opened and equipment restarted.

When used correctly, thermography is a valuable predictive maintenance and reliability tool. However, the derived benefits are directly proportional to how it is used. If it is limited to annual surveys of roofs or quarterly inspections of electrical systems, the resultant benefits are limited. When used to regularly monitor all critical process or production systems where surface temperature or temperature distribution is indicative of its reliability or operating condition, thermography can yield substantial benefits. To gain the maximum benefits from the investment in infrared systems, use it full power. Concentrate the program on those critical systems that generate capacity in the plant.

4.5. TRIBOLOGY

Tribology is the general term that refers to design and operating dynamics of the bearing-lubrication-rotor support structure of machinery. Two primary techniques are used for predictive maintenance: lubricating oil analysis and wear particle analysis.

Lubricating oil analysis, as the name implies, is an analysis technique that determines the condition of lubricating oils used in mechanical and electrical equipment. It is not a tool for determining the operating condition of machinery or detecting potential failure modes. Too many plants attempt to accomplish the latter and are disappointed in the benefits derived. Simply stated, lube oil analysis should be limited to a proactive program to conserve the use and extend the life of lubricants. While some forms of lubricating oil analysis may provide an accurate quantitative breakdown of individual chemical elements, both oil additives and contaminates, the technology cannot be used to identify the specific failure mode or root cause of incipient problems within the machines serviced by the lube oil system.

The primary applications for lubricating oil analysis are quality control, reduction of lubricating oil inventories, and determination of the most cost-effective interval for oil change. Lubricating, hydraulic, and dielectric oils can be analyzed periodically, using these techniques, to determine their condition. The results of this analysis can be used to determine if the oil meets the lubricating requirements of the machine or application. Based on the results of the analysis, lubricants can be changed or upgraded to meet the specific operating requirements.

In addition, detailed analysis of the chemical and physical properties of different oils used in the plant, in some cases, can allow consolidation or reduction of the number and types of lubricants required to maintain plant equipment. Elimination of unnecessary duplication can reduce required inventory levels and therefore maintenance costs.

As a predictive maintenance tool, lubricating oil analysis can be used to schedule oil change intervals based on the actual condition of the oil. In middle to large plants, a reduction in the number of oil changes can amount to a considerable annual reduction in maintenance costs. Relatively inexpensive sampling and testing can show when the oil in a machine has reached the point that warrants change.

Wear particle analysis is related to oil analysis only in that the particles to be studied are collected through drawing a sample of lubricating oil. Where lubricating oil analysis determines the actual condition of the oil sample, wear particle analysis provides direct information about the wearing condition of the machine train. Particles in the lubricant of a machine can provide significant information about the condition of the machine. This information is derived from the study of particle shape, composition, size, and quantity.

Two methods are used to prepare samples of wear particles. The first method, called *spectroscopy* or *spectrography*, uses graduated filters to separate solids by size. Normal spectrographic analysis is limited to particulate contamination with a size of 10 microns or less. Larger contaminants are ignored, which can limit the benefits that can be derived from the technique. The second method, called *ferrography*, separates wear particles using a magnet. Obviously, the limitation to this approach is that only magnetic particles are removed for analysis. Nonmagnetic materials, such as copper or aluminum, that constitute many of the wear materials in typical machinery therefore are excluded from the sample.

Wear particle analysis is an excellent failure analysis tool and can be used to understand the root cause of catastrophic failures. The unique wear patterns observed on failed parts, as well as those contained in the oil reservoir, provide a positive means of isolating the failure mode.

Three major issues affect using tribology analysis in a predictive maintenance program: equipment costs, acquiring accurate oil samples, and interpretation of data.

The capital cost of spectrographic analysis instrumentation normally is too high to justify in-plant testing. The typical cost for a microprocessor-based spectrographic system is between $30,000 and $60,000. Because of this, most predictive maintenance programs rely on third party analysis of oil samples. In addition to the labor cost associated with regular gathering of oil and grease samples, simple lubricating oil analysis by a testing laboratory ranges from about $20 to $50 per sample. Standard analysis normally includes viscosity, flash point, total insolubles, total acid number (TAN), total base number (TBN), fuel content, and water content. More detailed analysis, using spectrographic, ferrographic, or wear particle techniques, which include metal scans, particle distribution (size), and other data can range to well over $150 per sample.

A more severe limiting factor with any method of oil analysis is acquiring accurate samples of the true lubricating oil inventory in a machine. Sampling is not a matter of opening a port somewhere in the oil line and catching a pint sample. Extreme care must be taken to acquire samples that truly represent the lubricant that passes through the machine's bearings.

One recent example is an attempt to acquire oil samples from a bull gear compressor. The lubricating oil filter had a sample port on the clean (i.e., downstream) side. However, comparison of samples taken at this point and one taken directly from the compressor's oil reservoir indicated that more contaminates existed downstream from the filter than in the reservoir. Which location actually represented the oil's condition? Neither sample was truly representative of the oil condition. The oil filter had removed most of the suspended solids (i.e., metals and other insolubles) and therefore was not representative of the actual condition. The reservoir sample was not representative since most of the suspended solids had settled out in the sump.

Proper methods and frequency of sampling lubricating oil is critical to all predictive maintenance techniques that use lubricant samples. Sample points consistent with the objective of detecting large particles should be chosen. In a recirculation system, samples should be drawn as the lubricant returns to the reservoir and before any filtration. Do not draw oil from the bottom of a sump, where large quantities of material build up over time. Return lines are preferable to reservoir as the sample source, but good reservoir samples can be obtained if careful, consistent practices are used. Even equipment with high levels of filtration can be effectively monitored, as long as samples are drawn before oil enters the filters. Sampling techniques involve taking samples under uniform operating conditions. Samples should not be taken more than 30 minutes after the equipment has been shut down.

Sample frequency is a function of the mean time to failure from the onset of an abnormal wear mode to catastrophic failure. For machines in critical service, sampling every 25 hours of operation is appropriate. However, for most industrial equipment in continuous service, monthly sampling is adequate. The exception to monthly sampling is machines with extreme loads. In this instance, weekly sampling is recommended.

Understanding the meaning of analysis results is perhaps the most serious limiting factor. Most often results are expressed in terms totally alien to plant engineers or technicians. Therefore, it is difficult for them to understand the true meaning, in terms of oil or machine condition. A good background in quantitative and qualitative chemistry is beneficial. At a minimum, plant staff require training in basic chemistry and specific instruction on interpreting tribology results.

4.6. ULTRASONICS

Ultrasonics, like vibration analysis, is a subset of noise analysis. The only difference in the two techniques is the frequency band they monitor. In the case of vibration analysis, the monitored range is between 1 and 30,000 Hz; ultrasonics monitors noise frequencies above 30,000 Hz. These higher frequencies are useful for select applications, such as detecting leaks that generally create high-frequency noise caused by the expansion or compression of air, gases, or liquids as they flow through the orifice or leak in either pressure or vacuum vessels. These higher frequencies also are useful in measuring the ambient noise levels in various areas of the plant.

As applied as part of a predictive maintenance program, many companies are attempting to replace what is perceived as an expensive tool (i.e., vibration analysis) with ultrasonics. For example, many plants use ultrasonic meters to monitor the health of rolling-element bearings in the belief that this technology provides accurate results. Unfortunately, the perception is invalid. Since this technology is limited to a broadband (i.e., 30 kHz to 1 MHz), ultrasonics does not provide the ability to diagnose incipient bearing or machine problems. It certainly cannot specify the root cause of abnormal noise levels generated by either bearings or other machine-train components.

As part of a comprehensive predictive maintenance program, ultrasonics should be limited to the detection of abnormally high ambient noise levels and leaks. Attempting to replace vibration monitoring with ultrasonics simply does not work.

5

Reliability Processes

5.1. RELIABILITY SOFTWARE—MANAGING THE HEALTH OF THE ASSETS

Studies originally conducted in the airline industry and subsequently validated across other industries have shown that approximately 80% of all mechanical, electrical, and structural failures are random in nature and cannot be effectively correlated to time or operating hours. While, for decades, most industrial maintenance organizations have relied on time-based preventive maintenance (PM) tasks, the recognition that most failures are random has caused many companies to reevaluate their proactive maintenance programs. Invariably, the conclusion is that, if time-based maintenance represents the majority of proactive activity, then the wrong work is being done. And it is not just a question of using a tool such as Weibull analysis to fine-tune the timing of age-related PM tasks. Instead, leading companies are learning to develop the right kinds of proactive tasks to manage random failures.

Developing the right proactive work program is not simple, but many well-documented proactive methodologies—reliability centered maintenance (RCM), failure modes and effect analysis (FMEA), and maintenance task analysis (MTA)—can assist firms in identifying the right work. These approaches look at all the ways an asset can fail (failure modes) and take a different approach to failure management. Rather than attempting to use time-based tasks to manage the asset, the new approach to maintenance is focused on mitigating the consequence of failure at the failure mode level. This approach looks at each failure mode and determines the best proactive task or tasks to detect failure or prevent its consequences. Done properly, the result will be a high percentage (>80%) of tasks that require some form of condition monitoring and a much lower percentage (<20%) that rely on time-based tasks or tasks related to operating age. In addition, the failure analysis identifies the corrective work to be performed when early signs of failure are detected.

Condition monitoring tasks, driven by an understanding of failure modes, create a picture of equipment health from visual inspections, the appropriate use of predictive technology (thermography, vibration, nondestructive testing, etc.), and online equipment data (pressure, temperature, flow, amps, etc.). These condition monitoring activities generate massive amounts of data related to the health of the equipment. To be of real value to maintenance and operations, the data must be effectively centralized, analyzed, and compared against predefined "normal" states, allowing users to focus on just the abnormal data. Done effectively, this management of condition information will lead to dramatic improvements in asset reliability.*

Reliability software was developed to act as a single application with which to develop the complete asset reliability program—the list of all proactive tasks for optimal maintenance of each asset. In addition, reliability software is a tool to implement an asset reliability program, by creating a single place for monitoring the effectiveness of the program, collecting,

*The remainder of Section 5.1 was prepared in conjunction with Ivara Corporation of Burlington, Ontario, Canada, which provided the information and the graphics.

analyzing, and displaying all asset condition data. The software supports developing and managing the program and all the sources of data needed to manage the health of an asset.

5.1.1. Building an Effective Asset Reliability Program

Effective reliability software provides a tool to capture the entire process of identifying the right work to proactively maintain assets. Many environments today easily can make the transition from the current informal work identification (work ID) process (Figure 5.1) to a process that is more formalized. In a typical informal work ID process, work often is identified through operations-initiated work requests (signaling a failure that already occurred), manufacturer's generic time-based task suggestion, work that has always been done but for which the justification is nonexistent or unclear, as well as some level of predictive technology. If any RCM has been done, the results of the analysis usually ends up in a binder sitting on a shelf. These environments often have paper records of condition inspections being conducted; but they rarely lead to new work being identified, instead they pile up and serve only as an after-the-fact reminder that a failure could have been prevented.

Reliability software helps the development a more formal work identification process (Figure 5. 2), so that all proactive work is tied directly to a failure mode. Reliability software should be seen as a tool also used to systematically assign priorities to assets, document the work ID analysis, create the asset reliability program along with health indicators and alarm levels, and specify the recommended proactive corrective actions.

5.1.2. Using Reliability Software to Put the Program into Action

Whether the correct proactive work was identified through a formal reliability centered maintenance technique or some other methodology, such as FMEA

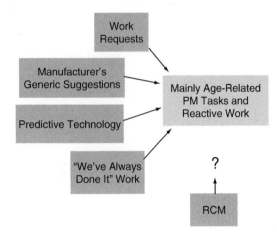

FIGURE 5.1. Typical informal work identification process. (Courtesy of Ivara Corporation.)

or MTA, software is available that has the technology to put a proactive asset reliability program into action. This is important because, typically, when a formal or informal analysis has been performed and a maintenance task identified, additional time and effort are needed to implement the new task. Reliability software records each task in the new program, linking all condition monitoring tasks directly to the indicators to be checked. From there, check sheets instruct the operations and maintenance personnel to conduct the inspections, and through handheld devices, current data are collected and fed to the reliability software. This software provides a single place from which to view and respond to any abnormal data that triggered alarms.

A sound failure analysis typically determines that about 33% of failure modes should be allowed to run to failure, while recommending redesign for about 4% of all failure modes. The remaining failure modes are the focus of the asset reliability program, which is managed within reliability software. Over 80% of the tasks in a well-defined asset reliability program are condition/state inspection tasks, while less than 20% of the tasks are age based (Figure 5.3). While all CMMS products can adequately handle trigger-

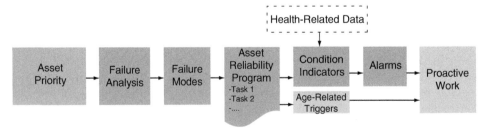

FIGURE 5.2. How reliability software supports the work identification process. (Courtesy of Ivara Corporation.)

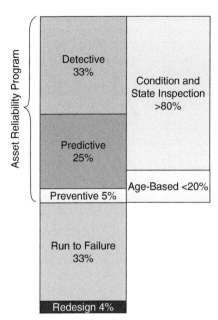

FIGURE 5.3. Breakdown of failure modes and asset reliability tasks. (Courtesy of Ivara Corporation.)

ing tasks based on operating age, only reliability software is designed to handle both age-based and condition inspection tasks.

As condition monitoring tasks are defined, appropriate indicators are set up for the inspections, along with defined normal and abnormal states. For a temperature reading, for example, the normal state may be defined by a range of temperature values. Various levels of abnormal states also may be defined to correspond to increasingly unacceptable temperature ranges. With each state, or temperature range in this case, a user-defined alarm level can be specified. As temperature readings are recorded, the reliability software compares the new value to the normal and abnormal states and triggers an alarm when abnormal values are recorded.

Reliability software collects condition and state data from a variety of sources, including visual inspections, predictive maintenance technologies, process controls, sensors, and data historians. Data from predictive technologies such as vibration analyses, thermography, and oil analysis also are utilized. Several kinds of condition indicators are tracked, including

- Simple numeric values.
- Qualitative or descriptive information, with user-defined value lists.
- Mathematical calculations.
- Rule-based configurations.

Reliability software uses single or multiple data points, applying rules and calculations to create a true picture of equipment health.

With the transition to an effective asset reliability program, maintenance and operations personnel collect and manage an increasing number of condition-based proactive tasks. In this environment, it is critical that the newly recorded readings be utilized immediately. The reliability software helps out with this data management challenge by sorting through the normal and abnormal data and displaying the results in ways that are easy to understand and utilize. Rather than requiring users to sift through piles of paper-based inspection readings, as abnormal values are recorded, alarms are triggered and displayed, drawing attention to only the few data points that currently signal the potential for equipment failure.

The plant, all of its assets, and failure-mode-specific health indicators can be displayed in two ways. The first display method uses an indicator panel, a two-panel screen showing the entire plant hierarchy and all assets on the left side (Figure 5.4) and relevant health indicators on the right side. The reliability software equipment hierarchy must align with the current enterprise asset management and computerized maintenance management systems (EAM/CMMS).

The indicator panel in reliability software allows monitoring asset condition and, at a glance, seeing any indications of impending failures, before the failures occur. Flashing alarms are displayed as assets move closer to functional failure and alarm severities are readily understood based on the type of icon displayed. Corrective maintenance decisions can be made based on asset health and risk to the business, so that the right work can be performed at the right time.

The second way to display asset health indicators is through graphics, photographs, and diagrams. As abnormal data trigger alarms, the indicators begin to flash on the drawing. And, since the alarms roll up the equipment hierarchy, a user could easily drill down from a higher level picture to the compressor, in this case, to quickly zero in on the flashing alarm and abnormal data. This capability makes it very easy for users to respond to data immediately as it is updated, without the need to review all data.

5.1.3. Using Handheld Devices to Collect and Upload Condition Inspection Data

The software replaces the manual paper-based approach to collecting, storing, and analyzing condition data (Figure 5.5), allowing inspection routes to be

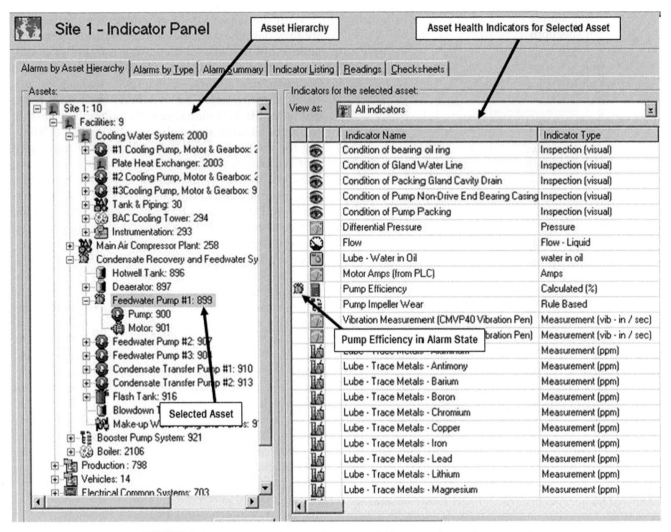

FIGURE 5.4.   Reliability software indicator panel. (Courtesy of Ivara Corporation.)

FIGURE 5.5   Tradesman performing manual calculations, which sometimes takes days. (Courtesy of Ivara Corporation.)

automated using simple handheld devices. In the case of subjective inspections, operators are presented with a predefined list of observation values from which to choose, making the condition data more quantifiable and useful (Figure 5.6).

5.1.4. Plotting Asset Health Trends

Many companies have collected condition readings for years but lacked the tools to manage the data properly. Charting asset health indicators allows trends in asset condition to be easily noticed. Bands of color (shading here) graphically show alarm severity ranges (Figure 5.7).

This graphing capability dramatically improves management's ability to proactively intervene to ensure that asset health is maintained and reliability is maximized.

5.1.5. Capturing the Experts' Knowledge about Asset Condition

Reliability software captures the knowledge of the equipment experts, the operators and maintainers who know the equipment best. In many companies, these employees have worked with the equipment daily for decades, so their knowledge is invaluable. The challenge is to find a way to store this information, so that all employees can take advantage of it for their daily work. Reliability software captures this knowledge and makes it available.

During the failure analysis, the maintainers and operators for the target asset are asked to contribute their knowledge of the ways the asset fails and the ways it found for detecting or preventing failure. The condition monitoring detail previously carried around in heads and personal pocket books becomes some of the critical knowledge stored in software, in the new asset reliability program. These employees usually know exactly how the equipment operates and how best to perform the required condition checks. In the context of a well-defined failure analysis, this knowledge is captured and formalized by linking the proactive tasks to specific failure modes and gaining agreement between operations and maintenance that the work being done is correct. For example, the reliability software screen shown in Figure 5.8 captures the calculation to determine the effectiveness of a heat exchanger. No longer must someone remember the engineering calculation, since the reliability software stores the expression, making it permanently available for all to use.

A reliability software program combines data from various indicators to determine the overall effectiveness of the heat exchanger and, when an abnormal value is found, prompts the user with the predetermined corrective action. Visual or other sensory inspections are logged via handheld data recorders (PDAs). Abnormal readings trigger alarms and follow-up work tasks to suggest more rigorous inspections or corrective work.

FIGURE 5.6. Tradesman collecting readings using the reliability software on a pocket PC, with instantaneous feedback on alarm conditions. (Courtesy of Ivara Corporation.)

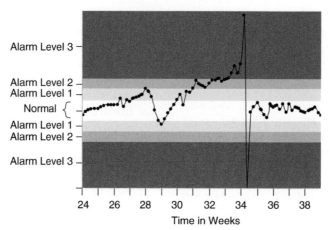

FIGURE 5.7.   Condition indicator graph. (Courtesy of Ivara Corporation.)

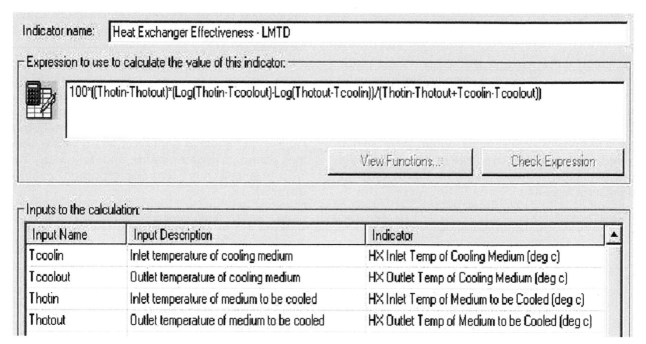

FIGURE 5.8.   Reliability software displays maintenance experts' knowledge and eliminates manual calculations. (Courtesy of Ivara Corporation.)

5.1.6. Integration to Enterprise Asset Management and Computerized Maintenance Management Systems

Asset reliability software helps companies to leverage and extend the benefits they receive from current investments in any EAM/CMMS. Maintenance organizations need to incorporate predictive and proactive, condition-based maintenance capabilities up front to monitor asset conditions and help maintenance personnel keep reliability and productivity on target. Reliability software complements EAM systems to make them more effective because they address up front the need for work identification and processes not addressed by the EAM (Figure 5.9).

While EAM/CMMS supports the maintenance control process (work planning, scheduling, and execution), reliability software supports the complete equipment reliability process.

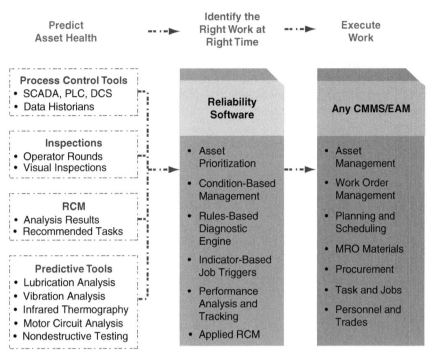

FIGURE 5.9. Data collection, dissemination, and execution. (Courtesy of Ivara Corporation.)

5.1.7. The Bottom Line

Achieving and sustaining improved levels of reliability through the monitoring and managing asset health helps manage maintenance costs and asset reliability more effectively. Asset reliability improves output and uptime, creates higher service levels, creates a safe environment, and enables companies to comply with environmental regulations. Using reliability technology for managing the health of a company's assets and protecting these assets ensures business goals are met and companies stay competitive.

5.2. SEVEN QUESTIONS ADDRESSED BY RELIABILITY CENTERED MAINTENANCE

An RCM process systematically identifies all of the asset's functions and functional failures and all reasonably likely failures and their causes. It then proceeds to identify the effects of these likely failure modes and in what way those effects matter. Once it has gathered this information, the RCM process then selects the most appropriate asset management policy.

RCM considers all asset management options: on-condition task, scheduled restoration task, scheduled discard task, failure-finding task, and one-time change (to hardware design, operating procedures, personnel training, or other aspects of the asset outside the strict world of maintenance). This consideration is unlike other maintenance development processes. Fundamentally, the RCM process seeks to answer the following seven questions in sequential order.

Functions

What functions and standards of performance are expected of the asset in its present operating context? The specific criteria that the process must satisfy are

- The operating context of the asset should be defined.
- All the functions of the asset or system should be identified (all primary and secondary functions, including the functions of all protective devices).
- All function statements should contain a verb, an object, and a performance standard (quantified in every case where this can be done).
- Performance standards incorporated in function statements should be the level of performance desired by the owner or user of the asset or system in its operating context.

The operating context is the circumstance in which the asset is operated. The same hardware does not always require same failure management policy in all installations. For example, a single pump in a system usually needs a different failure management policy from a pump that is one of several redundant units in a system. A pump moving corrosive fluids usually needs a different policy from a pump moving benign fluids. Protective devices often are overlooked; an RCM process should ensure that their functions are identified. Finally, the owner or user should dictate the level of performance the maintenance program is designed to sustain.

Functional Failures

In what ways can the asset fail to fulfill its functions? This question has only one specific criterion: All the failed states associated with each function should be identified. If functions are well defined, listing functional failures is relatively easy. For example, if a function is to "keep system temperature between 50 and 70°C," then functional failures might include inability to raise the system temperature above ambient or keep it above 50°C or below 70°C.

Failure Modes

What causes each functional failure? In failure modes, effects, and criticality analysis (FMECA), the term *failure mode* is used in the way that RCM uses the term *functional failure*. However, the RCM community uses the term *failure mode* to refer to the event that causes functional failure.

As a criterion, all failure modes reasonably probable to cause each functional failure should be identified. The method used to decide what constitutes a "reasonably probable" failure mode should be acceptable to the owner or user of the asset. Also, failure modes should be identified at a level of causation that makes it possible to identify an appropriate failure management policy. That is, lists of failure modes should include

- Failure modes that happened before, currently are being prevented by existing maintenance programs, and have yet to happen but are thought reasonably likely (credible) in the operating context.
- Any event or process likely to cause a functional failure, including deterioration, human error whether caused by operators or maintainers, and design defects.

RCM is the most thorough of the analytic processes that develop maintenance programs and manage physical assets. It therefore is appropriate for RCM to identify every reasonably likely failure mode.

Failure Effects

What happens when each of the failures occur? The criteria for identifying failure effects are

- Failure effects should describe what would happen if no specific task were done to anticipate, prevent, or detect the failure.
- Failure effects include all the information needed to support the evaluation of the consequences of the failure, such as
  - What evidence (if any) is there that the failure has occurred (in the case of hidden functions, what would happen if a multiple failure occurred)?
  - What (if anything) does the failure do to kill or injure someone or have an adverse effect on the environment?
  - What (if anything) does the failure do to have an adverse effect on production or operations?
  - What physical damage (if any) is caused by the failure?
  - What (if anything) must be done to restore the function of the system after the failure?

FMECA or FMEA usually describes failure effects in terms of the effects at the local level, the subsystem level, and the system level.

Failure Consequences

In what way does each failure matter? The standard's criteria for a process that identifies failure consequences are

- The assessment of failure consequences should be carried out as if no specific task currently is being performed to anticipate, prevent, or detect the failure.
- The consequences of every failure mode is formally categorized as follows. The consequence categorization separates hidden failure modes from evident failure modes. The process should clearly distinguish events (failure modes and multiple failures) that have safety or environmental consequences from those that have only economic consequences (operational and nonoperational consequences).

RCM assesses failure consequences as if nothing is being done about the failure. Some people are tempted to say, "Oh, that failure doesn't matter because we always do (something), which protects us from it." However, RCM is thorough: It checks the assumption that this action that "we always do" actually does protect the assets from failure, and it checks the assumption that this action is worth the effort.

RCM assesses failure consequences by formally assigning each failure mode into one of four categories: hidden, evident safety or environmental, evident operational, and evident nonoperational failures. The explicit distinction between hidden and evident failures, performed at the outset of consequence assessment, is one characteristic that clearly distinguishes RCM, as defined by Stan Nowlan and Howard Heap, from MSG-2 and earlier U.S. civil aviation processes.

Proactive Tasks

What should be done proactively to predict or prevent each failure? This is a complex topic, and so its criteria are presented in two groups. The first group pertains to the overall topic of selecting failure management policies. The second group of criteria pertains to scheduled tasks and intervals, which comprise proactive tasks as well as one default action (failure-finding task). The criteria for selecting failure management policies are

- The selection of failure management policies is carried out as if no specific task currently is being done to anticipate, prevent, or detect the failure.
- The selection process takes into account that the conditional probability of some failure modes increases with age (or exposure to stress), the conditional probability of others does not change with age, and the conditional probability of still others decreases with age.
- All scheduled tasks are technically feasible and worth doing (applicable and effective), and the means by which this requirement is satisfied are set out under scheduled tasks in the failure management section.
- If two or more proposed failure management policies are technically feasible and worth doing (applicable and effective), the policy that is most cost effective is selected.

Scheduled tasks are those performed at fixed, predetermined intervals, including continuous monitoring (where the interval is effectively zero). Scheduled tasks should be identified that fit the following criteria:

- In the case of an evident failure mode that has safety or environmental consequences, the task should reduce the probability of the failure mode to a level tolerable to the owner or user of the asset.
- In the case of a hidden failure mode where the associated multiple failure has safety or environmental consequences, the task should reduce the probability of the hidden failure mode to an extent that reduces the probability of the associated multiple failure to a level tolerable to the owner or user of the asset.
- In the case of an evident failure mode that has no safety or environmental consequences, the direct and indirect costs of doing the task should be less than the direct and indirect costs of the failure mode when measured over comparable periods of time. In this case, the direct and indirect costs of doing the task

also should be less than the direct and indirect costs of the multiple failure plus the cost of repairing the hidden failure mode when measured over comparable periods of time.

Three general categories of tasks are considered to be proactive in nature: on-condition tasks, scheduled discard tasks, and scheduled restoration tasks.

An on-condition task is a scheduled task used to detect a potential failure. Such a task has many other names in the maintenance community, such as *predictive* tasks (in contrast to preventive tasks, a name these people apply to scheduled discard and scheduled restoration tasks), *condition-based* tasks (referring to condition-based maintenance, again, in contrast to time-based maintenance or scheduled discard and scheduled restoration tasks), and *condition-monitoring* tasks (since the tasks monitor the condition of the asset).

The scheduled discard task is a scheduled task that entails discarding an item at or before a specified age limit regardless of its condition at the time. A scheduled discard task must be subjected to the following preconditions: a clearly defined (preferably a demonstrable) age at which there is an increase in the conditional probability of the failure mode under consideration or a sufficiently large proportion of the occurrences of this failure mode after this age to reduce the probability of premature failure to a level that is tolerable to the owner or user of the asset.

The scheduled restoration task is a scheduled task that restores the capability of an item at or before a specified interval (age limit), regardless of its condition at the time, to a level that provides a tolerable probability of survival to the end of another specified interval. In addition to the preconditions for a scheduled discard task, the following criterion must apply to a scheduled restoration task: The task should restore the resistance to failure (condition) of the component to a level that is acceptable to the owner or user of the asset.

Default Actions

What should be done if a suitable proactive task cannot be found? This question pertains to unscheduled failure management policies: the decision to let an asset run to failure and the decision to change something about the asset's operating context (such as its design or the way it is operated).

A failure-finding task is a scheduled task used to determine whether a specific hidden failure has occurred. Failure-finding tasks usually apply to protective devices that fail without notice. This task represents a transition from the question about proactive tasks to the one about default actions or actions taken in the absence of proactive tasks. Failure-finding tasks are scheduled like the proactive tasks. However, failure-finding tasks are not proactive. They do not predict or prevent failures. They detect failures that already happened, to reduce the chances of a multiple failure and the failure of a protected function while a protective device is already in a failed state.

If a process offers a decision to let an asset run to failure, the following criteria should be applied before accepting the decision. In cases where the failure, hidden or evident, has no appropr ate scheduled task, the associated multiple failure should have no safety or environmental consequences. In other words, the process must not allow its users to select "run to failure" if the failure mode or, in the case of a hidden failure, the associated multiple failure, has safety or environmental consequences.

5.3. FAILURE MODE AND EFFECTS ANALYSIS*

The FMEA (failure mode and effects analysis) is generally recognized as the most fundamental tool employed in reliability engineering. Because of its practical, qualitative approach, it is also the most widely understood and applied form of reliability analysis that we encounter throughout industry. Additionally, the FMEA forms the headwaters for virtually all subsequent reliability analyses and assessments because it forces an organization to systematically evaluate equipment and system weaknesses and their interrelationships that can lead to product unreliability.

But before we proceed to discuss the FMEA process, we feel it is important to address a semantics issue that often arises in this discussion. To put it most succinctly, failure to define failure can lead to some unfortunate misunderstandings.

For as long as we can recall, there have been varying degrees of confusion about what people mean when they use terminology that involves the word *failure*. Failure is an unpleasant word, and we often use substitute words such as anomaly, defect, discrepancy, irregularity, etc., because they tend to sound less threatening or less severe.

The spectrum of interpretations for failure runs from negligible glitch to catastrophe. Might we suggest that the meaning is really quite simple:

*Section 5.3 is taken from Anthony M. Smith and Glenn R. Hinchcliffe, *RCM—Gateway to World Class Maintenance* (Burlington, MA: Elsevier Butterworth–Heinemann, 2004), pp. 49–54.

Failure is the inability of a piece of equipment, a system, or a plant to meet its expected performance.

This expectation is always spelled out in a specification in our engineering world and, when properly written, leaves no doubt as to exactly where the limits of satisfactory performance reside. So, failure is the inability to meet specifications. Simple enough, we believe, to avoid much of the initial confusion.

Additionally, there are several important and frequently used phrases that include the word failure: failure symptom, failure mode, failure cause, and failure effect.

Failure symptom. This is a *tell-tale indicator* that alerts us (usually the operator) to the fact that a failure is about to exist. Our senses or instruments are the primary source of such an indication. Failure symptoms may or may not tell us exactly where the pending failure is located or how close to the full failure condition we might be. In many cases, there is no failure symptom (or warning) at all. Once the failure has occurred, any indication of its presence is no longer a symptom—we now observe its effect.

Failure mode. This is a brief description of *what is wrong*. It is extremely important for us to understand this simple definition because, in the maintenance world, it is the failure mode that we try to prevent, or, failing that, what we have to physically fix. There are hundreds of simple words that we use to develop appropriate failure mode descriptions: jammed, worn, frayed, cracked, bent, nicked, leaking, clogged, sheared, scored, ruptured, eroded, shorted, split, open, torn, and so forth. The main confusion here is clearly to distinguish between failure mode and failure cause—and understanding that failure mode is what we need to prevent or fix.

Failure cause. This is a brief word description of *why it went wrong*. Failure cause is often very difficult to fully diagnose or hypothesize. If we wish to attempt a permanent prevention of the failure mode, we usually need to understand its cause (thus the term, root cause failure analysis). Even though we may know the cause, we may not be able to totally prevent the failure mode—or it may cost too much to pursue such a path. As a simple illustration, a gate valve jams "closed" (failure mode), but why did this happen? Let's say that this valve sits in a very humid environment—so "humidity-induced corrosion" is the failure cause. We could opt to replace the valve with a high-grade stainless steel model that would resist (perhaps stop) the corrosion (a design fix), or, from a maintenance point of view, we could periodically lubricate and operate the valve to mitigate the corrosive effect, but there is nothing we can do to eliminate the natural humid environment. Thus, PM tasks cannot fix the cause—they can address only the mode. This is an important distinction to make, and many people do not clearly understand this distinction.

Failure effect. Finally, we briefly describe the *consequence of the failure mode* should it occur. To be complete, this is usually done at three levels of assembly—local, system, and plant. In describing the effect in this fashion, we clearly see the buildup of consequences. With our jammed gate valve, the local effect at the valve is "stops all flow." At the system level, "no fluid passes on to the next step in the process." And finally, at the plant level, "product production ceases (downtime) until the valve can be restored to operation."

Thus, without a clear understanding of failure terminology, reliability analyses not only becomes confusing but also can lead to decisions that are incorrect.

The FMEA embodies a process that is intended to identify equipment failure modes, their causes, and finally the effects that might result should these failure modes occur during product operation. Traditionally, the FMEA is thought of as a design tool whereby it is used extensively to assure a recognition and understanding of the weaknesses (i.e., failure modes) that are inherent to a given design in both its concept and detailed formulation. Armed with such information, design and management personnel are better prepared to determine what, if anything, could and should be done to avoid or mitigate the failure modes. This information also provides the basic input to a well-structured reliability model that can be used to predict and measure product reliability performance against specified targets and requirements.

The delineation of PM tasks is also based on a knowledge of equipment failure modes and their causes. It is at this level of definition that we must identify the proper PM actions that can prevent, mitigate, or detect onset of a failure condition. Specifying PM tasks without a good understanding of failure mode and cause information is, at best, nothing more than a guessing game.

How do we perform the FMEA? First, it should be clear by now that a fairly good understanding of the equipment design and operation is an essential starting point. The FMEA process itself then proceeds in an orderly fashion to qualitatively consider the ways in which the individual parts or assemblies in the equipment can fail.

These are the failure modes that we wish to list, and are physical states in which the equipment could be found. For example, a switch can be in a state where it cannot open or close. The failure modes thus describe necessary states within functions of the device, which have been lost. Alternatively, when suf-

ficient knowledge or detail is available, failure modes may be described in more specific terminology—such as "latch jammed" or "actuating spring broken." Clearly, the more precise the failure mode description, the more understanding we have for deciding how it may be eliminated, mitigated, or accommodated. Although it may be difficult to accurately assess, we also attempt to define a credible failure cause for every failure mode (maybe more than one if deemed appropriate to do so). For example, the failure mode "latch jammed" could be caused by contamination (dirt), and the "broken spring" could be the result of a material-load incompatibility (a poor design) or cyclic fatigue (an end-of-life situation).

Each failure mode is then evaluated for its effect. This is usually done by considering not only its local effect on the device directly involved, but also its effect at the next higher level of assembly (say, subsystem) and, finally, at the top level of assembly or product level (say, system or plant). It is usually most convenient to define two or three levels of assembly at which the failure effect will be evaluated in order to gain a full understanding of just how significant the failure mode might be if it should occur. In this way, the analyst gains a bottoms-up view of what devices and failure modes are important to the functional objectives of the overall system or product. A typical FMEA format is shown in Table 5.1.

By way of example, a filled-out FMEA is shown as Table 5.2, based on the simple lighting circuit schematic shown in Figure 5.10. In this instance, the FMEA is conducted at the system level, due to its simplicity, and we just move around the system circuit, device by device. In a more complex analysis, we might devote an entire FMEA to just one device, and break it into its major parts and assemblies for analysis. A pump or transformer are examples of where this might be done.

Frequently, FMEAs are extended to include other information for each failure mode, especially when the FMEA is conducted in support of a design effort. These additional items of information could include

- Failure symptoms.
- Failure detection and isolation steps.
- Failure mechanisms data (i.e., microscopic data on the failure mode and/or failure cause).
- Failure rate data on the failure mode (not always available with the required accuracy).
- Recommended corrective/mitigation actions.

When a well-executed FMEA is accomplished, a wealth of useful information is generated to assist in achieving the expected product reliability.

TABLE 5.1.   Failure Mode and Effects Analysis Format

	Equipment			Failure Effects		
ID #	Description	Failure Cause	Failure Mode	Local	System	Unit

TABLE 5.2.   Simple FMEA

Component	Mode	Effect	Comment
1. Switch Al	1.1 Fails open 1.2 Fails closed	1.1 System fails 1.2 None	1.1 Cannot turn on light. If A2 also fails closed, then system fails by premature battery depletion.
2. Switch A2	(Same as Al)	(Same as Al)	(Same as A1)
3. Light bulb C	3.1 Open filament 3.2 Shorted base	3.1 System fails 3.2 System fails; possible fire hazard	3.1 Cannot turn on light. 3.2 Cannot turn on light. May cause secondary damage to rest of system.
4. Battery B	4.1 Low charge 4.2 No charge 4.3 Overvoltage charge	4.1 System degraded; dim light bulb 4.2 System fails 4.3 System fails by secondary damage to light bulb C	4.1 May be precursor to no charge. 4.2 Cannot turn on light. 4.3 Secondary damage to light bulb C caused by overcurrent.

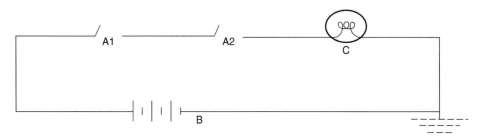

FIGURE 5.10.   Simple circuit schematic.

5.4. EQUIPMENT CRITICALITY ANALYSIS*

The proactive asset reliability process, shown in Figure 5.11, is used to nominate candidates and select equipment reliability improvement projects. It is an integral part of a larger manufacturing business process and focuses the maintenance of physical asset reliability as a contribution to the business goals of the company. The largest contributors are recognized as critical assets and specific performance targets are identified. The role of the maintenance function, accomplished through the six elements of the maintenance process, is to maintain the capability of critical equipment to meet its intended function at targeted performance levels.

*Section 5.4 was prepared in conjunction with Ivara Corporation of Burlington, Ontario, Canada, which provided the information and the graphics.

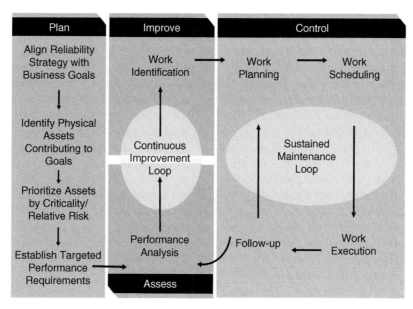

FIGURE 5.11. Proactive asset reliability process. (Courtesy of Ivara Corporation and *Plant Services Management*.)

The equipment criticality analysis is used to identify:

- Which equipment, if it fails, poses the most serious potential consequences on business performance? The resulting equipment criticality number is used to set a priority on the resources performing maintenance work.
- Which equipment is most likely to negatively affect business performance because both its failure poses serious consequences and it fails too often. The resulting relative risk number is used to identify candidates for reliability improvement.

The determination of critical equipment may vary from organization to organization. In fact, if the process is not formalized, there may be several interpretations of equipment criticality within a single organization. The assumptions used to assess what equipment is critical are not technically based. As a result, when different individuals are asked to identify their critical equipment, they will likely select different pieces of equipment. Often, they say, "All our equipment is critical!" Selections are based on individual opinions, lacking consensus. The potential for equipment failure having significant safety, environmental, or economic consequences may be overlooked.

Therefore, in the context of this document, critical equipment is that equipment whose failure has the highest potential impact on the business goals of the company. The relationship between equipment failure and business performance is an important factor in deciding where and when resources should be applied to maintain or improve equipment reliability.

To maintain reliable equipment performance requires the timely execution of maintenance work to proactively address causes of equipment failure. Large organizations normally manage a backlog of maintenance work. This maintenance work is made up of individual tasks that must be carried out over limited time periods, using limited resources to get the right work done at the right time. Effective maintenance scheduling requires an understanding of how critical the equipment is to which the task is applied, so that a priority can be assigned to each job, and the required time frame.

Equipment reliability improvement also requires human and financial resources. The business case for improvement justifies why the limited resources of the company should be applied to a project over the many possible alternatives that compete for the same resources. When justifying an improvement project, it is not sufficient to demonstrate benefits. It is necessary to demonstrate that the relative benefits of a project exceed the potential benefits of other projects.

Equipment reliability improvement projects benefit the organization by reducing the consequences of failure or the probability that the failure will occur. Equipment reliability improvement projects must focus on equipment that both matters a lot when it fails and fails a lot. The combination of failure consequence and failure probability is a measure of the risk posed to the organization by the specified equipment.

The discipline of risk management recognizes that failures with high consequence normally occur infrequently, while failures with low consequence occur more frequently. This is represented graphically in the risk spectrum of Figure 5.12. The consequence of a failure is plotted against the probability of the failure event. Probability is a measure of the number of events/time. The probability of an event like the nuclear accident at Chernobyl is very low but the consequence is very high. Alternately, many industrial organizations routinely experience failures within their plants. These failures affect business performance but their consequence is orders of magnitude less than the consequences of a Chernobyl-like incident. The majority of plant failures would fall to the right side of the risk spectrum.

The prerequisite to do Pareto analysis is to have failure data to analyze. This means that these failures must have occurred to be recognized. However, potential failures with very serious consequences are not considered because no failure data are associated with them. Therefore, it is necessary to manage events across the risk spectrum.

The criticality review process takes an integrated approach to setting project priority. The potential impact of equipment failure is assessed in each of the following categories: safety, environmental integrity, quality, throughput, customer service, and operating costs. The scales in each assessment category ensure that failure resulting in safety and environmental consequences is emphasized.

It also ensures that equipment whose failure affects the operational objectives of the organization are addressed. It is not possible to develop a separate maintenance strategy for each business driver. What is required is a comprehensive program that responds to the total needs of the organization. The equipment criticality analysis provides a ranked view of composite needs, which then become the focus of a suitable equipment reliability improvement strategy.

The equipment criticality evaluation provides a systematic, consistent approach to assessing equipment criticality. The relative risk rating is arrived at by consensus of the decision makers responsible for the nomination project, and the process can be completed in a short period of time. The focus is on business results managers already are accountable for achieving. The managers are committed to projects, which align with these objectives and are perceived as having the highest probability for success.

Finally, the use of a systematic process for focusing resource deployment supports a due diligence approach to physical asset management from a safety and environmental perspective. Projects having the largest

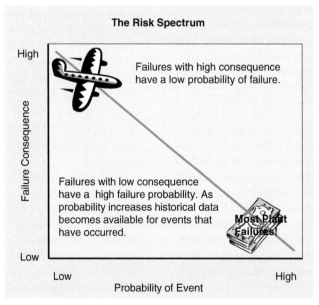

FIGURE 5.12.   Reliability must be managed across the risk spectrum. (Courtesy of Ivara Corporation.)

potential impact on the corporation, heavily weighted toward safety and environmental integrity, become the most critical. Projects with the potential to deliver the maximum benefit to the company by mitigating risk are identified to be the subject of equipment reliability improvement strategies.

5.4.1. Preparing for an Equipment Criticality Analysis

Equipment Hierarchy Review

Prior to performing an equipment criticality assessment, an equipment hierarchy must be produced and account for all equipment within the assessment area boundaries. This means that all maintainable components can be mapped and identified to an equipment or subequipment level. At the time the equipment criticality analysis is conducted, the equipment hierarchy might not be fully developed to the lowest level of detail desired. However, it is essential that the hierarchy be identified at least to a system level.

Registering the Equipment Criticality Analysis

All completed equipment criticality analyses should be consistently documented and recorded in an appropriate database. The analysis title should reflect the highest level in the equipment hierarchy to which the analysis applies; for example, XYZ Corporation, Port Operations, Sorting Plant, and Packaging Line Equipment Criticality Analysis. The date when the analysis is conducted should be recorded. Also identify the review team members and a description of their titles or positions.

The equipment criticality analysis should be reviewed and revised on an annual basis to reflect changes in business conditions and improvements in reliability and to identify new priorities for reliability improvement. Different review team members may be involved in the analysis review. The original team should be documented as well as the team members for the last revision.

Document a List of Equipment to be Assessed at the Appropriate Analysis Level

The level of analysis at which the assessment is completed is important. It is undesirable to evaluate the criticality of components. It also would be inappropriate to evaluate the criticality at the process or facility level. The level at which the analysis is done requires that the results of the analysis apply to all sublevel equipment not identified for analysis. Although somewhat imprecise, this provides a good definition for the first pass. In the evaluation process, it

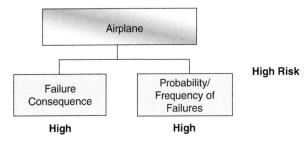

FIGURE 5.13. Airplane risk assessment. (Courtesy of Ivara Corporation.)

quickly becomes apparent if the equipment should be further divided into sublevels.

The two factors used in the risk assessment are the potential consequence of the failure and the probability that it will occur. If the level of the analysis is conducted too high, the resulting estimate of risk associated with the equipment may be misleading. This is illustrated in an airplane example (Figure 5.13). If a risk assessment is done at the airplane level, the result will be that flying in airplanes is high risk.

However, by simply moving the analysis down to a system level (see Figure 5.14), a much different perspective is achieved. The structural systems of the airplane have a high consequence if they fail but are extremely reliable, having a low failure rate. The airplane propulsion systems have a medium consequence when they fail, perhaps, because of built-in redundancy. The failure rate likely is higher than the failure rate for the structural systems. The relative risk therefore is greater. The comfort systems of the aircraft (such as seats, lights, entertainment plugs) have failure consequences much lower and likely failure rates much higher. Again overall risk is low. This result seems more reasonable. The list of equipment to be analyzed needs to be recorded at the desired level in the hierarchy, with specified parent and children relationships included in the analysis line item.

The facilitator prepares this list in advance of the analysis review meetings. It can also be revised during the review meetings. During the analysis, items and levels of detail omitted in the hierarchy are sometimes identified.

Specify the Equipment Criticality Assessment Criteria

In the default criteria, company goals are categorized under the themes of safety, environmental integrity, product quality, throughput, customer service, and total cost. An evaluation scale for consequence of failure potential is specified for each theme. If an equipment failure has no impact on a goal area, a score of zero (0) is assigned. If an equipment failure has an impact on a goal area, the rating is assigned that most closely fits the consequence description.

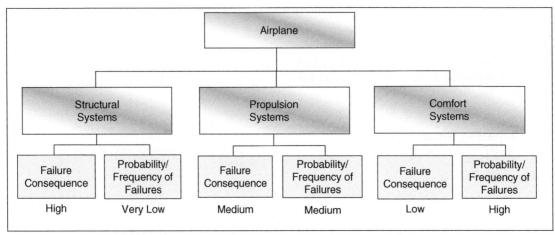

FIGURE 5.14.   Airplane risk assessment. (Courtesy of Ivara Corporation.)

Safety and environmental issues have a maximum scale of 40. Operational consequences independently score a maximum value of 10. Most equipment failures affect operations in several different ways, and in the extreme case, a total operating consequence of 40 could be achieved. The default criteria are provided in Table 5.3.

Against each of the criteria it is possible to have an explanation. A set of qualitative descriptions is provided for the environmental rankings in Table 5.4. Similar explanations could be provided for each of the assessment review areas.

Similar assessment criteria are provided for reviewing how likely a failure will occur on the selected equipment. This assessment is made along with the consequence evaluation for the asset being reviewed.

It is important to clarify the meaning of failure. The definition of *failure* used in the equipment criticality analysis is the inability to perform any function at its required level of performance. As a result of failure, corrective intervention is required to restore equipment capability. One way of interpreting how often the equipment fails is to assess how often any form of corrective maintenance is performed on the equipment. Corrective maintenance must be differentiated from preventive maintenance. The frequency or probability of failure number is used in the calculation of relative risk to the business. If an effective PM program controls failures, the equipment is unlikely to negatively affect business performance.

The probability or frequency of failure is evaluated on a scale ranging from 1 to 10 with 10 representing the highest failure rate. A description of the default criteria for this (how often failures occur) follows. It is possible for an intermediate value to be selected, such as 8.5, signifying that failures are felt to occur between weekly and monthly.

10 = Failures occur daily.
 9 = Failures occur weekly.
 8 = Failures occur monthly.
 7 = Failures occur between monthly and yearly.
 6 = Failures occur yearly.
 5 = Failures occur approximately between yearly and 1 in 5 years.
 4 = Failures occur between 1 in 5 and 1 in 10 years.
 1 = Failures occur less frequently than 1 in 10 years.

Document Assumptions Used in the Analysis.

During the analysis the review group must evaluate failure consequences based on various assumptions. For example, the impact of equipment failure with operational downtime depends on market conditions, inventory levels, capacity, and utilization. The assumptions used for evaluating the criteria should be documented against the assessment criteria, as in Table 5.5.

5.4.2. Conducting the Review

Facilitating the Review Meetings

The equipment criticality assessment is designed to achieve consensus among key decision makers in the organization. Review team members are selected based on their ability to assess the consequence of equipment failure on the business, the frequency of individual equipment failure, and their responsibility for nominating or sponsoring equipment reliability improvement projects. The assessment process is designed to minimize the time that the review team must dedicate to attending the assessment review meetings.

The analysis is conducted by answering a series of structured questions about each equipment line item. These questions assess both the consequence of

TABLE 5.3. Default Criteria (Courtesy of Ivara Corporation)

Safety	Environment	Quality	Throughput	Customer Service	Operating Cost
40 = Multiple fatalities	40 = Potential for severe environmental damage	10 = Scrap cannot be reworked or sold as secondary product	10 = Unable to recoup loss to attain production quota—must reduce future order bookings	10 = Loss of customer or potential litigation	10 = Increased costs of more than $500,000
38 = Fatality	32 = Potential for major environmental damage	8 = Out of specification, with rework can be sold as second at little or no profit	8 = Cannot make up lost production at facilities—have to purchase outside material or service	8 = Customer experiences downtime or excessive scrap loss, costs charged back	8 = Increased costs of more than $100,000 but less than $500,000
34 = Disabling injury	28 = Potential for significant environmental damage	6 = Out of spec, with rework can be sold as prime	6 = Lost production can be recovered within facilities but at additional cost (e.g., overtime) since no excess capability readily available	6 = Late delivery of majority of order quantity or customer rejects product as received	6 = Increased costs of more than $50,000 but less than $100,000
30 = Lost-time injury	20 = Minor or no environmental impact	5 = Out of spec, can be sold as seconds	4 = Can recover lost production through readily available excess capacity but has a significant impact on buffer inventory levels, putting other operations at risk of delay in supply	4 = Partial late delivery	4 = Increased costs of more than $10,000 but less than $50,000
20 = Minor injury such as contusions or lacerations	0 = No accidental release or emission	4 = Out of spec, can be reapplied to other prime order	2 = Lost production has no significant impact on buffer inventory levels	2 = On time delivery, but minor impact on order quality or quantity that the customer is willing to accept	2 = Increased costs of less than $10,000
0 = No injury		2 = Production within spec but process out of control	0 = No lost production	0 = Quality, quantity, and delivery date as promised to the customer at time of order placement	0 = No increased operating costs are incurred
		0 = Process remains in control			

equipment failure and the frequency and probability of failure against the predefined assessment criteria. A total consequence evaluation is compiled from the group's responses to the following questions using the assessment criteria for severity determination:

1. If the identified equipment fails, could it result in a safety consequence? If yes, how serious would the potential consequence be?
2. If the identified equipment fails, could it result in an environmental consequence? If yes, how serious would the potential consequence be?
3. If the identified equipment fails, could it result in a consequence affecting the quality of the product? If yes, how serious would the potential consequence be?
4. If the identified equipment fails, could it result in a consequence affecting the throughput capability of the plant? If yes, how serious would the potential consequence be?
5. If the identified equipment fails, could it result in a consequence affecting the service provided to the customer? If yes, how serious would the potential consequence be?

TABLE 5.4. Qualitative Descriptions for the Environmental Rankings (Courtesy of Ivara Corporation)

Environmental Consequence	Explanation or Example
40 = Potential for severe environmental damage	1. An environmental release causing death, injury, or evacuation of the surrounding community.
	2. Cost of clean up, damage to property, or interruption of production/business in excess of $1 million.
	3. Major kill of wildlife—generally fish or birds in the local area.
	4. Releases large quantities (>500 gallons) of toxic or environmentally persistent materials to the environment external to company property (ammonia, light oil, PCBs, etc.).
32 = Potential for major environmental damage	1. Discharges to storm sewers, sanitary sewers, or directly to water exceeding environmental regulations.
	2. Discharges to atmosphere causing property damage—particulate fall out, corrosion, etc.
	3. Discharges to atmosphere exceeding regulations and can cause health effects—particulate, sulfur dioxide, etc.
	4. Releases of large quantities of toxic materials to the ground (>500 gallons).
	5. Cost of clean up, damage to property, or interruption of production/business in excess of $100,000.
28 = Potential for significant environmental impact	1. Discharges to storm sewer, sanitary sewer, or water exceeding regulations.
	2. Discharges to atmosphere exceeding regulations, e.g., opacity.
	3. Operation of process equipment without environmental equipment even if there is no immediate or short-term impact.
	4. Releases to ground.
20 = Minor or no environmental impact	1. Accidental releases of process fluids to containment areas or treatment plant; e.g., tank leaks to containment pad.
	2. Use of containment areas for temporary storage.
	3. Releases inside buildings, which do not get to the natural environment.
	4. Events that cause upsets to treatment plants but do not necessarily result in excessive discharges.
	5. Operation of a process at production rates higher than specified in a certificate of approval.
	6. Operation of a process with feed materials not specified in a certificate of approval; e.g., feeding rubber tires to a coke plant.
0 = No accidental release or emission	1. Normal process and environmental-control equipment operation—within operating specifications and in compliance with regulations and environmental certificates of approval.

TABLE 5.5. Equipment Criticality Analysis Assumptions (Courtesy of Ivara Corporation)

Assessment Criteria	Assumption Description
Throughput	1. If equipment failure results in more than four hours of downtime, the buffer inventory stock will be depleted, interrupting customer supply.

6. If the identified equipment fails, could it result in a consequence affecting total operating costs? This includes the cost of maintenance to restore the equipment to full operational capability. If yes, how serious would the potential consequence be?

The frequency or probability of equipment failure assessment is made along with the consequence evaluation for the equipment line item being reviewed. In addition to the series of consequence questions asked of the review group, they are asked, How often do failures of the specified equipment occur? They choose their response from the predefined criteria.

Answers to all these questions should be recorded in the spreadsheet during the review team meetings.

Recognizing Capital Equipment Upgrade Requirements

In some cases, there is a preconceived belief that the equipment being assessed needs to be upgraded, consuming capital funds. Where physical redesign is the apparent solution, it is useful to capture this data during the assessment review. This can be done by placing an asterisk in front of the equipment line item under assessment.

5.4.3. Analyzing the Assessment Results

Calculating the Equipment Criticality Number

The criticality of equipment is a function of the impact of its failure on the business, regardless of how often it fails. Not all failures matter equally. The equipment criticality number assigned to a piece of equipment in the hierarchy is influenced by the severity of its failure and the consequence. Equipment criticality numbers are assigned between 1 and 9, where 9 is the highest and 1 is the lowest criticality.

During the review, the consequence of equipment failure is assessed against key company goal areas. The default criteria includes the potential impact of failure on the enterprise's safety and environmental integrity, considered fundamental to continued operation. Other key business goal areas are assessed, such as product quality, throughput, customer service, and operating costs. The assessment criteria may have been redefined, as previously discussed.

The default logic of Table 5.6 is used to calculate and assign the equipment criticality number, and Figure 5.15 shows an example. (This logic may need to be redefined by the organization if the consequence evaluation criteria are modified.)

Cascading the Equipment Criticality Number to Its Applicable Level in the Hierarchy

The equipment criticality analysis usually is performed at an intermediate level in the hierarchy. The equipment criticality number therefore applies to all children of the analysis level, except those children identified for analysis as well. Any parent level not analyzed adopts the equipment criticality value of its highest child. This is illustrated in Figure 5.15.

Determining Which Equipment Has the Greatest Potential Impact on Business Goals by Calculating Its Relative Risk

Risk incorporates the notion of severity of consequence when failure occurs and the likelihood that failure will occur. For example, being struck by lightning has a life-threatening consequence for an individual. The probability of being struck by lightning is low under normal circumstances. Therefore, the risk of being struck by lightning is low. Most people are not concerned about being struck by lightning. However, suppose a job involved working from heights where a fall could result in fatality, again a life-threatening consequence. If the probability of falling were great (perhaps the work platform is a crane runway), the

TABLE 5.6. Equipment Criticality Analysis Consequence Rating Worksheet
(Courtesy of Ivara Corporation)

Equipment Criticality Analysis Consequence Rating	Equipment Criticality Number
Safety, environmental, or total operational consequence ≥ 38	9
Safety, environmental, or total operational consequence ≥ 28 or any single operational consequence = 10	8
Safety, environmental, or total operational consequence ≥ 20 or any single operational consequence = 8	7
Safety, environmental, or total operational consequence ≥ 16 or any single operational consequence = 6	6
Total operational consequence ≥ 14 or any single operational consequence = 5	5
Total operational consequence ≥ 10	4
Total operational consequence ≥ 8	3
Total operational consequence ≥ 4	2
Total operational consequence < 4	1

For Asset B: Selected Levels of Analysis

Parent Level: B Hierarchy Level 2

Parent Level Adopts Equipment
Criticality Value of Highest Child: **9**

Analysis Level: **B1 Hierarchy Level 3**
    Includes: B1.1 Hierarchy Level 4

Analysis Level Criticality: **7**
    Child Level Criticality: **7** (Default)

Parent Level: B Hierarchy Level 2
                        B2 Hierarchy Level 3

Parent Criticality (Highest Child): **9**

Analysis Level: **B2.1 Hierarchy Level 4**
    Includes: Hierarchy Level 5

Analysis Level Criticality: **5**
    Child Level Criticality: **5** (Default)

Analysis Level: **B2.2 Hierarchy Level 4**
    Includes: B2.2.1 Hierarchy Level 5
                    B2.2.3 Hierarchy Level 5

Analysis Level Criticality: **8**
    Child Level Criticality: **8** (Default)
    (Analyzed Below)
    Child Level Criticality: **8** (Default)

Analysis Level: **B2.2.2 Hierarchy Level 5**

Analysis Level Criticality: **9**

FIGURE 5.15. Equipment criticality value. (Courtesy of Ivara Corporation.)

risk would be high as well. As a result, action to reduce the risk of falling is needed, perhaps by having each worker wear a safety harness.

The equipment criticality assessment uses the concept of risk to identify which equipment has the greatest potential impact on the business goals of the enterprise. This, in turn, is the equipment most likely to fail and have significant impact when the failure occurs. The relative risk (RR) number for the equipment is evaluated by calculating the product of the total consequence number and the frequency/probability (F/P) number. It is called *relative risk* because it has meaning only relative to the other equipment evaluated by the same method. Total consequence (TC) is the summation of the values assigned to each of the individual areas of consequence evaluation: safety (S), environmental (E), quality (Q), throughput (T), customer service (CS), and operating cost (OC).

$$TC = S + E + Q + T + CS + OC$$

$$RR = TC \times F/P$$

If different criteria are deemed integral to a firm's operation, then the total consequence would be the summation of scores applied in each area so defined.

Communicating Criticality Assessment Recommendations to All Stakeholders

The results of the equipment criticality assessment should be communicated and understood by every-one affected by the nominated equipment reliability improvement projects. This includes

- Senior and intermediate managers who sponsor or expect results from the project.
- Coaches and team leaders responsible for the assets that the project addresses.
- People assigned to the assets that the project addresses.
- Individuals who must commit time to the project or are directly affected by its outcome.
- People who are not immediately affected.

Often the last group demonstrates the greatest opposition because they believe that the selected projects "hog" the financial and human resources needed to address their high priorities.

The goal of this communication is to develop stakeholder understanding why each equipment reliability project is selected, its potential impact on business performance, and the resource expectations it is to deliver.

The initial output of the equipment criticality analysis should be a report suitable for binding and presenting. Some typical sections to include in the report follow:

- Recommendations.
- Analysis description, review team, date.
- Analysis assumptions.
- Equipment summary sorted by relative risk.
- Equipment summary sorted by criticality number.
- Detailed assessment results.
- Equipment failure consequence evaluation criteria.
- Probability evaluation criteria.
- Equipment criticality number conversion criteria.

5.4.4. Using the Output of the Equipment Criticality Assessment

The relative risk ranking provides a means of identifying which equipment poses the highest potential impact on the organization. The equipment with the highest relative risk ratings should be initially targeted for some reliability improvement strategy. In many applications, this method of establishing priority is sufficient for project nomination. The top 10 equipment items evaluated then would be subject to a project selection validation.

However, the ranking developed using relative risk alone does not consider the difficulty of improving the reliability of the critical equipment. Suppose this could be achieved only with a large commitment of human resources, over an extended time, and at high cost. In assessing the business case for proceeding with the reliability improvement project, each of these factors plays a role.

An alternate method assesses the human resource effort for an equipment reliability intervention. Alternatively, the cost of the intervention, the resulting redesign, or equipment replacement can be evaluated. The following subsections describe the process used to evaluate priority considering effort and cost. This approach normally is applied to the top 20% of "relative risk" equipment items to minimize the analysis effort. It is worthwhile estimating effort and cost for those reliability interventions generally providing the greatest potential impact on the business.

Estimating the Effort Required to Reduce the Risk to a Tolerable Level

The human resource effort required to proceed with the proposed equipment reliability improvement strategy is assessed. For example, the number of meetings to complete a reliability centered maintenance analysis is estimated. This effort provides an indication of the degree of difficulty required to overcome the performance gap.

Reliability improvement options include the application of reliability centered maintenance, predictive maintenance needs assessment, use of the reliability assessment, equipment maintenance program development, planning, and scheduling practice interventions. These can be estimated using an Excel spreadsheet.

The relative risk value is plotted on the vertical axis of a graph and the effort on the horizontal axis. Initially, work on projects with high potential impact that can be done quickly and work on projects with low impact, requiring large effort last.

To assign priorities to the proposed interventions, a diagonal line is drawn from the upper left corner of the risk/effort graph to the lower right. The slope of this line is calculated by summing all the relative risk values for each equipment item evaluated and dividing the total relative risk by the total effort, calculated by summing the effort values estimated for each equipment item. The downward slope of this line from the upper right to the lower left represents reduction in risk per unit effort. Consider a series of lines, drawn perpendicular to this diagonal completely covering the graph. Adjacent lines represent bands of relative priority.

Equipment reliability improvement projects addressing assets closest to the upper left corner of the plot should be addressed first, while those projects addressing assets in the lower right of the plot should come last. Each project can be assigned a specific priority. A sample plot is represented in Figure 5.16.

Note: This graph is a focusing tool only. The exact value and position on the graph is an indication of relative priority. Individual circumstances could require specific projects to proceed irrespective of their position on the graph. For example, a piece of equipment whose failure has serious safety implications and a high frequency or probability of failure, resulting in a high relative risk number, may require a large expenditure to improve its overall reliability. Legislation or a safety ruling may dictate that this project take precedence over another asset scoring equivalent relative risk and requiring much less cost. Nonetheless, the concept can be used successfully in most situations to develop a defensible position for assigning resources to address equipment reliability issues.

The priority of the proposed reliability intervention is identified mathematically by calculating the y-axis intercept of a perpendicular line passing through a point with the individual project (relative risk, effort) coordinates. A number 1 priority is assigned to the reliability intervention with the highest relative risk intercept. Lower priority is assigned to reliability interventions with successively lower relative risk intercepts.

Alternative to estimating the human resource effort is to estimate the cost to proceed with the chosen equipment reliability improvement strategy or equipment modification or replacement. This is an estimate of the cost required to overcome the performance gap.

Identifying Equipment Reliability Improvement Projects

The preceding subsection described several approaches for assigning priorities to equipment. The next step is to nominate a series of candidates for equipment reliability improvement projects. The equipment criticality evaluation process was designed as a focusing tool. It allows the organization to quickly

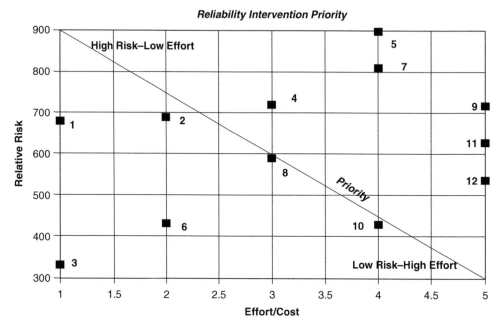

FIGURE 5.16.   Reliability intervention priority. (Courtesy of Ivara Corporation.)

understand where significant benefits may be achieved by improving equipment reliability. The methodology is not precise. The top 10% of the ranked items almost assuredly include the equipment where the organization wants to focus its attention. The exact ordering within the 10% may be imprecise. The next step is to develop a business case for each proposed candidate to justify proceeding with the project and validate the order in which the projects should proceed.

This is done using a cost/benefit analysis to validate nominated projects by evaluating their current performance or state against that desired. Quantify the potential benefits and estimate the costs to proceed.

The criticality assessment provides a means of identifying the equipment most likely to affect business performance by improving reliability and indicating what areas of performance are likely to be affected. In each category, safety, environmental integrity, quality, throughput, customer service, and operating cost, the current performance should be established and a performance target set that is achievable as an outcome of the improvement. The difference between current performance and the desired end state should be quantified either in terms of costs for operational improvements or reduced incidents or level of risk for safety and environmental issues. This gap is important in creating the required tension for change to maintain management commitment throughout the project. Estimate the costs of the reliability improvement intervention and summarize the costs and benefits.

As soon as capital or human resources are deployed, expectations are created to produce tangible benefits.

The development of the business case solidifies what results can be expected from the project. However, it is still necessary to demonstrate the improvement. This is effectively done through performance measurement. It is crucial that each of the stated performance benefits be monitored on a routine basis to validate improvement. If the required measurements are not currently collected, the project scope should formalize their creation. This permits the quantification of improvement benefits, sustaining project commitment and the management of long-term change.

5.4.5. Conclusions

The equipment criticality evaluation provides a systematic, consistent approach to assessing equipment criticality and nominating equipment reliability improvements. Rankings are arrived at by a consensus of the decision makers responsible for the project. By design, the process can be completed in a short period of time.

The focus is on business results that managers already are accountable for achieving. These managers are committed to the projects that align with these objectives and are perceived as having the highest probability for success.

Finally, the application of systematic processes for focusing resource deployment supports a due diligence approach to physical asset management. Projects having the largest potential impact on the corporation weighted toward safety and environmental integrity become the most critical. Projects with the potential to deliver the maximum benefit to the company by mitigating risk are identified for equipment reliability improvement strategies.

5.5. ROOT CAUSE ANALYSIS

Root cause analysis (RCA), also called *root cause failure analysis* (RCFA), is the systematic evaluation of problems to find the basic causes that, when corrected, prevent or significantly reduce the likelihood of a recurrence. These basic causes are called *root causes*. It is important to realize that most problems have more than one contributing cause, and if one of these contributing causes were eliminated, the problem would not recur.

Unexpected equipment failures are not normal and should not be tolerated. Equipment problems not determined sufficiently and solved only well enough to get the equipment back up and running will return; therefore, the cause of the failure must be understood and fixed, not just the failed equipment.

A structured RCA process is needed to make sure the true root cause(s) of a problem are understood instead of the more obvious symptoms. If the solution to the problem addresses the symptoms, obviously, the problem will reappear at some point in the future or other problems will be created by the implemented solution.

Every failure affords an opportunity to learn. Ignoring these opportunities means missing the chance to extend equipment life, decrease repair frequency, and improve profitability. RCA has four fundamental steps:

1. Quantify the magnitude of the problem and decide on the resources required to resolve it.
2. Perform the analysis by selecting the appropriate technique.
3. Develop a list of options for solving the problem and implement the most cost-effective solution.
4. Document the results of the analysis in the appropriate format.

Four methods generally are used, either singly or in combination, for conducting a root cause analysis: plan, do, check, act (PDCA).

5.5.1. Plan

Clearly Define the Problem (P1)

"A problem clearly stated is a problem half solved." Although it seems like a trivial step, do not take this step lightly. It is important to begin this problem-solving journey with a clear, concise problem statement. If this is not done properly, it could lead to excessive time spent identifying the cause due to a broad problem statement, or solution implementation rather than

root cause identification and remedy. A problem can occur in one of two contexts, and the context should be reflected in the problem statement: A specific set of conditions *prevents* a desired result or a specific set of conditions *causes* an undesired result.

The format for a problem statement is this:

| Problem stated in the context of an obstacle to or deviation from desired result; quantified. | + | Major consequence of problem. |

For example,

| Dilution is out of control 35% of the time | + | Resulting in product rejects, consumer complaints, and delays in processing, attempting to make adjustments. |

The key points to keep in mind are

- Avoid solution statements (e.g., "Not enough backup instrumentation" or "No in-house support").
- Include two parts, a description of the undesirable condition, along with what is causing or preventing it from happening, and the consequence of the problem.
- If data are available at this point, quantify the size of the problem within the problem statement (e.g., "Process is available 45% of the time" as opposed to "Process is available a large amount of the time").
- Define the problem as narrowly as possible (e.g., Is the problem on one maker or across all makers? Are there obsolete or missing material specs for all materials? Which ones in particular cause problems? Is parameter availability across all shifts, all days or can it be more specifically defined?).

Table 5.7 shows some examples.

Collect Evidence of Problem (P2)

Obtain the information or data to clearly demonstrate that the problem *does* exist. In the case of team problem solving, this should be a quick exercise, since the reliability engineering function must have been looking at data to create the team. However, it is important that the team gathers and views such data to answer the following questions:

- Does the problem truly exist?
- Is the problem measurable?
- Is the problem chronic?
- Do the data show the problem existing over an extended period?

TABLE 5.7. Examples

Original	"Good" Problem Statement
Reduce waste	Weekly waste is out of control, high 25–40% of the time, resulting in excessive cost/1000 units
Have maintenance manuals more accessible	Variation in repair time is high, resulting in increased variation in downtime, run costs
Improve machine care scheduling	The scheduling of machine care is not based on equipment performance data and results in ineffective prioritizing of resources, decreased average levels in process TQMs
Spare parts are never available when machine breaks down	Critical spare parts for makers are not available or difficult to locate, often resulting in excessive downtime, increased expediting costs, and frustration

- Is the problem significant? If solved, will it result in significant improvement or savings in time, money, morale, or resources?

The output of this activity will be a list of evidence statements (or graphs) to illustrate that the problem exists, its size, and the chronic nature of it.

Identification of Impacts or Opportunities (P3)

Identify the benefits that occur from successful problem solving. This activity needs to be thought of in two different perspectives because the work can take the form of control (fixing a problem that stands in the way of expected results) or pure improvement (attempting to take results to a new level of performance.)

Impacts. For control work, what is the consequence of not solving this problem?

Opportunities. For pure improvement work, what is the lost opportunity if this work is not initiated?

In each case, the output of this activity is a list of statements. The impact and opportunity statements should be stated in terms of loss of dollars, time, "product," rework, processing time, or morale.

Measurement of Problem (P4)

Before problem solving proceeds, it is important to do a quick check on the issue of how valid or reliable the data are on which the decision is made to tackle the problem.

For the parameter being used as evidence of the problem, is there any information that would question the validity, accuracy, or reliability of the data? This question should be examined whether relying on an instrument, a recorder, or people to record information or data.

If significant issues "cloud" the data, then these measurement problems need to be addressed, fixed, and new measures obtained *before* proceeding with the other segments of PDCA.

Measures of Effectiveness (P5)

At this point, identify how to measure success of problem solving. This is one of the most important steps in PDCA and one that certainly differentiates it from "traditional" problem solving. The strategy is to agree on the *what* and *how*, obtain the benchmark "before" reading, perform the PDCA activities, and remeasure or obtain the "after" measure. At that point, decide whether to "recycle" through PDCA to achieve the prestated objective.

- Determine what appropriate measures would directly reflect improvement and do not worry about the how at this point.
- Look at TQMs, customer feedback results, costs, and off-line parameters as possible measures.
- Typically one to three measures are sufficient.
- Then decide how the measure should be obtained and expressed (i.e., Pareto form, control chart form, or survey and tabulation of results). Guideline: Get creative!
- Obtain the "before" measure. Maybe the data exist and just need to be researched. Perhaps the vehicle to begin capturing the data needs to be developed or implemented.

The key point to remember is this: Only *after* the "before snapshot" has been taken and the data reviewed should the step of setting the objective occur.

Objective Setting (P6)

When the measure of success has been determined and the "before" or current level of performance of those measures obtained, the objective for improvement can be effectively set. Knowing the amount of effort and resources that will be utilized on this problem, what amount of improvement in the measure of effectiveness would provide a good return on investment? To do this activity productively requires discussion with the QC, PMT, SIT, or DMT, and the ability to fairly represent their expectations and setting the general time frame for problem solving activities.

The key points are these:

- The objective should be to significantly reduce the problem, not necessarily to totally eliminate it.
- Be aware of the effort/return ratio; that is, ensure that the expected benefit level is a meaningful return on the time, energy, and resources to be expended in solving the problem.
- The objective should be stated in terms of percentage of reduction in average level or percentage of change in measure of effectiveness, percentage of reduction in variation, percentage of reduction in cost or time.
- Objectives should be set so they can be achieved in a reasonable period of time.

Rough PDCA Timetable (P7)

For resource planning and with enough information to make the task possible, a rough timetable is projected for completing each segment of PDCA. This is a preliminary estimate based on the information currently available and will be revised as the work progresses. The format is simple:

Segment	Estimated Completion Date
Plan	
Do	
Check	
Act	

The information that the team should use to make these estimates is the

- Size of the problem.
- Amount of the problem to be solved (objective).
- Complexity of the problem.
- Other conflicts for time.
- QC, PMT, SIT, or DMT priority for resolution of the problem.

This timetable will be used in several ways:

- Objectives for getting work done.
- Planning purposes, so other activities can be scheduled with this timing objective in mind.

- QC, PMT, SIT, or DMT planning. With this information management can better direct the plan by knowing when other work must begin, the resources available, and the like.

Management Approval and Review (P8)

During the sequence of work, it will be necessary to maintain high-quality communications with management, whether it be line managers, SIT, DMT, PMT, or quality control. The nature of this communication could be to inform the management of progress or results or to review plans or obtain approval to carry out changes deemed necessary. Management should be concerned with the process (how the work is done) as well as the results. The more informed management is, the better it can set priorities and coordinate efforts, optimizing the allocation of a limited number of problem solving resources.

Who

In team work, these sessions with management could be done by the team leader, a rotating representative of the team, or in some cases, it might be appropriate for the entire team to be present.

How

The standardized forms to capture output should be the basis for the presentations. Talking from these forms should give management a good sense of the process (quality of problem-solving efforts) as well as the key conclusions. In the check segment especially, it might be necessary to use additional exhibits to demonstrate how the conclusions were drawn.

What

The content of these sessions vary depending on the stage of PDCA. For this first management approval, the major items to be covered are

- Problem statement.
- Evidence.
- Impact information.
- Measures of effectiveness with "before" measurements.
- Objective.
- Rough PDCA timetable.

5.5.2. Do

Generate Possible Causes (D1)

To avoid falling into the mode of solution implementation or trial-and-error problem solving, start with a "blank slate." From a fresh perspective, lay out all possible causes of the problem. From this point, use the data as well as collective knowledge and experience to sort through the most feasible or likely major causes. Proceeding in this manner helps ensure ultimately reaching the root causes of problems and not stopping at the treatment of symptoms. The best tool to facilitate this thinking is a cause and effect diagram done by those people most knowledgeable of and closest to the problem. To summarize this process,

Step 1. Construct causes and effects diagram.

Step 2. Obtain any currently available data on causes and effects, any available "clean" data that illustrate relationships between possible causes and the effects or dependent variables. If no data are available, go to step 3 immediately.

Step 3. Set priorities on major causes, based on data, knowledge, and experience. Suspected major causes are identified and initially investigated, indicated by circling those areas on the cause and effect diagram. As a guideline, target two to six bones, the first and second level causes that, if real, collectively reduce the problem to the level stated in the objective. In the absence of existing data (step 2), step 3 often becomes an exercise to determine which variables need to be included in a designed experiment. Keep in mind, if these PDCA activities do not meet the objective, come back to this cause and effect diagram and identify other suspected causes.

Step 4. Write an action plan or experimental test plan.

Identify Broken-Need-Fixing Causes and Work on Them (D2)

Before carrying out either an action plan (to remedy causes) or an experimental test plan, check whether parts of the process are "broken." This could take on many different forms. For example,

- Mechanical part known to be defective or incorrect.
- Piece of equipment not functioning as intended or designed.

- Erratic behavior of a piece of equipment.
- Temporary replacement of a part or piece of equipment that is not equivalent to the requirement.
- Method or procedural change made temporarily to "get around" a problem.

In most of these cases, the items are obvious as something "to just live with." These items, if not fixed, might obscure any experimental results or limit the amount of improvement realized in the action plan. A few key guidelines to remember in performing this activity are

- Focus on obvious items.
- Do not work on items without clear consensus.
- Address only those items that can be fixed in a short period of time (weeks not months).

Write Experimental Test or Action Plan (D3 and D4)

The PDCA strategy will take one of two directions at this point, depending on whether the problem is data based or data limited. Shown in Table 5.8 is the distinction between these two strategies and, in particular, the difference between an action plan and an experimental test plan. Note that, in some cases, it is necessary to use a combination of action and experimental test plans. That is, for some cause areas an action plan is appropriate, and for other causes within the same problem, an experimental test plan is the best route.

Write Action Plan for Cause Remedies (D3)

To be able to write the action plan, brainstorm possible solutions or remedies for each of the cause areas (circled bones) and reach consensus on the priorities of the solutions. In team work, this work can be carried out by a team or subteams; either way, agreement must be reached on proposed remedies and the action plan. The action plan will be implemented in the check segment. Who, when, and what to be done should be spelled out in the action plan. The format for this plan is shown in Table 5.9.

TABLE 5.8.   Action Plan or Experimental Test Plan?

Approach	When to Use	What
Action plan (for cause remedies)	Suspected cause(s) cannot be changed or undone easily once made; dependent variable (other than measure of effectiveness) not obvious; lack of data to study causes	Brainstorm solutions to major causes Identify solution areas for major causes Write action plan to describe what, who, how of solutions
Experimental test plan	Suspected causes(s) can operate at two or more levels; the levels can be deliberately and easily altered; the effects can be measured through dependent variables	Write experimental design test plan to test and verify all major causes, use other techniques or experimental design techniques

TABLE 5.9. Format for Action Plan

Function	Activities	Who	Time Frame
Cause area	Specific solution parts to be worked on related to this function	List all involved in this function	Beginning and ending date for each activity

As part of the action plan, clearly identify what the *dependent variable(s)* will be: After performing the activities associated with the cause area, what measurement can measure the effects (if any) of that solution? Underneath the action plan table, the following information needs to be provided—without this, the action plan should be regarded as incomplete:

- Cause area
- Measure (dependent variable)
- Measurement defined

Write Experimental Test Plan (D4)

The experimental test plan is a document that shows the experimental tests to be carried out. The tests verify whether the identified root cause really affects the dependent variable of interest. Sometimes, one test will test all causes at once, or a series of tests may be needed. Note: If there is a suspicion of an interaction among causes, those causes should be included in the same test.

The experimental test plan should reflect

- Time or length of test.
- How the cause factors are altered during the trials.
- Dependent variable (variable interested in affecting) of interest.
- Any noise variables that must be tracked.
- Items to be kept constant.

Everyone involved in the experimental test plan should be informed before the test is run. This should include

- Purpose of the test.
- Experimental test plan details.
- How that person will be involved.
- Key factors to ensure good results.

When solutions have been worked up, coordinate a trial implementation of the solutions and the "switch on/off" data analysis technique. (See activity C1.)

Identify Resources (D5)

Once the experimental test plan or the action plan is written, it will be fairly obvious what resources are needed to conduct the work. Construct a list of which resource people are needed, for what reason, the time frame, and the approximate amount of time needed. This information is given to management.

Revise the PDCA Timetable (D6)

At this point, there is a much better feel for what is involved in the remainder of the PDCA activities. Adjust the rough timetables projected in the plan segment. This information should be updated on the plan as well as taken to management.

Management Review and Approval (D7)

A critical point in the PDCA cycle has been reached. The activities about to be carried out have an obvious impact and consequences to the department. For this reason, it is crucial to make a presentation to management before proceeding. The content and purpose of this presentation is

- Present the output to date.
- Explain the logic leading up to the work completed to date.
- Present and get management approval for the measure of effectiveness with the "before" measure, high-priority causes, action plan (for cause remedies) or experimental test plan, and revised PDCA timetable.

5.5.3. Check

Carry out the Experimental Test or Action Plan (C1 and C2)

Depending on the nature of the problem, either conduct experimental test plan(s) to test and verify root causes or work through the details of the appropriate solutions for each cause area. Then, through data, verify to see if those solutions were effective.

On the following pages, we look at some general information and key points to remember for both strategies.

Carry Out Action Plan (C1)

In the case of action plans where solutions have been worked up and agreement reached, "switch on/ switch off" techniques need to be used to verify that the solutions are appropriate and effective (Table 5.10). To follow this strategy, the team needs to identify the dependent variable (the variable the team is trying to affect through changes in cause factors).

When using this strategy, remember these important points:

- Collect data on the dependent variable for a representative period before the test period. It should be comparable in length to the test period.
- Test for normality—develop control limits to define typical performance under the old system.
- During the test period, implement solutions. (Ensure the window of the test period is long enough to capture most sources of variation.)
- Compare test period data against already defined limits from the "before" data.
- Check to see if the level has shifted significantly—evidenced by OOCs.
- Switch off—undo the changes. See if the performance returns to "before" level.

Note: The purpose of the switch on/off technique is to guard against the situation in which the implemented changes had a positive effect, but the results did not show it because new causes (time related) entered the process and offset the positive effects of the planned changes. The data, in that case, would show no change. However, using the switch on/off technique helps overcome that phenomenon.

Carry Out the Experimental Test Plan (C2)

During the check segment, the experimental tests to check all the major high-priority causes are conducted, data analyzed, and conclusions drawn and agreed to. Remember a few key points in doing this series of activities:

- Confer with the TQI specialist to ensure appropriate data analysis techniques are used. The tools used depend on the nature of the test plan.
- Keep complete documentation while the test(s) are being run. This information will help the team decide if the results are valid.
- Construct clear, simple, concise, data recording sheets to ensure the right information is recorded and correct experimental conditions are set.
- Closely monitor test conditions to ensure the experimental test plan is followed as designed.

Analyze Data from Experimental or Action Plan (C3)

Typically, one person performs the analysis of the data from the test plan. When necessary, this person should use the department or plant resources available for guidance on the proper data analysis tools and the interpretation of output. The specific tools that should be used depend upon the nature of the test plan. Some of the most frequently used techniques include

- Analysis of variance—one way, multifactor.
- Tukey, Scheffe.
- Post-hoc technique.
- Significance testing (t test)—means, (F test)—variation.
- Regression fitting.
- Chi-square analysis.
- Fractional analysis of variance.
- Correlation analysis.
- Discriminate analysis.
- Nonparametric techniques (abnormal data).
- Switch on/off comparisons.
- Response surface.
- Stepwise regression.

In most cases, a combination of several techniques are used to analyze the data. The use of each of these techniques yields very specific outputs, which need to be interpreted and conclusions drawn from them. These conclusions need to be clearly documented and shared. It is important that everyone involved understand how these conclusions were reached, based on the raw data. Typically, these conclusions center around answering the following:

- Which (if any) causes demonstrated a significant impact (mean, variation) on the dependent variable?
- Were there any interactions? Did a combination of causes, not just one, create the difference?
- What is an accurate estimate of the expected impact on the dependent variable if the cause were eliminated?

TABLE 5.10.   **Flow of Activities for Implementing the Action Plan and "Switch on/off" Technique**

Switch On	Switch Off
Gather predata on dependent variable(s)	Undo changes (where feasible)
Implement solutions	Gather data again on dependent variable
Gather postdata on dependent variable	Analyze data for conclusions

Be careful not to "force" conclusions or try to creatively look at the data to create a difference. If the results of the technique applied indicate no significant impact, accept that conclusion and move on. Often times in data analysis, if carefully performed, evidence can indicate the presence of a cause variable that was not part of the design. This information then could be used when going back to the Do segment.

Decide—Back to "Do" Stage or Proceed (C4)

After reviewing the data analysis conclusions about the suspected causes or solutions that were tested, make a critical decision of what action to take based on this information. Table 5.11 shows some examples.

Write Implementation Plan to Make Change Permanent (C5)

The data analysis step could have been performed in either of the following contexts:

- After the action plan (solutions) was carried out, data analysis was performed to see if the dependent variable was affected. If the conclusions are favorable, go on to develop the implementation plan.
- The experimental test plan was conducted, data were analyzed to verify causes. If the conclusions were favorable (significant causes identified), develop solutions to overcome those causes *before* proceeding to develop the implementation plan (e.g., through the test plan, technician differences were found to contribute to measurement error).

Next, identify ways to eliminate these differences.

The implementation plan, to make the changes permanent, should cover the following areas. In each case, clear accountability for carrying out that function and activity should be identified.

- Changes needed to equipment, procedures, processes—what, who, when, how.
- Training needs.
- Communication needs.
- Approval steps to get changes made.

To write an implementation plan, the team should ask the following critical questions:

- What procedures need to be permanently modified?
- Who needs to be trained and in what to make this permanent?
- Who will do the training?
- What equipment needs to be modified, altered, or added?
- What job responsibilities need to be modified, added, and deleted?
- What work processes need to be altered; how can these changes be documented?
- Who needs to approve these changes?
- How will the changes be permanently implemented? When?
- How will they be phased in?
- Who needs to know that these changes are taking place? Who will do the communication?

Once these questions are answered thoroughly, construct an implementation plan to make the necessary changes permanent.

It is absolutely critical that this plan be carefully and thoroughly prepared to ensure that the *proven remedies* can be implemented smoothly, as intended, and with the support of those involved.

Perform a Force Field Analysis on Implementation (C6)

Once the implementation plan is written, do a force field analysis listing the factors pulling for and against a successful implementation—success in the sense that

TABLE 5.11 Decision (Action) Based on Conclusions from Data Analysis

Conclusion	Decision (Action)
Data indicates, cause(s) had a significant impact on the dependent variable and the effects are estimated to be large enough to affect the measure of effectiveness	Proceed to make changes permanent (Act segment)
Data indicated cause(s) had a significant impact on the dependent variable but the effects may not be large enough to affect the measure of effectiveness	Proceed to make changes permanent for causes that have significant effects but go back to the Do segment to study other suspected causes
Data indicates no significant impact on the dependent variable for any of the suspected causes	Go back to Do segment and recycle through PDCA
Data indicates conflicting conclusions of evidence that test plan was not run as prescribed	Draw no conclusions. Diagnose why problem occurred and rerun the test plan

the results seen in the test situation will be realized on a permanent basis once the solutions are implemented. As a result of this activity, the team should ask two questions:

1. Given these results, is the probability of success high enough to proceed?
2. Looking at the factors pulling against implementation, what (if anything) can be added to the implementation plan to minimize the effects of these negative factors?

The implementation plan should then be revised as needed and finalized.

Management Review and Approval (C7)

Once again, a critical point in the PDCA cycle has been reached, and management approval is needed before proceeding. This meeting is extremely important, because permanent changes need to be made to operations. Management not only needs to approve these changes but also the way in which they will be implemented.

The purpose of the meeting with management is to

- Provide details of the solutions developed as part of the action plan.
- Present the data and logic involved in the conclusions drawn from the data analysis.
- In the case of test plans, present solutions developed to overcome significant causes.
- Obtain approval to make the necessary changes permanent by carrying out the implementation plan or obtain approval to return to the Do segment.

The key output or information to be presented in this session should include the following:

- Experimental test plan data analysis and list of conclusions or action plan output, details of solutions for each cause area.
- In the case of action plans, switch on/off results after the trial.
- Implementation plan.
- Force field analysis on implementation.

5.5.4. Act

Carry out the Implementation Plan (A1)

If a complete, clear, and well-thought-through implementation plan has been written, it will be very obvious what work needs to be done, by whom, and when to carry out the Act segment of the PDCA cycle. Pay significant attention to assuring communications and training are carried out thoroughly, so department members know what is changing, why the change is being made, and what they need to do specifically to make implementation a success.

Determine Postmeasure of Effectiveness (A2)

After all changes have been made and sufficient time has passed for the results of these changes to have an effect, gather data on all the measures of effectiveness. The data then need to be analyzed to see if a *significant shift has occurred*. To accomplish this, do the following:

- Establish control limits for the measure based on the "before" data.
- Extend the limits.
- Plot the postmeasures on the same graph. Check to see if the chart goes OOC on the favorable side. See Figure 5.17.

Analyze Results versus Objectives (A3)

The previous step looked at whether the measure(s) of effectiveness had been affected in any significant way by the permanent implementation of the changes. Do not stop there. If the answer to that question is favorable, then verify if the *amount* of improvement was large enough to meet the objective.

To answer the question, use the tools such as hypotheses testing or confidence intervals. Note: The only evidence that should be accepted as proof that the change has "done its work" is a significant shift in the measure of effectiveness. Until this happens, "do" should not close out.

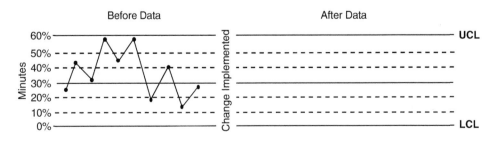

FIGURE 5.17.   Measure of effectiveness.

Gather Feedback (A4)

Once the decision has been made that the PDCA cycle has been successfully completed (based on measure of effectiveness change), present this information to management. Before this is done, gather feedback from those involved. This feedback can be in the form of a questionnaire that all should fill out. The results are tallied and recorded.

Then, call a meeting to review the results as part of the closeout. Also, share the results with management at the management closeout meeting.

The feedback questionnaire will attempt to assess perceptions in the following areas:

- How well was the PDCA format followed and used?
- How efficient was the process?
- How effective and efficient were meetings?
- How much did those involved learn or grow in PDCA and use of the tools and techniques?

Not only can management use these results, over time, as a measure of PDCA progress, the results also can provide valuable information so that appropriate steps may be taken to ensure the mastery of PDCA. There is a strong correlation between the degree to which employees can effectively use PDCA and the continuous improvement of processes, products, work life, and costs.

Hold the Management Closeout Meeting (A5)

Conduct a close-out meeting with management. The major areas to be covered in this meeting are

- Wrap up any implementation loose ends.
- Review the measure of effectiveness results, compare them to the objective.
- Ensure doc umentation is complete and in order.
- Share feedback on the experiences (standardized forms and informal discussion).

Note: A composite picture of the feedback should be given to management.

6

Key Performance Indicators

When identified and aligned properly, key performance indicators (KPIs) can save a plant, a job, a career. If management truly understood the power of KPIs, things would change quickly. In fact, managing without KPIs gives one the feeling of being lost with no hope (a reactive environment). Think of a car with the windshield painted black. The driver cannot see where the car is going but has a glimpse of where it has been, through the rearview mirror. The driver cannot tell if the trip is successful or not until either it's too late—or disaster strikes. The car could go into the ditch (high cost or worse) or never reach the destination (business goals not met). So, too, for a company blind to KPIs. This is a serious problem and it costs companies around the world billions of dollars as a result of what, we consider, a lack of management control. Peter Drucker, the industrial revolutionary, stated: "You cannot manage something you cannot control and you cannot control something you cannot measure."

6.1. DEFINING AND UNDERSTANDING KPIs

Let us first get down to basics and define KPIs. Within maintenance, we first define the performance we want to measure. Is it the performance of the equipment? Is it the performance of the spare parts warehouse? Is it the performance of the maintenance function? This may seem like a simple question, but often I see companies that do not understand their KPIs as they have not defined the specific area of business for which the performance is being measured.

Assume we want to measure the performance of the maintenance function. One of two kinds of KPIs must be chosen for measuring any particular function of a business: leading indicators or lagging indicators (also referred here as *leading* and *lagging KPIs*). Leading KPIs lead to results, such as scheduled compliance; lagging KPIs are the results, such as maintenance cost (effected if scheduling is not working).

We use leading indicators to manage a part of the business, while lagging indicators measure how well we have managed. With leading indicators, therefore, it is possible to directly and immediately respond when a poor result is found. With lagging indicators, we get value from knowing how well we performed but have little opportunity to immediately affect underperformance. Instead, when we see an unacceptable lagging indicator, we typically must drill down to the leading indicators to uncover the cause of the underperformance; and from there, we can implement appropriate changes.

Leading indicators for the maintenance function are those that measure how well we conduct each of the steps in the maintenance process. For example, a leading indicator for the work planning element of maintenance process could be the percentage of planned jobs executed using the specified amount of labor. If the planner is estimating labor correctly, we see a high percentage of jobs completed using the planned amount of labor hours. A maintenance manager who finds that the value of the KPI is lower than expected can speak with the planner about how best to improve the results immediately, possibly for the remainder of that day. With all KPIs, by definition, we measure past performance, so we do not suggest that leading indicators can be tweaked to improve past performance. But, if we are managing using leading indicators, we can respond immediately when needed.

So leading indicators measure how well we perform our jobs, while lagging indicators measure results. We manage using leading indicators, and we react to results using lagging indicators.

In the maintenance example, a lagging indicator measures the results of how well we managed the maintenance function. In a situation where the maintenance function is well managed, we expect an appropriate balance between the cost of maintenance and the plant availability. A lagging indicator therefore could be the actual maintenance cost for a month, as a percentage of the budgeted maintenance cost for that month. If the actual maintenance cost for last month is found to be 110% of budget, we can do very little to directly influence the performance of this KPI today. Instead, we look at all the leading indicators, probably including those that measure the performance of the maintenance process, to determine whether those values give us a signal for managing the problem.

Unfortunately, in the quest for excellence, we often are attracted to outside consultants that offer "benchmarking" services, claiming to provide all the KPIs we need to effectively run the business. Be careful, when considering these services, that you are not signing up for a laundry list of lagging indicators, since they will not help you with managing; they'll just quantify the problem you already acknowledged when you sought outside help.

Figure 6.1 shows how leading indicators for the maintenance process can provide management capability, while the lagging indicators show us how well we managed the maintenance function. Leading indicators such as Percent of Rework and Percent of PMs Executed on Time affect the overall performance of the maintenance process, which results in a certain level of performance. The lagging indicators in this case, which are affected by these leading indicators, are Maintenance Cost as a Percent of Budget and Plant Availability. At least one of these lagging indicators will suffer if there is sufficient underperformance in the leading indicators. This example shows the alignment of the maintenance process as KPIs transition from leading to lagging.

Figure 6.1 does not show the specific KPIs used to manage the maintenance process. Instead, some of those are listed in Table 6.1, along with the world-class target level, where applicable.

In the same way as in the maintenance example, KPIs can be used in other areas of the business. This approach is particularly interesting where multiple functional areas play roles in a given goal, such as plant reliability. Plant reliability is a shared responsibility of the maintenance, production, and engineering. Leading indicators for each departmental process would feed the lagging indicators for the department function, which would then summarize to the plant level, as shown in Figure 6. 2.

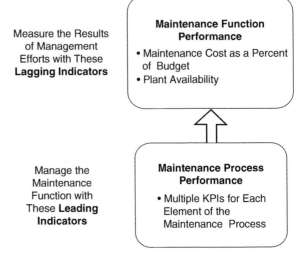

FIGURE 6.1.   Leading and lagging indicators.

6.1.1. The Problem

The problem is that management should learn to manage operations through KPIs (both leading and lagging). In over 30 years, we have seen many plants shut their doors forever. The closings were blamed on many reasons, but the one thing all had in common was that *none* had properly managed with the KPIs. The metrics or indicators they managed with were ones like cost, asset availability, equipment downtime, and overall equipment effectiveness.

All these measurements or indicators, while useful for measuring performance, cannot be used to manage the maintenance and reliability process. They are simply the results of all the previous actions in the maintenance and reliability process. Again, one cannot manage results. One can manage only the processes leading to the results. A company that uses any of these metrics to manage its operation, without leading indicators, is in a reactive mode.

Companies must ask themselves some very basic questions:

- Does the company differentiate between those KPIs that can be used to manage (leading indicators) from those that measure results (lagging indicators)?
- Does the company measure the performance of the maintenance process, which it easily can manage when needed?

If leading indicators show underperformance, then that underperformance will affect the lagging indicator, which could be reliability, cost, capacity, and the like. People must understand the relationship between a leading and lagging indicator and their effects on the maintenance and reliability function.

TABLE 6.1.   Specific Examples of Leading and Lagging Indicators

KPI Type	Measure	Key Performance Indicator	World Class Target Level
Result/lagging	Cost	Maintenance cost	Context specific
Result/lagging	Cost	Maintenance cost/replacement asset value of plant and equipment	2–3%
Result/lagging	Cost	Maintenance cost/manufacturing cost	<10–15%
Result/lagging	Cost	Maintenance cost/unit output	Context specific
Result/lagging	Cost	Maintenance cost/total sales	6–8%
Result/lagging	Failures	Mean time between failure	Context specific
Result/lagging	Failures	Failure frequency	Context specific
Result/lagging	Downtime	Unscheduled maintenance related downtime (hours)	Context specific
Result/lagging	Downtime	Scheduled maintenance related downtime (hours)	Context specific
Result/lagging	Downtime	Maintenance related shutdown overrun (hours)	Context specific
Process/leading	Maintenance strategy	Percentage of work requests in "request" status for less than 5 days, over the specified time period	80% of all work requests should be processed in 5 days or less
Process/leading Planning element/lagging	Planning	Percentage of work orders with work-hour estimates within 10% of actual, over the specified time period.	Accuracy of greater than 90%
Process/leading	Planning	Percentage of work orders, over the specified time period, with all planning fields completed	95% +
Process/leading	Planning	Percentage of work orders assigned "rework" status (due to a need for additional planning) over the last month.	Should not exceed 2–3%
Process/leading	Planning	Percenage of work orders in "new" or "planning" status less than 5 days, over the last month	80% of all work orders should be possible to process in 5 days or less; some work orders require more time to plan but attention must be paid to late finish date
Process/leading Scheduling element/lagging	Scheduling	Percentage of work orders, over the specified time period, having a scheduled date earlier or equal to the late finish or required by date	95%+ should be expected to ensure the majority of work orders are completed before their late finish date
Process/leading	Scheduling	Percentage of scheduled available work hours to total available over the specified time period	Target 80% of work hours applied to scheduled work
Process/leading	Scheduling	Percentage of work orders assigned "delay" status due to unavailability of personnel, equipment, space, or services over the specified time period	Number should not exceed 3–5%
Process/leading	Execution	Percentage of work orders completed during the schedule period before the late finish or required-by date	Schedule compliance of 90%+ should be achieved
Process/leading Execution element/lagging	Execution	Percentage of maintenance work orders requiring rework	Rework should be less than 3%
Process/leading	Follow up	Percentage of work orders closed within 3 days, over the specified time period	Should achieve 95%+; expectation is that work orders are reviewed and closed promptly

Most maintenance managers are told to control costs, improve reliability, and increase asset availability, with no idea where the problem may be in the maintenance process. Unfortunately, many managers lose their job as a result. No one can control cost, reliability, or availability without managing the maintenance process.

6.1.2. John Day

Since 1999, Alumax has been a leader in all alloys of aluminum. The Mt. Holly, SC, plant was rated as one of the best maintained plants in the world for over 20 years. John Day, the company's former engineering and maintenance manager comments on how

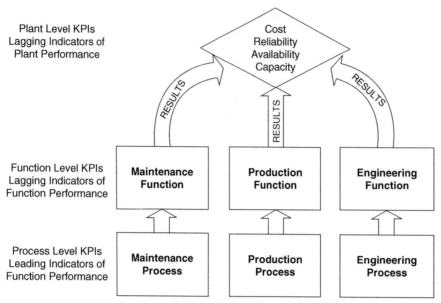

FIGURE 6.2.   The use of leading and lagging indicators across functional areas.

he managed using KPI's, "Hundreds of companies visited our plant, paying $1000 each to see our maintenance program up close, but only a few learned from their visit." Day feels they missed out on how Alumax managed with aligned KPIs. He also was invited to visit over 500 plants in the United States, Canada, and Australia and says, "The one thing over 90% of them had in common was they could not effectively manage their plants because they had no leading KPIs in place. Many of these companies were crying for help but did not know which way to go." Most managed and made decisions with only lagging indicators such as cost and reliability.

"For over 20 years, I could see problems brewing long before they would become a serious issue. Alumax had a system in place where we could measure everything in our maintenance process—from leading indicators such as the identification of potential failures through to the lagging financial results of all actions performed by maintenance." This separation of leading and lagging KPIs allowed him to make management decisions when leading KPI underperformance was identified before cost and reliability (the lagging indicators) were affected.

According to Day, most companies do not succeed because they do not know what information needs to be collected. In 1979, he worked with Alumax's accounting department to establish over 60 financial accounts just for maintenance. These financial accounts were linked to leading KPIs in the maintenance process, which provided information needed to manage proactively. In turn, these KPIs were linked to equipment performance, also lagging indicators. Each of these lagging KPIs had established benchmarks, which measured if the maintenance process was in or out of control. This approach may sound complex, but once you have it in place, management can truly manage the reliability of plant equipment.

Day shared 13 years of KPI data that was so impressive it would bring tears to any maintenance and reliability professional's eyes. Describing the data, he stated, "Everyone from a maintenance person to the plant manager had KPIs they looked at on a daily or weekly basis in order for them to make basic and immediate management decisions. Each level in our organization utilized a small number of lagging KPIs, along with a bigger number of leading KPIs that were important to managing their part of the business." In reviewing Alumax's KPIs over the 13-year period, we found that their maintenance cost (a lagging KPI) did not increase but was constant. Maintenance cost as a percentage of return on asset value held at around 3% for all of those years. Equally impressive was that the controllable plant operating cost was very constant over this same time period. The lagging indicator data pointed to the obvious fact that the reliability of equipment directly correlates to operating cost.

By managing the maintenance and reliability process, element by element, using leading indicators, Alumax was able achieve these results. Day's experience validates that managing with both leading and lagging KPIs is the only way to effectively manage an operation to achieve the results expected to succeed in a business. By the way, over 26 years ago, one of us worked for John Day at Alumax and enjoyed every day working for him.

6.1.3. The Solution

How much money do corporations lose every year due to plants not managing with good leading and lagging KPIs? The costs may be too high to calculate, so we must stop these massive losses now by putting in place a plan to develop and align KPIs. This chapter may save someone's plant or job. But, do not look for shortcuts in the process, because there are none.

Step 1. Educate management, from executive level to floor level supervisors, on KPIs and how the leading and lagging indicators should be aligned to meet the business goals. Then, provide a similar education to the maintainers and operators.

Step 2. Define and assess the current maintenance and reliability processes against a future state. A future state is formed of known maintenance and reliability "best practices." As part of this assessment, develop a business case with financial opportunities and the costs of change. This step continues the education process and creates an awareness of the opportunity at hand.

Step 3. Develop a plan based on the assessment to include financial opportunities and cost on a time line. This plan must include

- The definition of the elements of the maintenance and reliability process (work identification, planning, scheduling, work execution, etc.).
- The work flow process for each element in that process.
- The definition of roles and responsibilities for each task.
- The specification of leading and lagging KPIs in each element of the process.
- Targets and world-class benchmarks established against the defined KPIs.

Step 4. Implement the process and begin managing based on leading indicators. Begin by measuring only a few KPIs (maximum of three). Then, allow people at the lowest levels to make the decisions required to ensure the maintenance and reliability process is proactive and effective. The use of leading KPIs is a great awareness tool and brings everyone into the decision-making process.

This process is not easy, however; it is not magic either. Developing KPIs is time consuming but must be done for a company to survive.

6.2. KPI DASHBOARDS

KPI dashboards are the alignment of KPIs for specific positions that allows a specific individual or group of individuals the metrics required to manage the process for which they are responsible.

6.2.1. Plant Manager Dashboard

- Total cost per unit.
- Overall equipment effectiveness (OEE)—total plant (combination of all lines or production areas together providing a plant score).

OEE, for the plant manager, is defined in Figure 6.3 and includes *availability* (running 24/7, loss is measured by calculating any action or inaction that takes away from this time period, this identifies the "hidden plant"); *performance rate* (the best rate ever sustained in a production process or design rate, whichever is greater); and *quality rate* (calculated as "first-pass quality" only). Note: It is acceptable to have a low number in any of the categories. The objective of KPIs is to provide management with *real* numbers with which to manage and not to beat up people.

6.2.2. Plant Management Team Dashboard

The plant management team consists of maintenance, production, purchasing, accounting, and the like. The dashboard includes

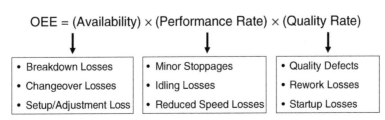

OEE = (Availability) × (Performance Rate) × (Quality Rate)

• Breakdown Losses	• Minor Stoppages	• Quality Defects
• Changeover Losses	• Idling Losses	• Rework Losses
• Setup/Adjustment Loss	• Reduced Speed Losses	• Startup Losses

FIGURE 6.3.   Definition of OEE for the plant manager.

- Total cost per unit by production area or line, including maintenance cost per unit and production cost per unit.
- Mean time between functional failures (MTBFF), by production area or line.
- OEE by production area or line.

OEE, for the plant management team, also is defined in Figure 6.3 and includes *availability* (running 24/7, loss is measured by calculating any action or inaction that takes away from this time period, this identifies the "hidden plant"); *performance rate* (the best rate ever sustained in a production process or design rate, which ever is greater); and *quality rate* (calculated as "first-pass quality" only). Note: It is acceptable to have a low number in any of the categories. The objective of KPIs is to provide management with *real* numbers with which to manage and not to beat up people.

As an example, consider a plant with one production line, operating eight hours a day, five days a week (40 hours per week). The current rate is 22,000 units per hour; the production design rate is 40,000 units per hour; quality losses are at 8% with second-quality product (which is sold) of 11%. The OEE for one week is

Availability (measured 24/7): 24 hours ×
7 days = 168 hours (100%)

Currently operates at 8 hours × 5 days = 40 hours

Availability: 40/168 = 23.8%

Performance: design rate = 40,000 units per hour

Current rate = 22,000 units per hour

Performance = 22,000/40,000 = 55%

Quality of first-pass quality only

Current quality losses including second quality = 19%

Quality = 100% − 19% = 81%

OEE = 23.8% × 55% × 81% = 10.6%

These numbers show the plant's current operating health, so smart decisions can be made to make the plant successful. Never use a modified OEE. Many companies fall into this trap and are led to believe a number that is not "real." Many companies do not like a number under 90%, so they develop a "modified" OEE. It does not work.

6.2.3. Production Manager (Supervisor) Dashboard

The production manager's dashboard has plant level KPIs specific to production and involve production cost per unit produced, which is reviewed and trended weekly.

- Total percentage of time that product changeover standard is met.
- Total percentage of time production equipment is operating at less than standard throughput: line speed, units per hour, and so forth.
- Total number of units produced (measured against the budget).
- Total percentage of off-quality (less than first-pass quality, anything less is not measured).
- MTBFF for the total plant.
- Percentage of overtime.
- Percentage of absenteeism.
- Number of discipline violations.
- Number of safety incidents.
- Number of days without loss-time accidents.

The production supervisor's dashboard has the same KPIs but at the level of that supervisor's production line or production area.

6.2.4. Production Operator Dashboard

- Total percentage of time product changeover standard is met.
- Total percentage of time production equipment is operating less than standard throughput: line speed, units per hour, and so forth.
- Total number of units produced (measured against the budget).
- Total percentage of off-quality (less than first-pass quality, anything less is not measured).
- MTBFF.
- Number of safety incidents.
- Number of days without loss-time accidents.
- Number of environmental incidents.

6.2.5. Maintenance Manager (Supervisor) Dashboard

The maintenance manager's dashboard has plant level KPIs specific to the total plant maintenance and involves the maintenance cost per unit produced. The following are measured as the maintenance cost of the return on asset value:

- MTBFF for the total plant.
- Maintenance labor cost (measured against a target).
- Maintenance material cost (measured against a target).
- Maintenance contractor cost (measured against a target).
- Percentage of overtime.
- Percentage of absenteeism.
- Number of discipline violations.
- Number of safety incidents.
- Number of days without loss-time accidents.

- Number of environmental incidents.
- Capital maintenance cost per unit produced.

The maintenance supervisor's dashboard has the same KPIs but at the level of that supervisor's area of responsibility only and does not include the capital maintenance cost per unit produced.

6.2.6. Maintenance Staff Dashboard

The maintenance staff's dashboard is at the level of that person's area of responsibility only and is measured as the maintenance cost per unit produced.

- MTBFF.
- Total percentage of time production equipment is operating less than standard throughput: line speed, units per hour, and so forth.
- Total number of units produced (measured against the budget).
- Total percentage of off-quality (less than first-pass quality, anything less is not measured).
- Number of safety incidents.
- Number of days without loss-time accidents.
- Number of environmental incidents.

6.2.7. Reliability Engineer Dashboard

- Bad actors report of top five critical assets with the highest total maintenance cost and worst reliability.
- MTBFF by production line and area.
- Percentage of critical assets with a maintenance strategy developed using an RCM methodology.
- Percentage of new assets with a maintenance strategy developed using an RCM methodology and ranked based on risk to the business.

6.2.8. Engineering Manager Dashboard

- Percentage of total projects completed on budget.
- Percentage of total projects completed within 10% of the budget.
- Dollar value of overspent budget for total projects by month and by year.
- Percentage of projects started up on time and with full capacity.
- Dollar value of change orders by month, project, and year.
- Percentage of projects started up on schedule with a maintenance strategy developed with an RCM methodology.
- Percentage of projects started up on schedule with all prints updated and in the native language.

- Downtime percentage of new project for one year.
- Throughput losses for a new project for one year.
- Percentage of projects with 80% of operating procedures developed prior to scheduled startup.
- Total MTBFF of all new projects for one year and two years.

6.2.9. Purchasing Manager Dashboard

- Percentage of time vendors deliver on time (standard = 98%).
- Percentage of time vendors deliver what was requested (type and quantity).
- Mean time to order emergency parts (standard = 30 minutes).
- Mean time to order normal parts (standard = 1 hour).

6.2.10. Maintenance Stores Manager

- Number of stockouts.
- Number of times an employee waits longer than five minutes for parts (24/7).
- Inventory accuracy (standard = 99%).
- Maintenance material as a percentage of return on asset value (world class = 0.25 to 0.75%).
- Percentage of stores preventive maintenance completed on time by month.
- Stores value by month.

6.2.11. Conclusion

Remove the black windshield and manage with leading indicators and not with lagging indicators. Leading KPIs should be used to drive the decision-making process. Remember leading indicators (KPIs) are manageable, while lagging indicators just offer information on how well the plant was managed. To be the best, step up to the plate and manage in the most efficient manner by following these recommendations.

6.3. MEASURING AND MANAGING THE MAINTENANCE FUNCTION*

Performance measurement is a fundamental principle of management. The measurement of performance is important because it identifies current performance

*Source: "Key Performance Indicators: Measuring and Managing the Maintenance Function" (Burlington, ONT, Canada: Ivara Corporation, January 2006).

gaps between current and desired performance and provides indication of progress toward closing the gaps. Carefully selected key performance indicators identify precisely where to take action to improve performance.

This section deals with the identification of key performance indicators for the maintenance function, by first looking at the ways that maintenance performance metrics relate to manufacturing metrics. Since performance measurements for maintenance must include both results metrics and metrics for the process that produces the results, this section presents a representation for the business process for maintenance. The section then identifies typical business process and results metrics that can be used as key performance indicators for the maintenance function.

6.3.1. Physical Asset Management

The purpose of most equipment in manufacturing is to support the production of product destined to downstream customers. Ultimately the focus is on meeting customer needs. This is illustrated in Figure 6.4. Customer expectations are normally defined in terms of product quality, on-time delivery and competitive pricing. By reviewing the composite requirements of all current customers and potential customers in those markets we wish to penetrate, the performance requirements of our physical assets can be defined. Manufacturing performance requirements can be associated with quality, availability, customer service, operating costs, safety and environmental integrity.

To achieve this performance there are three inputs to be managed. The first requirement is *Design Practices*. Design practices provide capable equipment "by design" (inherent capability), to meet the manufacturing performance requirements.

The second requirement is *Operating Practices* that make use of the inherent capability of process equipment. The documentation of standard operating practices assures the consistent and correct operation of equipment to maximize performance.

The third requirement is *Maintenance Practices* that maintain the inherent capability of the equipment. Deterioration begins to take place as soon as equipment is commissioned. In addition to normal wear and deterioration, other failures may also occur. This happens when equipment is pushed beyond the limitations of its design or operational errors occur. Degradation in equipment condition results in reduced equipment capability. Equipment downtime, quality problems or the potential for accidents and/or environmental excursions are the visible outcome. All of these can negatively impact operating cost.

Manufacturing key performance indicators provide information on the current state of manufacturing. Asset capability, operating practices and the maintenance of asset condition all contribute to the ability to meet these performance requirements.

Some typical key performance indicators for manufacturing include operating cost; asset availability, lost time injuries, number of environmental incidents, OEE and asset utilization.

Consider asset utilization, as depicted in Figure 6.5. Asset utilization is a manufacturing level key performance indicator. It is a function of many variables. For example, asset utilization is impacted by both maintenance and non-maintenance-related downtime. Non-maintenance related downtime may be attributed to

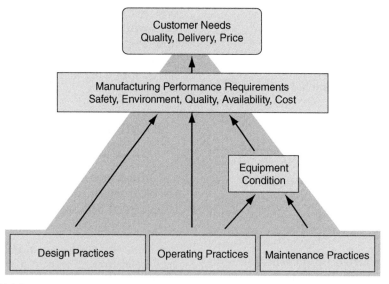

FIGURE 6.4. Managing manufacturing performance requirements to meet customer needs.

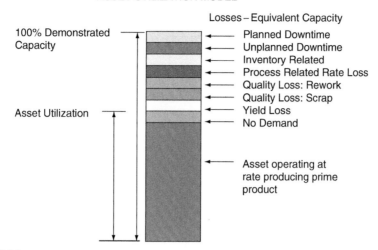

FIGURE 6.5. Asset utilization is an example of a manufacturing level key performance indicator.

lack of demand, an interruption in raw material supply or production scheduling delays beyond the control of the maintenance function. Asset utilization is also a function of operating rate, quality and yield losses, etc. In each of these areas maintenance may be a factor but it is not the only contributor. In order to maintain and improve performance each function in the organization must focus on the portion of the indicators that they influence.

Similarly, other manufacturing level performance indicators are not only a function of maintenance. They are affected by causes beyond the control of the maintenance function. Asset capability, operating practices and the maintenance of asset condition all contribute to the ability to meet performance requirements. If a manufacturing level indicator is used to measure maintenance performance, improved maintenance may not result in a proportional improvement in the manufacturing metric. For instance, in the asset utilization example, cited above, the maintenance contributors may all be positive and yet the resulting asset utilization may not improve due to other causes.

A key principle of performance management is to measure what you can manage. In order to maintain and improve manufacturing performance each function in the organization must focus on the portion of the indicators that they influence. Maintenance performance contributes to manufacturing performance. The key performance indicators for maintenance are children of the manufacturing key performance indicators.

Key performance indicators for maintenance are selected ensuring a direct correlation between the maintenance activity and the key performance indicator measuring it. When defining a key performance indica-

tor for maintenance a good test of the metric validity is to seek an affirmative response to the question; "If the maintenance function does 'everything right,' will the suggested metric always reflect a result proportional to the change; or are there other factors, external to maintenance, that could mask the improvement?"

This section focuses on defining key performance indicators for the maintenance function, not the maintenance organization. The maintenance function can involve other departments beyond the maintenance organization. Similarly, the maintenance department has added responsibilities beyond the maintenance function and, as such, will have additional key performance indicators to report. The key performance indicators for the maintenance organization may include key performance indicators for other areas of accountability such as health and safety performance, employee performance management, training and development, etc.

6.3.2. The Asset Reliability Process

The management of physical asset performance is integral to business success. What we manage are the business processes required to produce results. One of these business processes is responsible for the maintenance of physical asset reliability. The Asset Reliability Process is shown in Figure 6.6. It is an integral part of a much larger business process responsible for managing the total enterprise.

A proactive Asset Reliability Process, represented by the seven elements in the model, aims to deliver the performance required by the enterprise to meet all of its corporate objectives. Each element within the

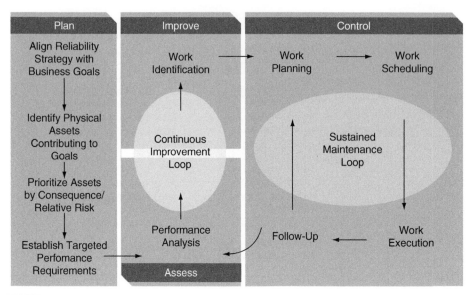

FIGURE 6.6.　The Asset Reliability Process identifies what's required to manage the maintenance function.

maintenance process is in itself a sub-process. A brief description of each element follows:

Business Focus, represented by the box on the left, focuses the maintenance of physical asset reliability on the business goals of the company. The potential contribution of the asset base to these goals is evaluated. The largest contributors are recognized as critical assets and specific performance targets identified.

Work Identification, as a process, produces technically based Asset Reliability Programs. Program activities identify and control failure modes impacting on the equipment's ability to perform the intended function at the required performance level. Activities are evaluated to judge if they are worth doing based on the consequences of failure.

Planning develops procedures and work orders for these work activities. The procedures identify resource requirements, safety precautions and special work instructions required to execute the work.

Scheduling evaluates the availability of all resources required for work "due" in a specified time frame. Often this work requires the equipment to be shut down. A review of production schedules is required. Resources are attached to a specific work schedule. The use of resources is balanced out.

In the *Execution* process, trained, competent personnel carry out the required work.

The *Follow-up* process responds to information collected in the execution process. Work order completion comments outline what was done and what was found. Actual time and manpower, to complete the job, is documented. Job status is updated as complete or incomplete. Corrective work requests, resulting from

the analysis of inspection data, are created. Requests are made for changes to drawings and procedures.

The process of *Performance Analysis* evaluates maintenance program effectiveness. Gaps between actual process performance and the required performance are identified. Historical maintenance data is compared to the current process performance. Maintenance activity costs are reviewed. Significant performance gaps are addressed by revisiting the Work Identification function.

Each element is important to provide an effective maintenance strategy. Omitting any element will result in poor equipment performance, increased maintenance costs or both.

For example, Work Identification systematically identifies the Right Work to be performed at the Right Time. Without proper Work Identification, maintenance resources may be wasted. Unnecessary or incorrect work will be planned. Once executed, this work may not achieve the desired performance results, despite significant maintenance costs. Without Planning the correct and efficient execution of the work is left to chance. The Planned Maintenance Process is a cycle. Maintenance work is targeted to achieve required asset performance. Its effectiveness is reviewed and improvement opportunities identified. This guarantees continuous improvement in process performance impacted by Maintenance.

Within the Planned Maintenance Process two internal loops exist. Planning, Scheduling, Execution and Follow Up make up the first loop. Once maintenance activities are initially identified, an asset maintenance program, based on current knowledge and requirements, is initiated. The selected maintenance activities will be enacted upon at the designed frequency and maintenance tolerance limits. The process is self-sustaining.

The second loop consists of the Work Identification and Performance Analysis elements. This is the continuous improvement loop. Actual asset performance is monitored relative to the required performance (driven by business needs). Performance gaps are identified. The cause of these gaps is established and corrective action recommended.

6.3.3. Performance Metrics for the Maintenance Function

The Asset Reliability Process represents the collection of all tasks required to support the maintenance function. The process is a supply chain. If a step in the process is skipped, or performed at a substandard level, the process creates defects known as failures. The output of a healthy reliability process is optimal asset reliability at optimal cost.

Asset Reliability Process measures are leading indicators. They monitor if the tasks are being performed that will lead to results. For example a leading process indicator would monitor if the planning function was taking place. If people are doing all the right things then the results will follow. The leading process indicators are more immediate than result measures.

Result measures monitor the products of the Asset Reliability Process. Result measures include maintenance cost (as a contributor to total operating cost), asset downtime due to planned and unplanned maintenance (as a contributor to availability) and number of failures on assets (the measure of reliability—this can then be translated into mean time between failures). Result measures lag. Failure is a good example. Typically the same piece of equipment doesn't fail day after day. Take a pump for example. Say the pump fails on average once every eight months. If we improve its reliability by 50% it will now fail every 12 months. You have to wait at least 12 months to see the improvement.

Key performance indicators for the maintenance function need to include both leading (maintenance process) measures and lagging (result) measures. This section focuses on identifying both leading and lagging measures of maintenance performance. Collectively, these measurements are the key performance indicators for the maintenance function.

6.3.4. Reliability Process Key Performance Indicators—Leading Measures

The maintenance process is made up of elements. All elements are required to complete the supply chain.

Key performance indicators of the maintenance process are process assurance measures. They answer the question "how do I know that this maintenance process element is being performed well?" The day-to-day execution of maintenance is addressed through the seven elements of the Reliability Process: Business Focus, Work Identification, Work Planning, Work Scheduling, Work Execution, Follow-up and Performance Analysis. Key performance indicators for each element are recommended.

It should be noted that variations of these metrics may be defined or additional performance metrics may be used. The metrics presented here provide a clear indication if the requirements of each element are being satisfied and, if not, what action should be taken to correct the lack of maintenance process adherence.

6.3.5. Work Identification

The function of work identification is to identify "the right work at the right time."

Work Requests

Initiating a work request is one method of identifying work. Once a work request is submitted it must be reviewed, validated and approved before it becomes an actual work order ready to be planned. If the work request process is performing well, the validation and approval/rejection of work requests should occur promptly.

A suggested measure for the work request process is:

- The percentage of work requests remaining in "Request" status for less than 5 days, over a specified time period (for example the last 30 days). The world class maintenance expectation is that most work (>80%) requests would be reviewed and validated within a maximum of five days.

Work requests rely on the random identification of problems or potential problems and bringing them to the attention of maintenance to address them. In a world class organization, work identification is not left to chance.

Proactive Work

The Asset Maintenance Program is designed to identify potential failure conditions, changes in state of hidden functions and known age related failure causes. The development of the Asset Maintenance Program defines the routine maintenance tasks that must be executed to achieve the performance levels required to meet business requirements. If the Asset Maintenance Program is effective, it will successfully identify and address most maintenance preventable causes of failure.

If the Work Identification function is working well, the majority of work performed by maintenance would consist of executing the Asset Maintenance Program (AMP) tasks and the corrective work originating from it.

The key performance indicator for the work identification element is:

- The percentage of available work hours used for proactive work (AMP + AMP initiated corrective work) over a specified time period. The world class maintenance target for proactive work is 75 to 80%. Recognizing that 5–10% of available work hours should be attributed to improvement work (non-maintenance) this would leave approximately 10–15% reactive work.

6.3.6. Work Planning

The primary function of the Work Planning element of the maintenance process is to prepare the work to achieve maximum efficiency in execution.

Amount of Planned Work

In general terms, planning defines how to do the job and identifies all the required resources and any special requirements to execute the work. A properly planned work order would include all this information. Maximizing maintenance efficiency requires a high percentage of planned work.

A measure of whether planning is taking place is:

- The percentage of all work orders, over a specified time period, with all the planning fields completed (e.g., labor assignments, task durations, work priority, required by date, etc.). The world class expectation is that >95% of all jobs should be planned.

Responsiveness of Planning

Another key performance indicator for planning is the time it takes a work order to be planned. A suggested measure of this is:

- The percentage of work orders in planning status for less than five days, over a specified time period. A world class performance level of at least 80% of all work orders processed in five days or less should be possible. Some work orders will require more time to plan but attention must be paid to late finish or required by date.

Quality of Planning

These key performance indicators for planning do not reflect the quality of the planning being done. A critical aspect of planning is estimating resources. The

quality of planning can be measured by monitoring the accuracy of estimating. Labor and material resources are the dominant resources specified on a work order.

The accuracy of estimating labor can be measured by:

- The percentage of work orders with work hour estimates within 10% of actual over the specified time period. Estimating accuracy of greater than 90% would be the expected level of world class maintenance performance.

A second metric of planning quality, addressing material estimates, would be:

- The percentage of planned, scheduled and assigned work orders, where execution is delayed due to the need for materials (spare parts) over the specified time period. The world class maintenance expectation is that less than 2% of all work assigned will have a material deficiency (due to planning). Note: this assumes the job should not have been scheduled if the materials were not available. Therefore, the problem is that the work order did not account for all the required materials.

6.3.7. Work Scheduling

Good planning is a prerequisite to scheduling. The primary function of scheduling is to coordinate the availability of the asset(s) to be maintained with all the required resources: labor, material and services creating a schedule to execute "the right work at the right time." The schedule is a contract between operations and maintenance. The "right work at the right time" implies that this work must be executed within the specified time period to achieve the desired level of performance. Failure to execute within the schedule period will increase the risk of failure.

With good work identification, planning and scheduling in place, the weekly maintenance schedule should be produced several days in advance of the beginning of the schedule period. There should be confidence that this schedule reflects the work that will be completed through the schedule period.

Quality of Scheduling

A key performance indicator for the scheduling function is:

- The percentage of work orders, over the specified time period, that have a scheduled date earlier or equal to the late finish or required by date. A world class maintenance target of >95% should be expected in order to ensure the majority of the work orders are completed before their late finish or required-by date.

A second measure of the quality of scheduling is:

• The percentage of work orders assigned "Delay" status due to unavailability of manpower, equipment, space or services over the specified time period.

Volume of Scheduled Work

The scheduling of properly planned work is also important to maximize maintenance efficiency. We would anticipate that a high percentage of the available maintenance work hours would be committed to a schedule. A second scheduling key performance indicator measures:

• The percentage of scheduled available work hours to total available work hours over the specified time period. A world class target of >80% of work hours should be applied to scheduled work.

It is not desirable to schedule 100% of available work hours within a schedule period, because we recognize that additional work will arise after the schedule has been cast. This includes both emergency work and other schedule write-ins that *must* be accommodated during the schedule period.

6.3.8. Work Execution

Work execution begins with the assignment of work to the people responsible for executing it and ends when the individuals charged with responsibility for execution provide feedback on the completed work.

Schedule Compliance

With a high quality of work identification, planning and scheduling, maintenance resources should execute according to the plan and schedule. Therefore, a key performance indicator of execution is schedule compliance. Schedule compliance is defined as:

• The percentage of work orders completed during the schedule period before the late finish or required by date. World class maintenance should achieve >90% schedule compliance during execution.

Quality of Work Execution

Work execution quality is measured by:

• The percentage of rework. World class levels of maintenance rework are less than 3%.

Work Order Completion

The purpose of identifying maintenance process key performance indicators is to help manage the maintenance process. The ability to successfully monitor and manage the process and measure the results of the process is highly dependent on gathering correct information during work execution. The vehicle for collecting this information is the work order. Work orders should account for all work performed on assets. This is necessary to gather accurate maintenance cost and history data, enabling the management of the physical asset through its life cycle.

A returned work order should indicate the status of the job (complete, incomplete), the actual labor and material consumed, an indication of what was done and/or what was found and recommendations for additional work. In addition, information about process and equipment downtime and an indication of whether the maintenance conducted was in response to a failure should be provided.

The idea that the job is not done until the work order is completed and returned is a significant challenge to many organizations. For this reason it is also important to have a key performance indicator on work order completion. This metric should look at:

• The percentage of work orders turned in with all the data fields completed. World class maintenance organizations achieve 95% compliance.

6.3.9. Follow-Up

In the follow-up element of the maintenance process, actions are initiated to address the information identified during execution. Some key follow-up tasks include reviewing work order comments and closing out completed work orders, initiating corrective work and initiating part and procedural updates as required.

Work Order Closure

Timely follow-up and closure of completed work orders is essential to maintenance success. A key performance indicator for follow-up is:

• The percentage of work orders closed within a maximum of three days, over the specified time period. The expectation is that >95% of all completed work orders should be reviewed and closed within three days.

6.3.10. Performance Analysis

The performance analysis element of the maintenance process evaluates maintenance effectiveness by focusing on key performance indicators of maintenance results. Gaps between the actual and required performance of the maintained asset are identified. Significant performance gaps are addressed by initiating work identification improvement actions to close the performance gap.

Presence of Performance Analysis

One indication that performance analysis is being executed is the existence of the maintenance result metrics described in "Key Performance Indicators of Maintenance Effectiveness" below.

Quality of Performance Analysis

From a maintenance process perspective it is important that these results are driving action. Therefore, a key performance indicator for performance analysis is a measure of:

- The number of reliability improvement actions initiated through performance analysis during the specified period. No absolute number is correct but no number suggests inaction.
- A second measure is the number of asset reliability actions resolved over the last month. In other words, a measure of how successful the organization is in performance gap closure.

6.3.11. Key Performance Indicators of Maintenance Effectiveness (Result Measures)

The product of maintenance is reliability. A reliable asset is an asset that functions at the level of performance that satisfies the needs of the user. Reliability is assessed by measuring failure.

Failures

The primary function of maintenance is to reduce or eliminate the consequences of physical asset failures. The definition of functional failure is any time that asset performance falls below its required performance. Therefore a key performance indicator for *maintenance effectiveness* is some measurement of failure on the asset(s). If the maintenance function is effective, failures on critical assets and thus their consequences should be reduced or eliminated.

Failure consequence impacts manufacturing level key performance indicators. Failure classification by consequence identifies the contribution of maintenance function to manufacturing level performance.

Failure consequences are classified into the following categories:

1. Hidden Consequence—there is no direct consequence of a single point failure other than exposure to the increased risk of a multiple failure (a second failure has to occur to experience a consequence).
2. Safety Consequence—a single point failure results in a loss of function or other damage which could injure or kill someone.
3. Environmental Consequence—a single point failure results in a loss of function or other damage which breaches any known environmental standard or regulation.
4. Operational Consequence —a single point failure has a direct adverse effect on operational capability (output, product quality, customer service or operating costs in addition to the direct cost of repair).
5. Non-Operational Consequence—a single point failure involving only the cost of repair.

Therefore, it is important to track the number and frequency of asset failures by area of consequence. There is no universal standard for this metric because of the diversity of industries and even of plants within industry segments. It is however reasonable to expect a downward trend and to set reduction targets based on current performance levels and business needs.

Maintenance Costs

Maintenance costs are another direct measure of maintenance performance. Maintenance costs are impacted by both maintenance effectiveness and the efficiency with which maintenance is performed.

Maintenance maximizes its effectiveness by ensuring that it performs "the right work at the right time." Proactive maintenance means intervening before the failure event occurs. The impact of proactive maintenance is not only to minimize the safety, environmental and operational consequences of failure but also to reduce the cost of maintenance by reducing secondary damage. For example, if the potential failure of a pump bearing was detected proactively, the catastrophic failure of the bearing could be prevented. The catastrophic failure of the pump bearing would likely result in damage to the casing, wear rings, impeller, mechanical seals, etc. The corrective repair would require an extensive pump rebuild. Utilizing a proactive task such as vibration monitoring to detect the bearing deterioration permits the scheduled replacement of the bearing prior to the occurrencev of secondary damage. Less secondary damage means that it takes less time to repair (labor savings) and consumes fewer parts (material savings). The overall effect is the repair costs much less.

Maintenance costs are also impacted by increasing the efficiency of maintenance. These efficiency gains are achieved through improved planning and scheduling of "the right work at the right time." Published data suggests that companies with estimated wrench times of 25% to 30% can increase wrench time to between 40% and 60% through better planning and scheduling.

There are several useful maintenance cost related measures:

- Maintenance Cost: The target maintenance cost depends on the asset and its operating context (how the asset is applied and used).
- Maintenance Cost/Unit Output: The target maintenance cost depends on the asset and its operating context (how the asset is applied and used).
- Maintenance Cost/Replacement Asset Value of Plant and Equipment: This metric is a useful benchmark at a plant and corporate level. The world class benchmark is between 2% and 3%.
- Total Maintenance Cost/Total Manufacturing Cost: This metric is a useful benchmark at a plant and corporate level. The world class benchmark is <10% to 15%.
- Total Maintenance Cost/Total Sales: This metric is a useful benchmark at a plant and corporate level. The world class benchmark is between 6% and 8%.

Maintenance Related Downtime

The maintenance function's impact on asset availability is through minimizing downtime attributed to maintenance. This includes both scheduled and unscheduled maintenance related downtime. A key objective of proactive maintenance is to identify potential failures with sufficient lead-time to plan and schedule the corrective work before actual failure occurs. If the maintenance function is successful unscheduled maintenance related downtime will be reduced.

It is equally important to measure scheduled downtime. The work identification element of the maintenance process strives to eliminate unnecessary scheduled maintenance by focusing on only performing "the right work at the right time."

Through more formal work identification and enhanced planning and scheduling, shutdown overruns should be minimized.

Useful key performance indicators associated with asset downtime attributable to maintenance are:

- Unscheduled downtime (hours)
- Scheduled downtime (hours)
- Shutdown overrun (hours)

Note: It is useful to distinguish between "equipment down" where a specific piece of equipment is unavailable and "process down" where production has stopped.

6.3.12. The Importance of the Work Order

Implementation of the suggested key performance indicators for the maintenance function requires a reliable source of data on asset failures, maintenance costs and downtime. If any time maintenance is performed on an asset a record should be kept. The vehicle for collecting this data is the maintenance work order.

Whenever maintenance is performed against an asset, work order completion data should include the following information:

- Identification of the asset at the level in the asset hierarchy where the work was performed.
- Date, time and duration of the maintenance event.
- An indication if failure has occurred: yes or no (no if proactive)
- When failure has occurred, identification of the failure consequence: hidden, safety, environment, operational (product quality, throughput, customer service, operating costs) or non-operational involving only the cost of repair
- Actual costs (labor, materials, services, etc.)
- Process downtime (loss of production)
- Asset downtime (equipment out of service but process still able to produce)

Queries in your computerized maintenance management system can then be developed to track and report key performance indicators for asset failure, maintenance costs and downtime.

6.3.13. Reporting and Use of Key Performance Indicators

Key performance indicators should be aligned with defined roles and responsibilities for the maintenance function against the assets for which they apply. For example, a planner responsible for "Area A" would be responsible for the planning function key performance indicators for the "Area A" assets.

The manager responsible for "Area A" assets would monitor all process and result metrics for, "Area A." Each metric should roll up the asset hierarchy, in alignment with individual responsibility for the assets. Management action is directed at improving compliance with the requirements of Work Identification, Planning, Scheduling, Execution and Follow-up. In this way, the process is managed leading to world class results. This logic is repeated at each level of management in the organization. At the plant and/or corporate level, management is exercising accountability for plant-wide maintenance metrics, both process and results.

6.3.14. Conclusion

Maintenance and reliability business process metrics (leading indicators) provide a clear indication of compliance to the maintenance business process. They indicate where to take specific action because of a gap in the way maintenance is being performed.

This gap in the execution of the maintenance process will eventually lead to asset failure(s). The consequence of these failures translates into poor manufacturing performance.

Therefore, maintenance, reliability, engineering and operations need to work together to define and measure the leading indicators for the Asset Reliability Process (the seven elements required to support the maintenance function). The result will be optimal asset reliability at optimal cost—the output of a healthy Asset Reliability Process.

TABLE 6.2.   Summary of Maintenance Key Performance Indicators

	Type of Measure	Measuring	Key Performance Indicator	World Class Target Level
1	Result, Lagging	Cost	Maintenance Cost	Context specific
2	Result, Lagging	Cost	Maintenance Cost/Replacement Asset Value of Plant and Equipment	2–3%
3	Result, Lagging	Cost	Maintenance Cost/Manufacturing Cost	<10–15%
4	Result, Lagging	Cost	Maintenance Cost/Unit Output	Context specific
5	Result, Lagging	Cost	Maintenance Cost/Total Sales	6–8%
6	Result, Lagging	Failures	Mean Time Between Failure (MTBF)	Context specific
7	Result, Lagging	Failures	Failure Frequency	Context specific
8	Result, Lagging	Downtime	Unscheduled Maintenance Related Downtime (hours)	Context specific
9	Result, Lagging	Downtime	Scheduled Maintenance Related Downtime (hours)	Context specific
10	Result, Lagging	Downtime	Maintenance Related Shutdown Overrun (hours)	Context specific
11	Process, Leading	Work Identification	Percentage of work requests remaining in "Request" status for less than 5 days, over the specified time period.	80% of all work requests should be processed in 5 days or less. Some work requests will require more time to review but attention must be paid to "late finish date" or required by date.
12	Process, Leading	Work Identification	Percentage of available work hours used for proactive work (AMP + AMP initiated corrective work) over a specified time period.	Target for proactive work is 75–80%. Recognizing 5–10% of available work hours attributed to redesign or modification (improvement work) this would leave approximately 10–15% reactive.
13	Process, Leading	Work Identification	Percentage of available work hours used on modifications over the specified time period.	Expect a level of 5–10% of work hours spent on modification work.
14	Process, Leading	Work Planning	Percentage of work orders with work hour estimates within 10% of actual over the specified time period.	Estimating accuracy of greater than 90% would be the expected level of performance.
15	Process, Leading	Work Planning	Percentage of work orders, over the specified time period, with all planning fields completed.	95% + should be expected. Expect a high level of compliance for these fields to enable the scheduling function to work.
16	Process, Leading	Work Planning	Percentage of work orders assigned "Rework" status (due to a need for additional planning) over the last month.	This level should not exceed 2 to 3%.

TABLE 6.2. (*Continued*)

	Type of Measure	Measuring	Key Performance Indicator	World Class Target Level
17	Process, Leading	Work Planning	Percentage of Work Orders in "New" or "Planning" status less than 5 days over the last month.	80% of all work orders should be possible to process in 5 days or less. Some work orders will require more time to plan but attention must be paid to "late finish date."
18	Process, Leading	Work Scheduling	Percentage of work orders, over the specified time period, having a scheduled date earlier or equal to the late finish or required by date.	95%+ should be expected in order to ensure the majority of the work orders are completed before their "late finish date."
19	Process, Leading	Work Scheduling	Percentage of scheduled available work hours to total available work hours over the specified time period.	Target 80% of work hours applied to scheduled work.
20	Process, Leading	Work Scheduling	Percentage of work orders assigned "Delay" status due to unavailability of manpower, equipment, space or services over the specified time period.	This number should not exceed 3–5%.
21	Process, Leading	Work Execution	Percentage of work orders completed during the schedule period before the late finish or required by date.	Schedule compliance of 90%+ should be achieved.
22	Process, Leading	Work Execution	Percentage of maintenance work orders requiring rework.	Rework should be less than 3%.
23	Process, Leading	Work Execution	Percentage of work orders with all data fields completed over the specified time period.	Should achieve 95%+. Expectation is that work orders are completed properly.
24	Process, Leading	Work Follow-up	Percentage of work orders closed within 3 days, over the specified time period.	Should achieve 95%+. Expectation is that work orders are reviewed and closed promptly.
25	Process, Leading	Performance Analysis	Number of asset reliability improvement actions initiated by the performance analysis function, over the specified time period.	No number is correct but level of relative activity is important. No actions being initiated when lots of performance gaps exist is inappropriate.
26	Process, Leading	Performance Analysis	Number of equipment reliability improvement actions resolved, over the specified time period. (Did we achieve performance gap closure?)	This is a measure of project success.

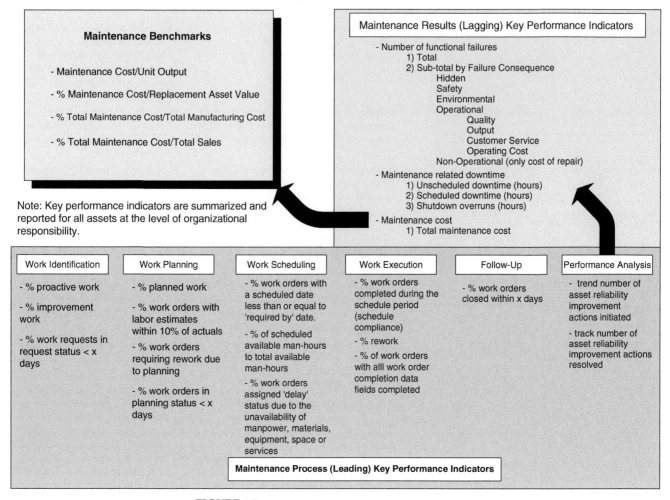

FIGURE 6.7.   Example of how maintenance KPIs are used.

7

Total Productive Maintenance

7.1. INTRODUCTION TO TOTAL PRODUCTIVE MAINTENANCE

Total productive maintenance (TPM) is a company-wide equipment maintenance system involving all employees, from top management to production line workers and the building custodians. It is but one of many approaches to maintenance. The word *total* in total productive maintenance is common to

- *Total equipment effectiveness* in pursuit of profitability, not simply maintenance cost reduction.
- *Total maintenance system*, including maintenance prevention (MP), maintainability improvement (MI), and good old-fashioned preventive maintenance (PM), is a maintenance plan for the equipment's entire life span. MP translates into maintenance-free design and is pursued during the equipment design stage. MI translates into repairing or modifying equipment to prevent breakdown and facilitate ease of maintenance.
- *Total participation* by all employees.

The idea behind TPM is not revolutionary. Simply stated, it is cooperation to get the important job of maintenance done reliably and effectively. It is supplementary to, rather than a replacement for, established principles of successful maintenance management. Built around five focal points, TPM combines concepts of continual improvement, total quality, and employee involvement:

1. Activities to optimize overall equipment effectiveness.
2. Elimination of breakdowns through a thorough system of maintenance throughout the equipment's entire life span.

3. Autonomous operator maintenance (this does not imply that operators perform all maintenance):
   - Use lower-skilled personnel to perform routine jobs that do not require skilled technicians.
   - Use operators to perform specific routine maintenance tasks on their equipment.
   - Use operators to assist technicians in the repair of equipment when it is down.
   - Use computerized technology enabling operators to calibrate selected instrumentation.
   - Use technicians to assist operators during shutdown and startup.
4. Day-to-day maintenance activities involving the total workforce (engineering, operations, custodians, maintenance, and management).
5. Company directed and motivated, yet autonomous, small group activities. Small group goals to coincide with company goals.
6. Continuous training: formal training, on job training, one-point lessons, and team members training each other.

7.1.1. The TPM Organization

In the TPM organization, operators and technicians are organized into zone teams. They are trained and certified to perform particular tasks in their zone as the need arises, such as one point lessons (one per week) and higher levels of maintenance by operators, which usually are voluntary. Individuals who spot problems pitch in to restore equipment, resulting in less downtime. Major maintenance work requiring high craft skills is performed by centralized (or less centralized) maintenance forces.

7.1.2. TPM Objectives

While the dual goals of TPM are zero breakdowns and zero defects, TPM is more about performance improvement, employee interaction, and positive reinforcement than maintenance-specific technology. Human resource factors and technical factors must be balanced. TPM works toward elimination of nine formidable obstacles to equipment effectiveness:

- Downtime
- Equipment failure from breakdown
- Setup and adjustment
- Speed losses
- Idling and minor stoppages due to abnormal operation of sensors, blockages of chutes, and the like
- Reduced speed due to discrepancies between designed and actual speed of equipment
- Defects
- Process defects—scrap, downgrades, rejects, returns, and so forth
- Reduced yields from all resources such as raw materials, packaging, energy, and labor

7.1.3. Autonomous Maintenance

Group activities are promoted throughout the organization to gain greater equipment effectiveness. Operators are trained to *share* with maintenance personnel the responsibility for routine maintenance. This is referred to as *autonomous maintenance*. Routine maintenance normally includes

- Housekeeping.
- Equipment cleaning.
- Protection of components from dirt.
- Lubrication by operators.
- Equipment inspection by operators and maintenance.
- Setups and adjustments.

Autonomous maintenance may or may not include minor equipment repairs. Each company decides how autonomous maintenance is to be. Operators must receive a large amount of skills training and be certified as they progress through various skill levels. As operators become trained, an organized transfer of tasks takes place.

7.1.4. Equipment Management

The second focus of TPM is on equipment management (EM), not simply on maintenance. EM is pursued through autonomous, small group (team) activity comprising operators, engineers, and technicians. Equipment improvement never stops.

The aim of team efforts is to optimize overall equipment effectiveness and eliminate breakdowns through a thorough system of maintenance throughout equipment's entire life span. Through involvement, operators develop ownership of and an affinity for "their equipment." Team members involved in problem solving develop a strong urge to see the problem fixed. They therefore participate in equipment management activities to make sure the problem does not recur. They take pride in their accomplishments.

The organization trains, solves problems, and works as teams. Peer support is important. Team competition is healthy and encouraged.

7.1.5. TPM Integration

The TPM approach does not preclude need of an integrated maintenance program, including computerized support, formalized planning and scheduling, and insightful equipment history.

7.1.6. TPM Is an Investment

TPM is not a short-lived, problem-solving, maintenance cost-reduction program. It is a process that changes the corporate culture and permanently improves and maintains the overall effectiveness of equipment through active involvement of operators and all other members of the organization. The required TPM investment, as well as the return, is very high. Systematic TPM development and full implementation requires a three-year program. Over time, the cooperative effort creates job enrichment and pride that dramatically increases productivity and quality, optimizes equipment life cycle cost, and broadens the base of every employee's knowledge and skill. Initially, the company must bear the additional expense of restoring equipment to its proper condition and educating personnel about their equipment.

While TPM is popularly viewed as a Japanese concept, more accurately it should be viewed as a program that effectively *integrates* a number of maintenance management concepts that did not originate in Japan. However, to the credit of the Japanese, they have a knack for turning good ideas into enormously successful practices.

The Japanese business culture is more receptive to TPM requirements than the dominant business culture in the United States (see Table 7.1).

While initial TPM successes occurred largely in automated machining and assembly operations, it has also been effectively applied to other types of organizations as well.

TABLE 7.1. Japanese and U.S. Business Cultures

Typical Conditions in Japan	Typical Conditions in the United States
Total corporate commitment to TPM	Lack of management involvement
Very long-range planning	Focus on quarterly results
Few cost constraints	Severe cost constraints
Pressure to succeed from top management	Less sustained pressure to succeed from top management
Practically no limit on training	Limited training time; however, plenty of time for meetings
Ability to absorb concurrent activities	Inability to absorb concurrent activities
Employees volunteer own time	Time constraints

7.1.7. Calculating Major Losses Is Key to TPM's Success

TPM activities should focus on results. One of the fundamental measures used in TPM is overall equipment effectiveness (OEE):

OEE = equipment availability × performance efficiency × first-pass quality

World-class levels of OEE start at 85% based on the following values:

90% equipment availability × 95% performance efficiency × 99% rate of quality = OEE of 84.6%

The OEE calculation factors in the major losses that TPM seeks to eliminate. The first focus of TPM should be on major equipment effectiveness losses, because this is where the largest gains can be realized in the shortest time. The 11 major areas of loss fall within four broad categories.

Planned shutdown losses:

1. No production, breaks, or shift changes.
2. Planned maintenance.

Downtime losses:

3. Equipment failure or breakdowns.
4. Setups and changeovers.
5. Tooling or part changes.
6. Startup and adjustment.

Performance efficiency losses:

7. Minor stops (less than six minutes).
8. Reduced speed or cycle time.

Quality losses:

9. Scrap product or output.
10. Defects or rework.
11. Yield or process transition losses.

Planned Shutdown Losses

Valuable operating time is lost when no production is planned. However, there are reasons that this is not often considered a "loss." It is probably best stated as a "hidden capacity to produce."

Loss 1. Production is not scheduled. The facility may be a three-shift operation where the third shift is the maintenance shift. It may be a plant that does not run on Saturday and Sunday. Employee shift changes, breaks, and mealtimes also fall into this category. During breaks, some equipment gets shut down. If an operation is a continuous process, obviously, that does not happen. But in plants making piece parts, some assembly lines and processes shut down for breaks and shift changes.

Loss 2. Planned maintenance includes periodic shutdowns of equipment, processes, and utilities for major maintenance. These shutdowns represent a period of time when no production occurs. Typically, OEE calculations factor planned maintenance out of the equation. It is assumed that it is planned maintenance—it must be done, it cannot be reduced or eliminated, so leave it in. Auto racing represents a different story about "planned maintenance." A planned-maintenance stop or "pit stop" on the NASCAR circuit in 1950 typically was nearly four minutes long. Now pit crews perform the same pit stops in 17.5 to 22.5 seconds. In that less-than-23-second period, the same things happen today that were happening in 1950 in four minutes: change four tires, dispense 22 gallons of fuel, make chassis adjustments, wipe off the windshield, give the driver some water, and wipe the rubber dust off the radiator. In about 20 seconds, the car is back on the track. At some point, somebody said, "We can do these pit stops in less than four minutes, can't we?" Planned maintenance is a pit stop. TPM advocates must ask, "How much time are we spending in the pits?"

A walk-through of almost any plant uncovers ways that the same amount of maintenance can be done in less time or more maintenance done in the same amount of time. And those things can be accomplished, not with more contractors and not with more work hours, but just by doing things differently and working smarter, not working harder.

Downtime Losses

Downtime is the second category of major equipment losses. This category includes the following.

Loss 3. Typically, when a company's personnel consider losses, they think of equipment failures or breakdowns. But there are other unplanned downtime losses.

Loss 4. This loss includes how long it takes to set up for production processing and how long it takes to shift from one product or lot to another. Determining these losses should take into account how long it takes to start up after a changeover and run a new product. In auto racing, this type of loss includes the preparation and setup for qualifying and racing.

Loss 5. This loss includes the time it takes to make tool changes and production-part changes. Industries in which certain tooling, machine devices, or parts have to be changed can learn from an auto racing pit stop. The techniques are the same. In an actual pit stop, every single action is taken into account.

Loss 6. This loss occurs when equipment or processes start up. It includes the warmup time and the run-in time that must be set aside to get everything in the process ready to produce high-quality output. Pit stops are in part successful if the driver optimizes the car's speed on the slowdown lap before the pit stop and the speed-up lap after the pit stop. If the driver brings the car up to speed too quickly, the drive train and tires may be damaged. A race car handles poorly until the new tires are properly conditioned on the track.

Performance Efficiency Losses

The third category of major equipment losses is performance loss, when machines operate at less than designed speed, capacity or output.

Loss 7. These types of loss—minor stops or "machine hiccups"—are the little things that companies usually do not track. Quite often, equipment downtime of less than six minutes is not tracked. However, consider the impact if this six-minute downtime occurs during each shift in a three-shift, five-day operation. That all adds up to 1½ hours of downtime per work week or 75 hours of lost production per 50-week year. Little losses add up.

Loss 8. Equipment and processes running at less than design speeds and cycle times result in lower output. As machines age and components wear, they tend to run slower. At times, machines are run at lower speeds because the people who operate and maintain them have compensated for problems and believe that running them slower is better and results in fewer breakdowns.

Quality Losses

The final category of major equipment losses is loss of quality—first-pass quality only. One of the fundamental truths of TPM is this: If equipment is available 24 hours a day, seven days a week—and if it performs at its highest design cycle rate—then if it is not producing the highest level of quality, it is just producing scrap at full capacity. So, quality is a very important element among the major losses.

Loss 9. Scrap loss is fairly straightforward. Every time equipment runs and produces unusable product or output, valuable operating time is lost.

Loss 10. Defective output, even if it can be reworked or recycled, is considered a major loss to be eliminated. As with scrap, every time equipment runs and produces unusable product or output, valuable operating time is lost.

Loss 11. Yield or transition losses often occur when equipment and processes require a warmup or run-in time. During that time, they often produce an off-quality output. This type of loss also includes the lost output that results from transitioning from one chemical product to another in a process. Equipment running and wasting raw materials create yield and transition losses.

TPM can yield results in two months—sometimes two weeks—when activities are focused on results and there is regular monitoring, recording, and trending of OEE data. It is not as important to focus on OEE percentage as on each factors of OEE and perform root cause failure analysis on the major losses of each—equipment availability, performance efficiency, and rate of quality. Use the OEE data to communicate how well the equipment performs and how well the TPM activities work.

As the foundation of lean maintenance, TPM is popularly viewed as a Japanese concept. To be more accurate it should be viewed as a program that effectively integrates a number of maintenance management concepts that did not originate in Japan. Yet, to the credit of the Japanese, they have a knack for turning good ideas into enormously successful practices. The Japanese business culture is more receptive to TPM requirements than is the dominant business culture in the United States.

7.2. LEAN RELIABILITY*

Lean thinking has been widely adopted in manufacturing operations. Toyota set the standard in discrete manufacturing industries for producing products in the quickest and most efficient way. Other leading companies have also adopted productivity-enhancing manufacturing techniques. Yet in highly capital-intensive industries, including mining, metals, pulp and paper, and power generation, equipment reliability plays a far more critical role in business success because degradation in equipment condition results in reduced equipment capability. Equipment downtime, quality problems, and the potential for safety or environmental incidents are the result of poor performing equipment. All these can negatively affect plant output.

This section explains how traditional lean thinking and principles can be applied specifically to the equipment reliability process to achieve breakthrough performance improvement. This business process focuses people on managing physical asset reliability to meet the business goals of the company. Lean reliability brings together maintenance and operations personnel to manage the equipment reliability process in a way that reduces waste and continuously improves the process. Lean reliability focuses on managing asset health to achieve optimal performance at optimal cost.

7.2.1. The Evolution from Lean Manufacturing to Lean Maintenance to Lean Reliability

Lean manufacturing has been implemented in many organizations to optimize the production process. Lean improvement efforts have successfully reduced manufacturing lead times, reduced work

* Section 7.2 was prepared in conjunction with Ivara Corporation of Burlington, Ontario, Canada, which provided the information and the graphics.

in process, and generally improved the work environment and manufacturing operation. Yet many companies continue to face major issues due to poor performing capital assets (either not meeting capacity requirements or overspending in maintenance to achieve required performance levels). These companies have not seen the full benefits of lean. The reason for this underachievement is the lack of integration of maintenance as a true partner of manufacturing in achieving optimal performance of the physical assets that contribute to achieving company goals. Maintenance needs to "join the lean team" and contribute as a partner to the operation.

For some companies, the old way of thinking about lean within maintenance simply meant cutting the maintenance budget by 20 to 30%, sometimes by even 50%, then demanding that maintenance do more with less. This approach was known as the "slash and dash" approach. As a result, plant performance suffered while management struggled to cope with acute reliability issues and a disgruntled workforce.

For other companies, maintenance was simply a necessary evil and a creeping maintenance budget was the result—spending too much money for the required uptime or equipment availability. As Ron Moore, author of *Making Common Sense Common Practice* defined it, companies have "too much maintenance in their reliability." If the maintenance budget continues to rise and reliability either stays the same or decreases, the maintenance program is the problem.

And, for yet others, reliability and maintenance issues remain a hidden problem—even though they are the root cause of the company losing large amounts of money by the minute.

When we compare maintenance costs within a typical company to world class (see Table 7.2), we recognize that there is a huge opportunity to improve. In the United States, alone, billions of dollars are spent unnecessarily on maintenance expenditures due to lack of control, lack of process.

TABLE 7.2.   Maintenance Costs in Typical and World-Class Companies

Metric	Typical	World Class
Maintenance cost/replacement asset value		
Maintenance cost must include labor (including overtime), materials, contract maintenance, and capital replacements, and maintenance (replacing worn-out assets because they were never properly maintained)	3.5–9%	2.0–3.0%
Maintenance materials cost/replacement asset value		
Maintenance materials cost must include material in storeroom stock plus material in other locations (maintenance shop, plant floor, etc.)	1.0–3.5%	0.25–0.75%

Many companies have shut their doors and used many excuses, including not being able to compete with cheaper labor overseas. We, in maintenance and operations, can either improve our plant operational performance and control our destiny or someone else will, by closing the plant.

7.2.2. Managing Asset Performance to Meet Customer Needs

Customer expectations normally are defined in terms of product quality, competitive pricing, service levels, on-time delivery, and overall solution delivery. Within the organization, we can determine, measure, and control the performance requirements of the physical assets to meet business goals and market demand (e.g., quality, availability and overall equipment effectiveness, cost/unit, and safety and environmental integrity). To achieve the asset performance requirements, we must manage three inputs: process technology, standard operating practices, and asset care practices (see Figure 7.1).

The first input is *process technology*, which simply delivers the inherent capability of the equipment "by design" to meet the equipment performance requirements. The second input is the *operating practices*, which make use of the inherent capability of process equipment. The documentation of standard operating practices assures the consistent and correct operation of equipment to maximize performance. The third input is the *asset care practices*, which maintain the inherent capability of the equipment. Deterioration begins to take place as soon as equipment is commissioned. In addition to normal wear and deterioration, other failures occur. Failures happen when equipment is pushed beyond the limitations of its design or operational errors occur. Degradation in equipment condition results in reduced equipment capability, equipment downtime, quality problems, or the potential for accidents and environmental incidents. In lean, one goal is to reduce the losses associated with product quality, plant capacity, and safety (Figure 7.2).

Plant capacity, asset availability, and asset utilization losses are affected by scheduled and unscheduled downtime. This downtime includes equipment breakdowns, scheduled outages, product changeovers, breaks and lunch, shift changeovers, and equipment startup. Product quality losses can be affected by both maintenance and operations and are the responsibility of both parties. Quality losses include operator adjustments and errors that are a result of equipment reliability issues; raw material problems could be caused by the supplier (low chance), storage, or transfer issues, which could be operations or equipment reliability problems. Safety and environmental integrity losses can affect the health of the plant workers and the community. This category is difficult to place a specific hard value to but the results nonetheless can be measured.

Everyone in an operation from senior management to floor level personnel must understand that these losses can cause a plant to lay off employees or, worse,

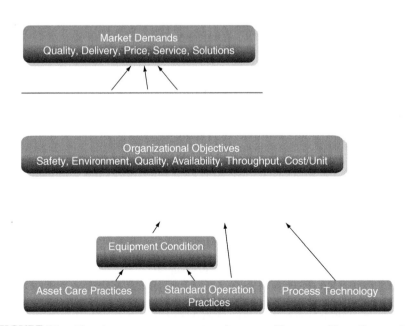

FIGURE 7.1.   Three inputs to manage asset performance. (Courtesy of Ivara Corporation.)

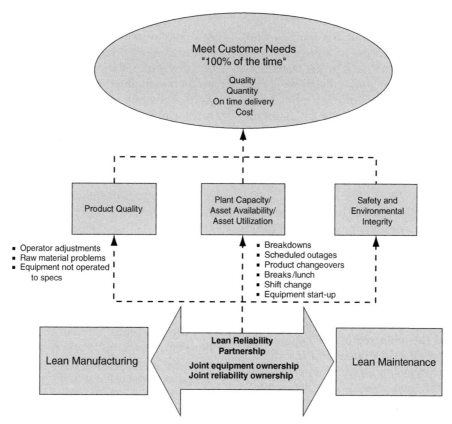

FIGURE 7.2. A lean reliability partnership is key to meeting customer needs. (Courtesy of Ivara Corporation.)

shut down. In most cases, plant layoffs and shutdowns could have been avoided if addressed jointly in a lean reliability initiative (because of the huge financial impact increased reliability provides). The specific financial value of the initiative must be determined in terms of addressing the major losses. Jointly, the plant leadership team, along with the comptroller, can determine the opportunity in the short and long term.

The New Lean Team

To achieve and sustain lean reliability requires joint responsibility between maintenance and operations. Ownership of equipment and reliability is a shared responsibility, which must be demonstrated and proven through reduction in cost and risk to the business. Working together at all levels, maintenance and operations are the new "lean team," providing a solution to address the major losses that can be caused by equipment reliability issues. This new team needs to drive reliability from the floor level and monitor progress of reliability by establishing targets for improvements and measuring progress with KPIs (key performance indicators). Only then can this team leverage a sense of shared ownership in lean reliability to achieve breakthrough performance.

Lean reliability requires a business process to manage asset reliability. This process needs to be jointly managed by maintenance and operations, working together to optimize asset reliability at optimal cost.

A Proactive Asset Reliability Process

Evolving from lean manufacturing to lean reliability requires the development of a lean asset reliability process. The output of this proactive process is optimal asset reliability at optimal cost. The core concept in a lean asset reliability process is providing a sound technical basis to focus on the right work at the right time, regardless of who is conducting the work (the operator or trades person).

In the majority of plants with high maintenance costs (excluding perhaps nuclear power generation plants), we find reactive activities. By our very nature, people resist change and are trapped in this reactive environment, We become such good firefighters, that we are rewarded for our reactive behavior—so we believe this reactive culture is a good thing, that is until the next time the equipment breaks down, causing a major downtime incident (hero one day, enemy the next).

Changing from a reactive culture to a proactive culture cannot happen overnight. It is a journey that starts with implementing a process for achieving reliability. All the companies successful in the transition to proactive asset reliability focused culture have one ingredient in common—they follow a formal business process to govern the work done to maintain their assets. The reason for this is that the reliability of assets is related far more to the things people do than to anything else. With the right process in place, we can ensure that people are doing the right things to maintain plant assets.

While most plants have in place a process to govern the work done in an asset reliability process (refer back to Figure 5.12), maintenance, the typical process includes only the planning, scheduling, work execution, and follow-up elements of this process. These process elements, shown in the right-hand box of the process diagram, are necessary elements but they do not represent an effective, proactive process.

Before we discuss the other required elements of a reliability process, we should understand the definition of *optimal asset reliability*. Optimal asset reliability means that, for the least possible cost, we achieve the level of performance we need from equipment to meet business goals (plant or company goals). Equipment performance, in this case, is not production output but the required level of uptime (such as mean time between failure) or the maintenance cost needed to assure the desired performance. Given that the business goals must be supported by the equipment, the asset reliability process must include this connection to business goals, as shown in the left-hand box of the process diagram. Next, determine the assets that are most critical when they fail and the highest risk in terms of impact on business performance. For these assets, establish specific performance targets. This stage focuses improvements on the performance targets of critical assets that contribute most to the company's success. The assess stage then compares the asset performance targets to the maintained asset's actual performance, which is learned in the center box as we execute work. This stage identifies and sets priorities on gaps in performance by performing specific performance analyses. Because functional failure is the inability to meet performance requirements, a performance gap is really a functional failure. In the improve stage, the team selects an appropriate work identification strategy to understand and address all causes of failure for the asset under consideration. One of the toughest challenges on the road to improved asset reliability is to determine the prescription of proactive work that should be done to maintain the assets so that they deliver the reliability (at an optimal cost). This

topic, also known as *work identification*, represents the cornerstone of an effective asset reliability process. The resulting asset reliability program includes some mix of preventive maintenance, detective maintenance, predictive maintenance, and some run-to-failure decisions. The outcome of the work identification element is the right work at the right time (the right work in terms of the tasks and the timing for conducting them). The process is self-sustaining, with opportunities to continuously improve and evaluate the overall effectiveness of the asset reliability process as well as revisit reliability programs and continuously improve. These activities optimize the effectiveness of the lean team.

Supporting Reliability Practices and Technology

The new asset reliability process creates an opportunity for maintainers, operators, and management to learn new things and grow into new roles and responsibilities. Implementing one system at a time, one can gradually transform maintenance and operations employees into experts in reliability. Much of the knowledge of the equipment and how to maintain it to ensure optimal reliability lies within the existing workforce. The knowledge from these equipment experts can be formalized and made available through the reliability process. This knowledge can be leveraged across all employees, assets, and even other plants.

The transition to a reliability focused approach to asset care results in the need to manage an enormous amount of equipment health data. Successful lean reliability initiatives use technology to their advantage to capture the knowledge of equipment experts then manage plant assets from an on-line picture of the health. Rather than reacting to equipment failures, operators and maintainers proactively maintain optimal equipment health.

7.2.3. The Basic Principles of Lean Reliability

The three basic principles of lean reliability are eliminate waste, continuously improve, and team maintenance and operations.

Eliminate Waste

A huge opportunity to eliminate waste occurs when we implement a proactive asset reliability process. Fundamentally, for the most part, people are doing the wrong work. The wrong work is a combination of work done that is too much too early or too little too late. The classic example of the wrong work is justified as "we have always done it this way." There is a very significant financial opportunity associated with identifying the right work to do.

In a reactive environment, the focus of maintenance work is repairing failed equipment. In a fully proactive environment, the focus on maintenance work becomes inspections of asset heath to enable proactive intervention prior to failure (Figure 7.3). To make the transition to a proactive asset reliability program, we need effective work identification capabilities.

The right work is the minimum amount of work necessary to ensure the asset provides the necessary level of performance. Since the majority of failures occur randomly and are not related to age, other techniques such as asset heath monitoring are needed to allow intervening prior to loss of asset function. Work identification represents the fundamental shift from conventional time-based maintenance to an asset reliability approach to maintenance. Yet, the answer is not as simple as solely identifying the right work. In a reactive environment, a binder filled with proactive tasks is not a solution. The solution must be a proactive process for guiding people's activity. In addition, to successfully execute the process, employees must be trained in the latest reliability thinking and practices and equipped with tools to manage the data inherent in a proactive environment. Finally, an appropriate change management implementation approach must be used. To implement a proactive solution in a reactive culture requires a unique approach, one that transforms the culture one asset at a time.

Determining the right reliability program for an asset is no easy task. It might seem that, the longer a company has been around, the more effective the asset maintenance programs would be. Unfortunately, this is not always true. The effectiveness of a reliability program has little to do with the number of years a company has been doing maintenance. Most companies do too much maintenance too soon or too little too late, either of which has cost consequences to the organization. To begin, we must understand how equipment really fails.

Understand the New Definition of Failure

Reliability studies over the last 30 years say that 80% of asset failures are random. This is quite a departure from what we would expect, yet research has proven that, for the majority of components, there is no correlation between age and how likely they are to fail. However, with the right practices and technologies, early signs of random failure can be detected by monitoring the health indicators to determine whether asset health is degrading. As discussed in Chapter 1, Section 1.3.3, the PF interval is the time between the detection of a potential failure (P) and functional failure (F); see Figure 1.4. Proactive corrective action is scheduled before functional failure occurs.

To summarize the discussion in Chapter 1, take a step back and review the way equipment performance is managed. If equipment continues to fail after preventive maintenance or overhauls, then something must change. If a preventive maintenance task is not tied to a failure mode (root cause), question why it is done.

Understand the function of an asset and its operating context; otherwise, efforts are wasted. Understanding this function and its performance requirements can have a profound effect on how that equipment is operated and maintained and, hence, can affect the overall reliability of the asset. This can be a challenging task and needs to involve the operators and maintainers of the asset because they know best what the asset must do to achieve the operating targets.

Capture the Knowledge of an Aging Workforce before They Retire

The imminent aging workforce issue has made asset reliability a hot topic at the executive level of most capital-intensive companies. Executives have recognized the challenge in retaining valuable maintenance and equipment reliability knowledge as their

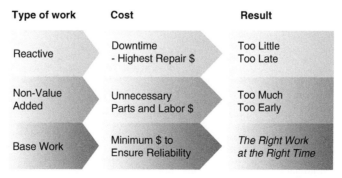

FIGURE 7.3. "The right work at the right time." (Courtesy of Ivara Corporation.)

workforce retires. During their tenure, well-seasoned operators and maintenance veterans became intimate with their equipment and could quickly repair it to avoid downtime. But this acquired knowledge is rarely documented or transferred to others and will be lost if companies do not systematically collect this important information on how employees perform their jobs.

The reliability program captures the knowledge of the equipment experts, the operators and maintainers who know the equipment best. In many companies, these employees have worked with the equipment daily for decades, so their knowledge is invaluable. The challenge is to find a way to store this information so that all employees can take advantage of it in their daily work. Reliability software captures this knowledge and makes it available.

During the development of the asset reliability program, these maintainers and operators are asked to contribute their knowledge of the ways the asset fails and the ways that have found to detect or prevent failure. In the context of a well-defined failure analysis, their knowledge is captured, formalized by linking proactive tasks to specific failure modes. The details previously carried around in personal pocket books becomes readily available as the new asset reliability program is defined and deployed.

It's not a matter of *if* the aging workforce issue will affect an organization but *when*. Do not waste this expertise, do not let it walk out the door. Act now, or the problem will worsen.

By capturing the knowledge of experts within a system, the amount of time wasted each day manually calculating condition data is reduced. For example, the screen in Figure 5.9 captures the calculation to determine the effectiveness of a heat exchanger.

No one needs to remember the engineering calculation, since software can store the expression, making it permanently available for all to use. Combined data from various indicators to determine the overall effectiveness of the heat exchanger and, when an abnormal value is found, the user is prompted with the predetermined corrective action (which once was known only by the expert). Islands of data from multiple sources within the plant can be consolidated to evaluate current asset health. This ensures an accurate picture of asset health in a timely manner.

The majority of inspections involve rather subjective assessments of the equipment condition. Visual or other sensory inspections can be logged via handheld data recorders (PDAs, see Figure 7.4). Abnormal readings trigger alarms and follow-up work tasks to suggest more rigorous inspections or corrective work.

This also removes the subjectivity of the inspection by providing the inspector with the ability to choose a statement that corresponds to the observed condition, from a predefined list of such statements. This approach enables each employee that conducts the inspection to do so in a consistent manner.

There are many more opportunities to eliminate waste when a proactive asset reliability process is implemented. These ideas are just the tip of the iceberg.

Continuously Improve

Continuous improvement in lean reliability has some very basic principles that must be met to be successful. Most companies never obtain true continuous improvement because one or more of these basic principles is not met; thus, the lean effort never

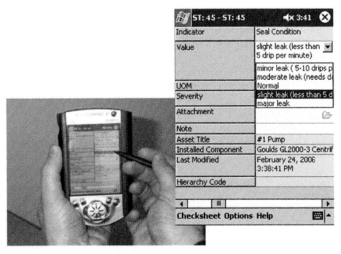

FIGURE 7.4. Automating paper check sheets improves consistency and quality of data. (Courtesy of Ivara Corporation.)

provides all the rewards available in the form of lower total cost, higher asset reliability, higher asset availability, and higher capacity. These principles include

1. Base work flow processes on known best practices and periodically review them.
2. Have the discipline to follow the established work flow processes. Defining clear roles and responsibilities as well as KPIs helps ensure adherence to process.
3. Develop preplanned job packages with defined procedures, specifications, materials, tools, safety, and other items required to ensure repeatability of the maintenance work. These preplanned work packages provide a means for maintenance work practices to be improved based on subject matter experts. With preplanned work packages, maintenance work can continue to improve and reduce the potential for lost time, lost knowledge, and maintenance induced failures.
4. Define the "right work at the right time" (preventive, predictive, detective maintenance), utilizing a technically sound process determined and managed by both maintenance and operations. Analyze known failures and improve the reliability program (learn from mistakes). An automated system makes this a simple process.
5. Do not let the magnitude of failure data prevent you from succeeding. Monitoring and dissemination of asset health using predictive maintenance, visual inspections, PLC (Programmable Logic Controller) data, and the like becomes more manageable once a system is in place.

All these principles need not be followed, but with every principle not followed, there comes a risk of lean not meeting the expected goals. Is it worth the risk? Most companies select the principles that are easiest to put in place or that they understand first.

Team Maintenance and Operations

The third, and most important, element of lean reliability is teaming maintenance and operations to manage the physical assets required to achieve the goals of the company. Asset management needs to be a shared responsibility. Together, this team can provide a solution to address the major losses caused by equipment reliability issues.

Lean reliability can be successful only when maintenance and operations become true partners in managing assets. With a proactive process focusing on the value-added functions required to produce optimal equipment reliability at optimal cost, lean reliability is about people and creating a permanent environment focused on reliability of equipment as a way of life in maintenance and operations. After all, it is the operators and maintainers that are the equipment experts. They know how the equipment can fail. They know how to detect early signs of failure. They need to understand whether or not it matters to the business if the asset fails.

This new team needs to drive reliability from the plant floor level and support it from the top to be successful.

7.2.4. How Lean Reliability Aligns with TPM, Kaizen, Five S, and Six Sigma

Over the years, there have been a multitude of manufacturing initiatives to improve operational effectives and efficiency. Lean reliability applies these concepts to the equipment reliability process.

Total Productive Maintenance

The basic principle of TPM is to empower employees to get involved with process improvement to prevent unplanned equipment downtime and minimize waste. With the objective to lower costs and improve return on assets, the basic asset care philosophy is about "autonomous maintenance" or "operator-driven maintenance." While this concept of basic care is a valuable starting point toward optimizing asset performance at optimal cost, it falls short in the technical validity of the asset reliability program.

One way to make the transition from reactive to proactive is to enhance the work identification process. Rather than relying on an informal program (Figure 7.5) of reactive work requests, mostly time-based equipment maintenance suggestions (potentially unjustified work) create a formal process (Figure 7.6) based on an understanding of the relative risk of the asset, followed by a technically sound failure analysis using a formal work identification methodology, to understand all the asset's failure modes. Using one or more of these work identification methodologies for those failure modes managed through the maintenance function, outline

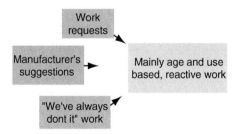

FIGURE 7.5. Informal work request process. (Courtesy of Ivara Corporation.)

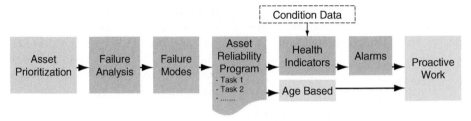

FIGURE 7.6. Formal work request process. (Courtesy of Ivara Corporation.)

the complete asset reliability program or the list of tasks to mitigate the consequences of those failures. About 80% of the asset reliability program tasks, on average, are health based, requiring definition of normal and abnormal values and corresponding alarms. This process maximizes the extent to which the maintenance function is proactive and, therefore, improves the bottom line through improved reliability.

Kaizen

Kaizen, like lean, focuses on eliminating waste and continuous improvement. *Kaizen* is about empowering employees to get involved with their own work organization. With a *kaizen* mindset, the employee that does a job is the expert of that job. Maintenance supervisors typically are encouraged to lead structured improvement efforts in their own work areas. *Kaizen* helps the supervisor stimulate employees' creativity. While this philosophy has merit on its own, the missing element that would enable a sustainable change is a proactive asset reliability process supported by appropriate tools and practices. Nevertheless, the philosophy is a good one and guides the people responsible for creating and implementing lean reliability.

Five S

Five S, in the traditional manufacturing world, is a methodology for organizing the workplace. It involves sorting, setting in order, shining, standardizing, and sustaining the environment. Five S can be applied to lean reliability as follows:

Sort. Organize assets based on risk, allow maintenance and operations efforts to be jointly focused on the right targets for improved asset reliability and plant performance. When "sorting" is well implemented, communication between workers is improved.

Set in order. Organize the work, create sound asset reliability programs. Ensure doing "the right work at the right time." Use a work identification methodology that delivers a technically sound program. Involve operators and maintainers and create and

implement the asset reliability programs, one asset at a time.

Shine. Take pride (and ownership) in the reliability of assets. Identify the health of an asset based on indicators.

Standardize. Focus on standardizing maintenance work. In preplanned job packages, document everything needed to perform the maintenance work to save time and avoid the potential for a self-induced failure. Maintenance labor loss and precious production downtime is kept to a minimum. Orderliness and control is the core of "standardization."

Sustain. Focus on discipline and commitment; without a focus on "sustaining" reliability, it is easy to revert back. The team must be trained in the process and equipped with supporting practices and tools. Employees are empowered to make decisions.

Six Sigma

Six Sigma focuses on removing defects (failures) and reducing variation in a process. Six Sigma uses a variety of statistical analysis tools to analyze reliability data.

The best application of six sigma in lean reliability is through Six Sigma's DMAIC process. We can overlay the DMAIC process on top of the lean reliability process, a perfect fit (Figure 7.7).

Step 1. *Define* a reliability strategy and plan aligned with the goals of the company. To define where to focus efforts, set asset priorities on assets according to their relative risk to the business (consequence of failure multiplied by frequency or probability of failure).

Step 2. *Measure* the level of the reliability and performance for the highest priority assets. To measure asset utilization, ensure that availability is consistently measured and includes both planned and unplanned downtime.

Step 3. Perform an *analysis* on the asset working with operators and maintenance. The goal in the early stages should be to put in place proactive asset reliability programs for critical assets (starting with highest priority ones). Statistical methods like Weibull can be used to analyze the performance.

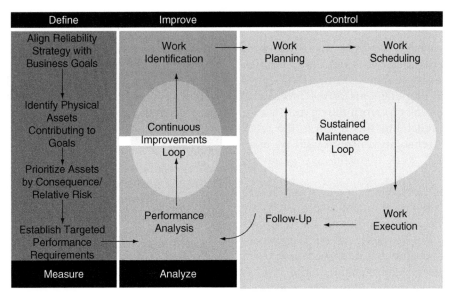

FIGURE 7.7. The DMAIC asset reliability process. (Courtesy of Ivara Corporation.)

Step 4. Continuously *improve* the asset's reliability program using a formal work identification methodology. For example, choose to use reliability centered maintenance or maintenance task analysis to get started quickly. The method of work identification used for an asset should be determined by the level of knowledge and definition (documentation) of the asset's current reliability program and the relative risk of the asset's failure to the business.

Step 5. Develop a *control* plan for maintenance and operations. An asset's proactive asset reliability program dictates a new way of life for trades people and operators, including a new sense of ownership and willingness to care for assets.

Six Sigma's DMAIC process can add quick value to a lean reliability initiative.

7.2.5. Key Elements to Implement and Sustain Lean Reliability

When implementing any lean initiative, there are key elements to successfully implementing and sustaining the improvement. The elements are the areas that bring true success to the initiative.

Education. Education is so important for personnel from senior management to floor level personnel to truly understand all aspects of how lean reliability affects the operation. Everyone in a plant should understand reliability and what it means to the success of the company. Operators and maintenance personnel

need to understand the reality of how equipment fails and learn to manage asset health.

Support from senior management. The easiest and fastest way to get support from senior management is to develop a business case. The business case must be developed with plant leadership identifying the opportunity based on the business goals of the company. It should include an assessment to determine gaps between the current and future state and identify specific opportunities for improvement, both financial and nonquantifiable. An action plan also should include costs and a return on investment to obtain senior management support. Without executive support, the entire project is a waste of time—it either will not get off the ground, will not succeed, or will not be sustainable. When assessing the organization and setting the plan, ensure employees are involved, so later they feel the ownership and are empowered to make decisions.

Effective change management. Implementing lean reliability is all about people. Changing the way people think—the culture of a company—is very difficult. Do not expect to change the culture within the plant overnight. Change happens gradually. With the right practices and tools to support a lean project, it can be implemented one asset at a time and a significant return on investment quickly realized. Remember the Five S and the DMAIC processes, and focus on one asset at a time. This will provide the needed and sustainable change.

Technology support. Technology can make a data-intensive process easier to manage. Ensure the systems chosen can be integrated and work effectively together.

Value stream mapping. This mapping process is critical to the success of lean. It is used to determine current work processes against a future state eliminating non-value-adding elements of the process. In manufacturing, this refers to work flow processes for product output. In reliability, this refers to the work flow processes for reliability output, such as asset criticality assessment, work identification, planning, scheduling, and the like.

Roles and responsibilities. Map the tasks required to manage and execute the asset reliability process to the roles required in the organization and the responsibilities of those people associated with equipment reliability. The goal is to ensure that improved equipment performance is achieved and sustained. The need is for clearly defined role descriptions with associated responsibilities. Everyone must focus on executing the asset reliability process. Start by analyzing the business process tasks, then identify the duties required for each role to ensure that optimal equipment performance is sustained.

Key performance indicators. KPIs in lean begins with the manufacturing KPIs (quality, throughput, OEE, asset utilization, safety) and all other KPIs must align with them. KPIs should be categorized as leading or lagging. The key lagging (or results) indicators for reliability are failures (MTBF is the recognized measure of reliability), downtime attributed to maintenance, and cost. These KPIs demonstrate whether everything is done right or wrong in a lean process. Lagging indica-

tors cannot be managed because they are the results of everything already done. Leading (or process) indicators provide an indication of where problems are occurring before they affect the lagging indicators. Leading indicators are what can be managed. Reliability process metrics are identified to drive specific actions in the process.

Many other valuable metrics can be identified for benchmarking, but we focus on the metrics that drive the execution of a successful reliability process and directly measure its impact. As an example, wrench time is a useful metric to benchmark maintenance efficiency, but it tells nothing about how to improve it. However, by acting on process metrics for planning, scheduling, and execution, wrench time can be improved.

7.2.6. Summary

In today's competitive world, lean reliability is key to a company's survival. Eliminating non-value-adding tasks and continuously improving must be a part of everyone's daily life. As lean reliability progresses, reactivity begins to vanish and a proactive lean asset reliability becomes the focus. The hidden plant finally is found, where the plant experiences more capacity and asset availability and lower cost than anyone ever imagined.

Do not think this journey is easy or without sacrifices, because it is not. However, once you have traveled this journey, you will never return.

EQUIPMENT AND PROCESSES

8

Chain Drives[*]

Chain drives are an important part of a conveyor system. They are used to transmit needed power from the drive unit to a portion of the conveyor system. This chapter will cover:

1. Various types of chains that are used to transmit power in a conveyor system.
2. The advantages and disadvantages of using chain drives.
3. The correct installation procedure for chain drives.
4. How to maintain chain drives.
5. How to calculate speeds and ratios that will enable you to make corrections or adjustments to conveyor speeds.
6. How to determine chain length and sprocket sizes when making speed adjustments.

Chain drives are used to transmit power between a drive unit and a driven unit. For example, if we have a gearbox and a contact roll on a conveyor, we need a way to transmit the power from the gearbox to the roll. This can be done easily and efficiently with a chain drive unit.

Chain drives can consist of one or multiple strand chains, depending on the load that the unit must transmit. The chains need to be matched with the sprocket type, and they must be tight enough to prevent slippage.

Chain is sized by the pitch or the center-to-center distance between the pins. This is done in $\frac{1}{8}''$ increments, and the pitch number is found on the side bars. Examples of the different chain and sprocket sizes can be seen in Figures 8.1 and 8.2.

*Source: Ricky Smith and Keith Mobley, *Industrial Machinery Repair: Best Maintenance Practices Pocket Guide* (Boston: Butterworth–Heinemann, 2003), pp. 120–134.

Sometimes chains are linked to form two multi-strand chains. The number designation for this chain would have the same pitch number as standard chain, but the pitch would be followed by the number of strands.

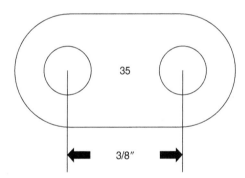

FIGURE 8.1. Chain size, $\frac{3}{8}''$.

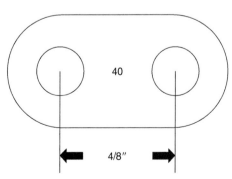

FIGURE 8.2. Chain size, $\frac{4}{8}''$.

8.1. CHAIN SELECTION

8.1.1. Plain or Detachable-Link Chain

Plain chains are usually used in slow speed applications like conveyors. They are rugged, designed to carry heavy loads, and when properly maintained, can offer years of reliable service. They are made up of a series of detachable links that do not have rollers.

The problem is that, if the direction of the chain is reversed, the chain can come apart. When replacing a motor, the rotation of the coupling must be the same before you connect the coupling to the driven unit.

8.1.2. Roller Chain

Roller chains are made up of roller links that are joined with pin links. The links are made up of two side bars, two rollers, and two bushings. The roller reduces the friction between the chain and the sprocket, thereby increasing the life of the unit.

Roller chains can operate at faster speeds than plain chains, and properly maintained, they will offer years of reliable service.

Some roller chains come with a double pitch, meaning that the pitch is double that of a standard chain, but the width and roller size remains the same. Double-pitch chain can be used on standard sprockets, but double-pitch sprockets are also available. The main advantage to the double-pitch chain is that it is cheaper than the standard pitch chain. So, they are often used for applications that require slow speeds, as in lifting pieces of equipment in a hot press application.

8.1.3. Sprockets

Sprockets are fabricated from a variety of materials depending on the application of the drive. Large fabricated steel sprockets are manufactured with holes to reduce the weight of the sprocket on the equipment. Because roller chain drives sometimes have restricted spaces for their installation or mounting, the hubs are made in several different styles (see Figure 8.3).

Type A sprockets are flat and have no hub at all. They are usually mounted on flanges or hubs of the device that they are driving. This is accomplished through a series of holes that are either plain or tapered.

Type B sprocket hubs are flush on one side and extend slightly on the other side. The hub is extended to one side to allow the sprocket to be fitted close to the machinery that it is being mounted on. This eliminates a large overhung load on the bearings of the equipment.

Type C sprockets are extended on both sides of the plate surface. They are usually used on the driven

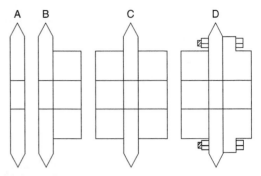

FIGURE 8.3.   Types of sprocket hubs.

sprocket where the pitch diameter is larger and hub classification where there is more weight to support on the shaft. Remember this: The larger the load is, the larger the hub should be.

Type D sprockets use an A sprocket mounted on a solid or split hub. The type A sprocket is split and bolted to the hub. This is done for ease of removal and not practicality. It allows the speed ratio to be changed easily by simply unbolting the sprocket and changing it without having to remove bearings or other equipment.

8.2. CHAIN INSTALLATION

When the proper procedures are followed for installing chains, they will yield years of trouble-free service. Use the following procedure to perlform this task:

1. The shafts must be parallel or the life of the chain will be shortened. The first step is to level the shafts. This is done by placing a level on each of the shafts, then shimming the low side until the shaft is level (see Figure 8.4).
2. The next step is to make sure that the shafts are parallel. This is done by measuring at different points

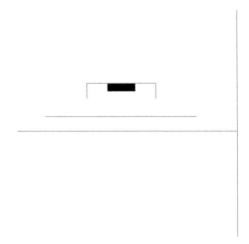

FIGURE 8.4.   Alignment, leveling the shafts.

on the shaft and adjusting the shafts until they are an equal distance apart. Make sure that the shafts are pulled in as close as possible before performing this procedure. The jacking bolts can be used to move the shafts apart evenly after the chain is installed (see Figure 8.5).

3. Before installing a set of used sprockets, verify the size and condition of the sprockets.

4. Install the sprockets on the shafts following the manufacturer's recommendations. Locate and install the first sprocket, then use a straightedge or a string to line the other one up with the one previously installed.

5. Install the chain on the sprockets, then begin increasing the distance between the sprockets by turning the jacking bolts; do this until the chain is snug but not tight. To set the proper chain sag, deflect the chain ¼″ per foot of span between the shafts. Use a string or straightedge and place it across the top of the chain. Then push down on the chain just enough to remove the slack. Use a tape measure to measure the amount of sag (see Figure 8.6).

6. Do a final check for parallel alignment. Remember: The closer the alignment, the longer the chain will run (see Figure 8.7).

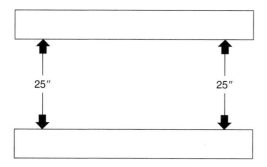

FIGURE 8.5. Alignment, making the shafts parallel and close.

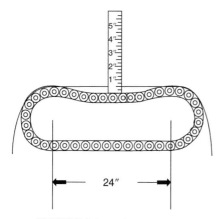

FIGURE 8.6. Adding tension.

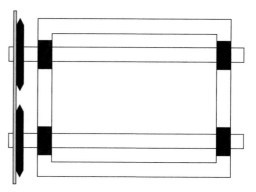

FIGURE 8.7. Final alignment.

8.3. POWER TRAIN FORMULAS

8.3.1. Shaft Speed

The size of the sprockets in a chain drive system determines the speed relationship between the drive and driven sprockets. For example, if the drive sprocket has the same size sprocket as the driven, then the speed will be equal (see Figure 8.8).

If we change the size of the driven sprocket, then the speed of the shaft will also change. If we know what the speed of the electric motor is, and the size of the sprockets, we can calculate the speed of the driven shaft by using the following formula (see Figure 8.9):

Driven shaft rpm =

$$\frac{\text{Drive shaft sprocket \# teeth} \times \text{drive shaft rpm}}{\text{Driven sprocket \# teeth}}$$

$$\text{Driven shaft rpm} = \frac{6 \times 1800}{12} \qquad 900 = \frac{6 \times 1800}{12}$$

Now we understand how changing the size of a sprocket will also change the shaft speed. Knowing this, we could also assume that to change the shaft rpm we must change the sprocket size.

The problem is how do we know the exact size sprocket that we need to reach the desired speed? Use

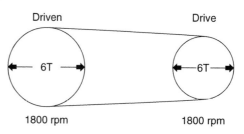

FIGURE 8.8. Ratio.

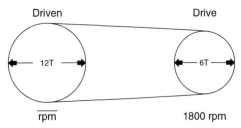

FIGURE 8.9.   Speed ratio change example.

the same formula that was used to calculate shaft speed, only switch the location of the driven shaft speed and the driven sprocket size:

Driven shaft rpm =

$$\frac{\text{Drive sprocket \# teeth} \times \text{drive shaft rpm}}{\text{Driven sprocket \# teeth}}$$

Let's change the problem to look like this:

Driven sprocket teeth =

$$\frac{\text{Drive sprocket \# teeth} \times \text{drive shaft rpm}}{\text{Driven shaft rpm}}$$

Let's say that we have a problem similar to the ones that we just did, but we want to change the shaft speed of the driven unit. If we know the speed we are looking for, we can use the formula above to calculate the sprocket size required.

$$\text{Driven shaft rpm} = \frac{6 \times 1800}{900} \quad 12 = \frac{6 \times 1800}{900}$$

Let's change the speed of the driven shaft to 900 rpm (see Figure 8.10).

8.4. CHAIN LENGTH

Many times when a mechanic has to change our chains there is no way of knowing how long the chain should be. One way is to lay the new chain down beside the old chain, but remember that the old chain has been stretched.

Or, maybe you are installing a new drive and you want to have the chain made up before you install it. So what do you do? One method is to take a tape measure and wrap it around the sprockets to get the chain length.

However, this is not a very accurate way to determine the length. Instead, let's take a couple of measurements, then use a simple formula to calculate the actual length that is needed.

First, move the sprockets together until they are as close as the adjustments will allow. Then move the motor or drive out ¼ of its travel. Now, we are ready to take our measurements. The following information is needed for an equation to find the chain length:

1. Number of teeth on the drive sprocket.
2. Center-to-center distance between the shafts.
3. The chain pitch in inches.

Now use the following formula to solve the equation (see Figure 8.11):

$$\text{Chain length} = \frac{\text{\#teeth drive} \times \text{pitch}}{2}$$
$$+ \frac{\text{\#teeth driven} \times \text{pitch}}{2} + \text{center to center} \times 2$$

Use the formula above to find the chain length.

$$\text{Chain length} = \frac{6 \times 0.5}{2} + \frac{12 \times 0.5}{2} \times 35'' \times 2$$

$$74.5 = \frac{6 \times 0.5}{2} + \frac{12 \times 0.5}{2} \times 35'' \times 2$$

8.5. MULTIPLE SPROCKETS

When calculating multiple sprocket systems, think of each set of sprockets as a two-sprocket system.

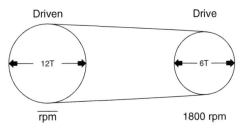

FIGURE 8.10.   Sprocket calculations.

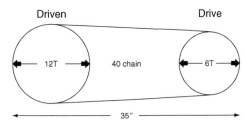

FIGURE 8.11.   Chain size calculations.

8.6. CHAIN SPEED

In order to calculate the speed of a chain in feet per minute (FPM), we need the following information:

1. The number of teeth on the sprocket.
2. The shaft rpm of the sprocket.
3. The pitch of the chain in inches.

With this information we can use the following

$$FPM = \frac{\# \text{ teeth} \times \text{pitch} \times \text{rpm}}{12}$$

formula:

$$FPM = \frac{\# \text{ teeth} \times \text{pitch} \times \text{rpm}}{12}$$

$$450 = \frac{6T \times 0.5 \times 1800}{12}$$

Use this formula to find the speed of the following chain (see Figure 8.12):

Chain drives are used to transmit power between a drive unit and a driven unit. For example, if we have a gearbox and a contact roll on a conveyor, we need a way to transmit the power from the gearbox to the roll. This can be done easily and efficiently with a chain drive unit.

Chain drives can consist of one or multiple strand chains, depending on the load that the unit must transmit. The chains need to be matched with the sprocket type, and they must be tight enough to prevent slippage.

An effective preventive maintenance program will provide extended life to a chain drive system, and through proper corrective maintenance procedures we can prevent premature failures.

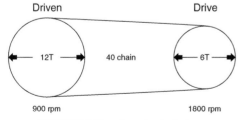

FIGURE 8.12.   Speed calculations.

8.7. PREVENTIVE MAINTENANCE PROCEDURES

Inspection (risk of failures for not following the procedures below is noted along with a rating): LOW: minimal risk/low chance of failure; MEDIUM: failure is possible but equipment not operation to specification is highly probable; HIGH: failure will happen prematurely.

- Inspect a chain for wear by inspecting the links for worn bushings. If worn bushings are noted, write a corrective maintenance work order so that the replacement can be planned and scheduled at a later time. *Risk if the procedure is not followed: high.* Chain breakage will occur.

- Lubricate chain with lightweight oil recommended by chain manufacturer. (Ask your chain supplier to visit your site and make recommendations based on documentation they can present to you.) *Risk if the procedure is not followed: high.* Chain breakage will occur.

- Check chain sag. Measure the chain sag using a straight edge or string and measure the specifications noted on this PM (preventive maintenance) task. (The chain sag specification can be provided by your chain supplier, or you can use the procedure noted earlier in this chapter.) Warning: The specification must be noted on the PM procedure.

- Set tension, and make a note at the bottom of the PM work order, if a deficiency is noted. *Risk if the procedure is not followed: medium.* Sprocket and chain wear will accelerate, thus causing equipment stoppage.

- Inspect sprockets for worn teeth and abnormal wear on the sides of the sprockets. (The question is: Can the sprockets and chain last for two more weeks without equipment stoppage?) If the sprockets and chain can last two weeks, then write a corrective maintenance work order in order for this job to be planned and scheduled with the correct parts. If the sprocket cannot last two weeks, then change all sprockets and the chain. Set and check sheave and chain alignment and tension. Warning: When changing a sprocket, all sprockets and the chain should be changed, because the difference between a worn and new sprocket in pitch diameter can be extreme, thus causing premature failure of the sprockets and chain. *Risk if the procedure is not followed: high.* Worn sprockets are an indication of the equipment being in a failure mode. Action must be taken.

8.7 PREVENTIVE MAINTENANCE PROCEDURES

8.6 CHAIN SPEED

In order to calculate the speed of a chain in feet per minute (FPM), we need the following information:

1. The number of teeth on the sprocket.
2. The shaft rpm of the sprocket.
3. The pitch of the chain in inches.

With this information, we can use the following formula:

$$FPM = \frac{\text{teeth} \times \text{pitch} \times \text{rpm}}{12}$$

C H A P T E R

9

Hydraulics[*]

9.1. HYDRAULIC KNOWLEDGE

People say knowledge is power. This is true in hydraulic maintenance. Many maintenance organizations do not know what their maintenance personnel should know. We believe in an industrial maintenance organization where we should divide the hydraulic skill necessary into two groups. One is the hydraulic troubleshooter; they must be your experts in maintenance, and this should be as a rule of thumb 10% or less of your maintenance workforce. The other 90% plus would be your general hydraulic maintenance personnel. They are the personnel that provide the preventive maintenance expertise. The percentages we give you are based on a company developing a true preventive/proactive maintenance approach to its hydraulic systems. Let's talk about what the hydraulic troubleshooter knowledge and skills should be.

9.2. HYDRAULIC TROUBLESHOOTER

Knowledge:

- Mechanical principles (force, work, rate, simple machines)
- Math (basic math, complex math equations)
- Hydraulic components (application and function of all hydraulic system components)

*Source: Ricky Smith and Keith Mobley, *Industrial Machinery Repair: Best Maintenance Practices Pocket Guide* (Boston: Butterworth–Heinemann, 2003), pp. 314–326.

- Hydraulic schematic symbols (understanding all symbols and their relationship to a hydraulic system)
- Calculating flow, pressure, and speed
- Calculating the system filtration necessary to achieve the system's proper ISO particulate code

Skill:

- Trace a hydraulic circuit to 100% proficiency
- Set the pressure on a pressure compensated pump
- Tune the voltage on an amplifier card
- Null a servo valve
- Troubleshoot a hydraulic system and utilize "root cause failure analysis"
- Replace any system component to manufacturer's specifications
- Develop a PM (preventive maintenance) program for a hydraulic system
- Flush a hydraulic system after a major component failure

9.3. GENERAL MAINTENANCE PERSON

Knowledge:

- Filters (function, application, installation techniques)
- Reservoirs (function, application)
- Basic hydraulic system operation
- Cleaning of hydraulic systems
- Hydraulic lubrication principles
- Proper PM techniques for hydraulics

Skill:

- Change a hydraulic filter and other system components
- Clean a hydraulic reservoir
- Perform PM on a hydraulic system
- Change a strainer on a hydraulic pump
- Add filtered fluid to a hydraulic system
- Identify potential problems on a hydraulic system
- Change a hydraulic hose, fitting, or tubing

9.4. BEST MAINTENANCE HYDRAULIC REPAIR PRACTICES

In order to maintain your hydraulic systems, you must have preventive maintenance procedures and you must have a good understanding and knowledge of "best maintenance practices" for hydraulic systems. We will convey these practices to you. See Table 9. 1.

9.5. ROOT CAUSE FAILURE ANALYSIS

As in any proactive maintenance organization you must perform root cause failure analysis in order to eliminate future component failures. Most maintenance problems or failures will repeat themselves without someone identifying what caused the failure and proactively eliminating it. A preferred method is to inspect and analyze all component failures. Identify the following:

- Component name and model number
- Location of component at the time of failure
- Sequence or activity the system was operating at when the failure occurred.
- What caused the failure?
- How will the failure be prevented from happening again?

Failures are not caused by an unknown factor like bad luck, or "it just happened," or "the manufacturer made a bad part." We have found most failures can be analyzed and prevention taken to prevent their recurrence. Establishing teams to review each failure can pay off in major ways.

9.6. PREVENTIVE MAINTENANCE

Preventive maintenance (PM) of a hydraulic system is very basic and simple and, if followed properly, can eliminate most hydraulic component failure. Preventive maintenance is a discipline and must be followed as such in order to obtain results. We must view PM programs as performance oriented and not activity oriented. Many organizations have good PM procedures but do not require maintenance personnel to follow them or hold personnel accountable for the proper execution of these procedures. In order to develop a preventive maintenance program for your system you must follow these steps:

First: Identify the system operating condition.

- Does the system operate 24 hours a day, seven days a week?
- Does the system operate at maximum flow and pressure 70% or better during operation?
- Is the system located in a dirty or hot environment?

Second: What requirements does the equipment manufacturer state for preventive maintenance on the hydraulic system?

Third: What requirements and operating parameters does the component manufacturer state concerning the hydraulic fluid ISO particulate?

Fourth: What requirements and operating parameters does the filter company state concerning its filters' ability to meet this requirement?

Fifth: What equipment history is available to verify the above procedures for the hydraulic system?

As in all preventive maintenance programs, we must write procedures required for each PM task. Steps or procedures must be written for each task, and they must be accurate and understandable by all maintenance personnel from entry level to master.

Preventive maintenance procedures (see Figure 9.1) must be a part of the PM job plan, which includes:

- Tools or special equipment required for performing the task.
- Parts or material required for performing the procedure, with the store room number.
- Safety precautions for this procedure.
- Environmental concerns or potential hazards.

A list of preventive maintenance tasks for a hydraulic system could be

- Change the hydraulic filter (could be the return or pressure filter).
- Obtain a hydraulic fluid sample.
- Filter hydraulic fluid.
- Check hydraulic actuators.
- Clean the inside of a hydraulic reservoir.
- Clean the outside of a hydraulic reservoir.
- Check and record hydraulic pressures.
- Check and record pump flow.
- Check hydraulic hoses, tubing, and fittings.
- Check and record voltage reading to proportional or servo valves.

TABLE 9. 1.   Best Maintenance Repair Practices: Hydraulics

Component	Component Knowledge	Best Practices	Frequency
Hydraulic fluid filter	There are two types of filters on a hydraulic system: 1. Pressure filter: Pressure filters come in collapsible and noncollapsible types. The preferred filter is the noncollapsible type. 2. Return filter: Typically has a bypass, which will allow contaminated oil to bypass the filter before indicating the filter needs to be changed.	Clean the filter cover or housing with a cleaning agent and clean rags. Remove the old filter with clean hands and install new filter into the filter housing or screw into place. CAUTION: NEVER allow your hand to touch a filter cartridge. Open the plastic bag and insert the filter without touching the filter with your hand.	Preferred: based on historical trending of oil samples. Least preferred: Based on equipment manufacture's recommendations.
Reservoir air breather	The typical screen breather should not be used in a contaminated environment. A filtered air breather with a rating of 10 micron is preferred because of the introduction of contaminants to a hydraulic system.	Remove and throw away the filter.	Preferred: Based on historical trending of oil samples. Least preferred: Based on equipment manufacturer's recommendations.
Hydraulic reservoir	A reservoir is used to: Remove contamination. Dissipate heat from the fluid. Store a volume of oil.	Clean the outside of the reservoir to include the area under and around the reservoir. Remove the oil by a filter pump into a clean container, which has not had other types of fluid in it before. Clean the insides of the reservoir by opening the reservoir and cleaning the reservoir with a lint-free rag. Afterward, spray clean hydraulic fluid into the reservoir and drain out of the system.	If any of the following conditions are met: A hydraulic pump fails. If the system has been opened for major work. If an oil analysis reveals excessive contamination.
Hydraulic pumps	A maintenance person needs to know the type of pump in the system and determine how it operates in the system. Example: What is the flow and pressure of the pump during a given operating cycle? This information allows a maintenance person to trending potential pump failure and troubleshooting a system problem quickly.	Check and record flow and pressure during specific operating cycles. Review graphs of pressure and flow. Check for excessive fluctuation of the hydraulic system. (Designate the fluctuation allowed.)	Pressure checks: Preferred: daily least Preferred: weekly Flow and pressure checks: Preferred: two weeks Least preferred: monthly

- Check and record vacuum on the suction side of the pump.
- Check and record amperage on the main pump motor.
- Check machine cycle time and record.

Preventive maintenance is the core support that a hydraulic system must have in order to maximize component life and reduce system failure. Preventive maintenance procedures that are properly written and followed properly will allow equipment to operate to

```
                        ABC COMPANY
              PREVENTIVE MAINTENANCE PROCEDURE

TASK DESCRIPTION:           P.M. – Inspect hydraulic oil reserve tank level
EQUIPMENT NUMBER:           311111
FILE NUMBER:                09
FREQUENCY:                  52
KEYWORD, QUALIFIER:         Unit, Hydraulic (Dynamic Press)
SKILL/CRAFT:                Production
PM TYPE:                    Inspection
SHUTDOWN REQUIRED:          No
REFERENCE MANUAL/DWGS:
1. See operator manual F-378
REQUIRED TOOLS/MATERIALS:
1. Oil, Texaco Rando 68 SDK #400310
2. Flashlight
3. Oil Filter09

SAFETY PRECAUTIONS:
1. Observe plant and area specific safe work practices.

MAINTENANCE PROCEDURE:
1. Inspect hydraulic oil reserve tank level as follows:
   a) If equipped with sight glass, verify oil level at the full mark. Add oil as required.
   b) If not equipped with sight glass, remove fill plug/cap.
   c) Using flashlight, verify that oil is at proper level in tank. Add oil as required.
2. Record discrepancies or unacceptable conditions in comments.
```

FIGURE 9.1.   Sample preventive maintenance procedure. (Courtesy of Life Cycle Engineering, Inc.)

its full potential and life cycle. Preventive maintenance allows a maintenance department to control a hydraulic system rather than the system controlling the maintenance department. We must control a hydraulic system by telling it when we will perform maintenance on it and how much money we will spend on the maintenance for the system. Most companies allow hydraulic systems to control the maintenance on them at a much higher cost.

9.7. MEASURING SUCCESS

In any program we must track success in order to have support from management and maintenance personnel. We must also understand that any action will have a reaction, negative or positive. We know successful maintenance programs will provide success, but we must have a checks and balances system to ensure we are on track.

In order to measure success of a hydraulic maintenance program we must have a way of tracking success but first we need to establish a benchmark. a benchmark is a method by which we will establish certain key measurement tools that will tell you the current status

of your hydraulic system and then tell you if you are succeeding in your maintenance program.

Before you begin the implementation of your new hydraulic maintenance program it would be helpful to identify and track the following information.

1. Track all downtime (in minutes) on the hydraulic system with these questions (tracked daily):
   - What component failed?
   - Cause of failure?
   - Was the problem resolved?
   - Could this failure have been prevented?
   - Track all costs associated with the downtime (tracked daily).
   - Parts and material cost?
   - Labor cost?
   - Production downtime cost?
   - Any other cost you may know that can be associated with a hydraulic system failure?
2. Track hydraulic system fluid analysis. Track the following from the results (taking samples once a month):
   - Copper content
   - Silicon content
   - H_2O

- Iron content
- ISO particulate count
- Fluid condition (viscosity, additives, and oxidation)

When the tracking process begins, you need to trend the information that can be trended. This allows management the ability to identify trends that can lead to positive or negative consequences. See Figure 9.2.

Fluid analysis will prove the need for better filtration. The addition of a 3-micron absolute return line filter to supplement the "kidney loop" filter can solve the problem.

Many organizations do not know where to find the method for tracking and trending the information you need accurately. A good computerized maintenance management system can track and trend most of this information for you.

9.8. RECOMMENDED MAINTENANCE MODIFICATIONS

Modifications to an existing hydraulic system need to be accomplished professionally. A modification to a hydraulic system in order to improve the maintenance efficiency is important to a company's goal of maximum equipment reliability and reduced maintenance cost.

First: Filtration Pump with Accessories

Objective: The objective of this pump and modification is to reduce contamination that is introduced into an existing hydraulic system through the addition of new fluid and the device used to add oil to the system.

Additional information: Hydraulic fluid from the distributor usually is not filtered to the requirements of an operating hydraulic system. Typically, this oil is strained to a mesh rating and not a micron rating. How clean is clean? Typically, hydraulic fluid must be filtered to 10 microns absolute or less for most hydraulic systems; 25 microns is the size of a white blood cell, and 40 microns is the lower limit of visibility with the unaided eye.

Many maintenance organizations add hydraulic fluid to a system through a contaminated funnel and may even, without cleaning it, use a bucket that has had other types of fluids and lubricants in it previously.

Recommended equipment and parts:

- Portable filter pump with a filter rating of 3 microns absolute.
- Quick disconnects that meet or exceed the flow rating of the portable filter pump.
- A ¾" pipe long enough to reach the bottom of the hydraulic container your fluids are delivered in from the distributor.
- A 2" reducer bushing to ¾" NPT to fit into the 55-gallon drum, if you receive your fluid by the drum. Otherwise, mount the filter buggy to the double wall "tote" tank supports if you receive larger quantities.
- Reservoir vent screens should be replaced with $\frac{3}{10}$ micron filters, and openings around piping entering the reservoir sealed.

Show a double wall tote tank, of about 300 gallons mounted on a frame for fork truck handling, with the pump mounted on the framework. Also show pumping from a drum mounted on a frame for fork truck handling, sitting in a catch pan, for secondary containment, with the filter buggy attached.

Regulations require that you have secondary containment, so make everything "leak" into the pan. See Figure 9.3.

Second: Modify the Hydraulic Reservoir (see Figure 9.4)

Objective: The objective is to eliminate the introduction of contamination through oil being added to the system or contaminants being added through the air

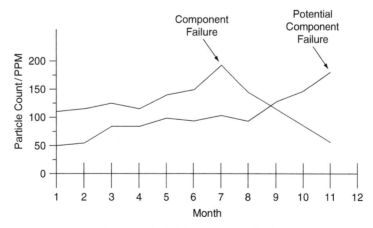

FIGURE 9.2. Hydraulic system fluid samples.

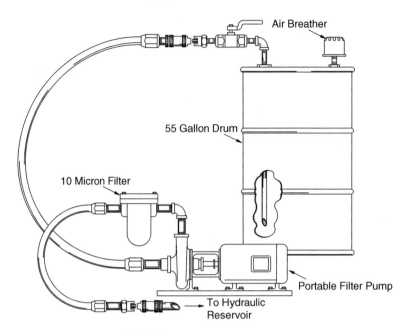

FIGURE 9.3.   Filter pumping unit.

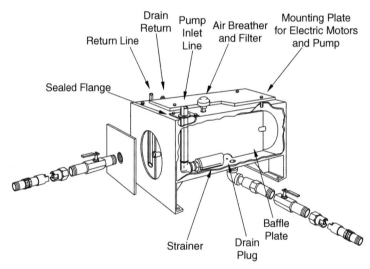

FIGURE 9.4.   Hydraulic reservoir modification.

intake of the reservoir. A valve needs to be installed for oil sampling.

Additional information: The air breather strainer should be replaced with a 10-micron filter if the hydraulic reservoir cycles. A quick disconnect should be installed on the bottom of the hydraulic unit and at the three-quarter level point on the reservoir with valves to isolate the quick disconnects in case of failure. This allows the oil to add from a filter pump as previously discussed and would allow for external filtering of the hydraulic reservoir oil if needed. Install a petcock valve on the front of the reservoir, which will be used for consistent oil sampling.

Equipment and parts needed:

- Quick disconnects that meet or exceed the flow rating of the portable filter pump.
- Two gate valves with pipe nipples.
- One 10-micron filter breather.

Warning: Do not weld on a hydraulic reservoir to install the quick disconnects or air filter.

To summarize, maintenance of a hydraulic system is the first line of defense to prevent component failure and thus improve equipment reliability. As discussed earlier, discipline is the key to the success of any proactive maintenance program.

10

Maintenance Welding*

An important use of arc welding is the repair of plant machinery and equipment. In this respect, welding is an indispensable tool without which production operations would soon shut down. Fortunately, welding machines and electrodes have been developed to the point where reliable welding can be accomplished under the most adverse circumstances. Frequently, welding must be done under something less than ideal conditions, and therefore, equipment and operators for maintenance welding should be the best.

Besides making quick, on-the-spot repairs of broken machinery parts, welding offers the maintenance department a means of making many items needed to meet a particular demand promptly. Broken castings, when new ones are no longer available, can be replaced with steel weldments fashioned out of standard shapes and plates (see Figure 10.1). Special machine tools required by production for specific operations often can be designed and made for a fraction of the cost of purchasing a standard machine and adapting it to the job. Material-handling devices can be made to fit the plant's physical dimensions. Individual jib cranes can be installed. Conveyors, either rolldown or pallet-type, can be tailor-made for specific applications. Tubs and containers can be made to fit products. Grabs, hooks, and other handling equipment can be made for shipping and receiving. Jigs and fixtures, as well as other simple tooling, can be fabricated in the maintenance department as either permanent tooling or as temporary tooling for a trial lot.

The almost infinite variety of this type of welding makes it impossible to do more than suggest what can be done. Figures 10.2 through 10.5 provide just a few examples of the imaginative applications of welding technology achieved by some maintenance technicians.

*Source: Ricky Smith and Keith Mobley, *Industrial Machinery Repair: Best Maintenance Practices Pocket Guide* (Boston: Butterworth–Heinemann, 2003), pp. 460–538.

The welding involved should present no particular problems if the operators have the necessary training and background to provide them with a knowledge of the many welding techniques that can be used.

A maintenance crew proficient in welding can fabricate and erect many of the structures required by a plant, even to the extent of making structural steel for a major expansion. Welding can be done either in the plant maintenance department or on the erection site. Structures must, of course, be adequately designed to withstand the loads to which they will be subjected. Such loads will vary from those of wind and snow in simple sheds to dynamic loads of several tons where a crane is involved. Materials and joint designs must be selected with a knowledge of what each can do. Then the design must be executed by properly trained and qualified welders. Structural welding involves out-of-position work, so a welder must be able to make good welds under all conditions. Typical joints that are used in welded structures are shown in Figures 10.6 through 10.9.

Standard structural shapes, including pipe, which makes an excellent structural shape, can be used. Electrodes such as the E6010-11 types are often the welder's first choice for this kind of fabrication welding because of their all-position characteristics. These electrodes, which are not low-hydrogen types, may be used providing the weldability of the steel is such that neither weld cracks nor severe porosity is likely to occur.

Scrap materials often can be put to good use. When using scrap, however, it is best to weld with a low-hydrogen E7016-18 type of electrode, since the analysis of the steel is unlikely to be known and some high-carbon steels may be encountered. Low-hydrogen electrodes minimize cracking tendencies. Structural scrap frequently comes from dismantled structures such as elevated railroads, which use rivet-quality steel that takes little or no account of the carbon content.

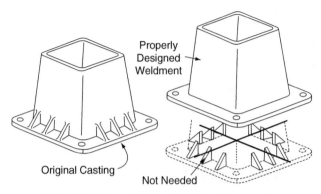

FIGURE 10.1.   Weldment fashioned for a repair.

FIGURE 10.4.

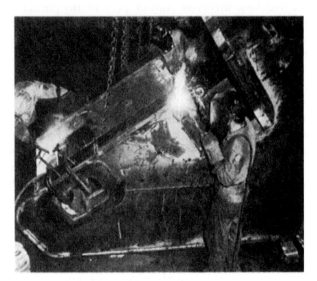

FIGURE 10.2.

FIGURE 10.5.

FIGURE 10.3.

10.1. SHIELDED METAL ARC WELDING (SMAW), "STICK WELDING"

Shielded metal arc welding is the most widely used method of arc welding. With SMAW, often called "stick welding," an electric arc is formed between a consumable metal electrode and the work. The intense heat of the arc, which has been measured at temperatures as high as 13,000°F, melts the electrode and the surface of the work adjacent to the arc. Tiny

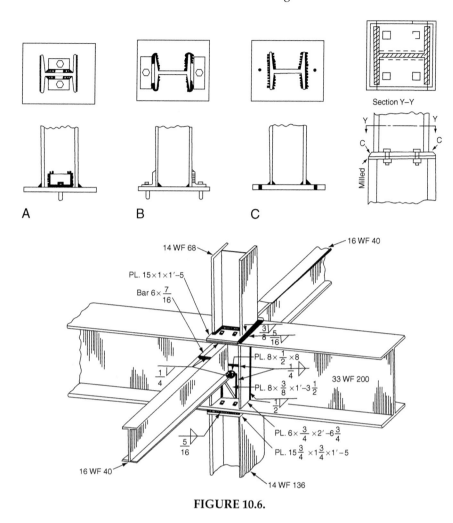

FIGURE 10.6.

globules of molten metal rapidly form on the tip of the electrode and transfer through the arc, in the "arc stream," and into the molten "weld pool" or "weld puddle" on the work's surface (see Figure 10.10).

Within the shielded metal arc welding process, electrodes are readily available in tensile strength ranges of 60,000 to 120,000 psi (see Table 10.1). In addition, if specific alloys are required to match the base metal, these, too, are readily available (see Table 10.2).

10.2. FLUX-CORED ARC WELDING (FCAW)

Flux-cored arc welding is generally applied as a semiautomatic process. It may be used with or without external shielding gas depending on the electrode selected. Either method utilizes a fabricated flux-

cored electrode containing elements within the core that perform a scavenging and deoxidizing action on the weld metal to improve the properties of the weld.

10.2.1. FCAW with Gas

If gas is required with a flux-cored electrode, it is usually CO_2 or a mixture of CO_2 and another gas. These electrodes are best suited to welding relatively thick plate (not sheet metal) and for fabricating and repairing heavy weldments.

10.2.2. FCAW Self-Shielded

Self-shielded flux-cored electrodes, better known as Innershield, are also available. In effect, these are stick electrodes turned inside out and made into a continuous coil of tubular wire. All shielding, slagging, and

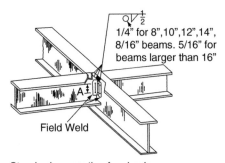

1/4" for 8",10",12",14",
8/16" beams. 5/16" for
beams larger than 16"

Field Weld

Standard connection for simple
beam-to-beam framing

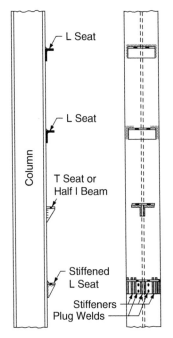

FIGURE 10.7.

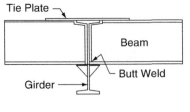

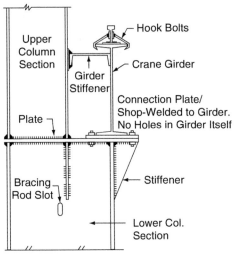

FIGURE 10.9.

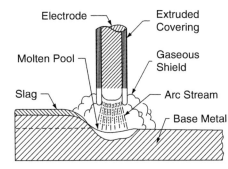

FIGURE 10.10.

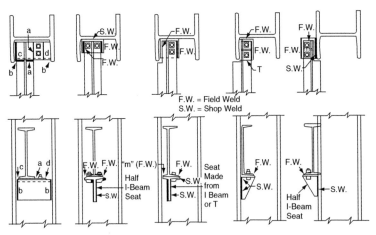

FIGURE 10.8.

TABLE 10.1.

AWS Classification	Tensile Strength, Min. psi	Yield Strength, Min. psi
E6010-11	62,000	50,000
E7010-11	70,000	57,000
E7016-18	70,000	57,000
E8016-18	80,000	67,000
E9016-18	90,000	77,000
E10016-18	100,000	87,000
E11016-18	110,000	97,000
E12016-18	120,000	107,000

Note: E6010-11 and E7010-11 are cellulosic electrodes. All others are low-hydrogen electrodes and are better suited to welding higher-strength steels.

deoxidizing materials are in the core of the tubular wire. No external gas or flux is required.

Innershield electrodes offer much of the simplicity, adaptability, and uniform weld quality that account for the continuing popularity of manual welding with stick electrodes, but as a semiautomatic process, they get the job done faster. This is an open-arc process that allows the operator to place the weld metal accurately and to visually control the weld puddle. These electrodes operate in all positions: flat, vertical, horizontal, and overhead.

The electrode used for semiautomatic and fully automatic flux-cored arc welding is mechanically fed through a welding gun or welding jaws into the arc from a continuously wound coil that weighs approximately 50 pounds.

TABLE 10.2.

AWS Classification[a]	Chemical Composition, Percent[b]								
	C	Mn	P	S	Si	Ni	Cr	Mo	V
Chemical composition, percent[b]									
E7010-Al	0.12	0.60	0.03	0.40	0.60	—	—	0.40–0.65	—
E7011-AI	0.12	0.60	0.03	0.40	0.60	—	—	0.40–0.65	—
E7015-AI	0.12	0.90	0.03	0.60	0.60	—	—	0.40–0.65	—
E7016-Al	0.12	0.90	0.03	0.04	0.60	—	—	0.40–0.65	—
E7018-Al	0.12	0.90	0.03	0.80	0.60	—	—	0.40–0.65	—
E7020-A1	0.12	0.60	0.03	0.40	0.60	—	—	0.40–0.65	—
E7027-Al	0.12	1.00	0.03	0.40	0.60	—	—	0.40–0.65	—
Chromium-molybdenum steel electrodes									
E8016-B1	0.05–0.12	0.90	0.03	0.04	0.60	—	0.40–0.65	0.40–0.65	—
E7017-B1	0.05–0.12	0.90	0.03	0.04	0.80	—	0.40–0.65	0.40–0.65	—
E8015-B2L	0.05	0.90	0.03	0.04	1.00	—	1.00–1.50	0.40–0.65	—
E8016-B2	0.05–0.12	0.90	0.03	0.04	0.60	—	1.00–1.50	0.40–0.65	—
E8018-B2	0.05–0.12	0.90	0.03	0.04	0.60	—	1.00–1.50	0.40–0.65	—
E8018-B2L	0.05	0.90	0.03	0.04	0.80	—	1.00–1.50	0.40–0.65	—
E9015-B3L	0.05	0.90	0.03	0.04	1.00	—	2.00–2.50	0.90–1.20	—
E9015-B3	0.05–0.12	0.90	0.03	0.04	0.60	—	2.00–2.50	0.90–1.20	—
E9016-B3	0.05–0.12	0.90	0.03	0.04	0.60	—	2.00–2.50	0.90–1.20	—
E9018-B3	0.05–0.12	0.90	0.03	0.04	0.80	—	2.00–2.50	0.90–1.20	—
E9018-B3L	0.05	0.09	0.03	0.04	0.80	—	2.00–2.50	0.90–1.20	—
E8015-B4L	0.05	0.90	0.03	0.04	1.00	—	1.75–2.25	0.40–0.65	—
B8016-B5	0.07–0.15	0.40–0.70	0.03	0.04	0.30–0.60	—	0.40–0.60	1.00–1.25	0.05
Nickel steel electrodes									
E8016-C1	0.12	1.25	0.03	0.04	0.06	2.00–2.75	—	—	—
E8018-C1	0.12	1.25	0.03	0.04	0.08	—	—	—	—
E7015-C1L	0.05	1.25	0.03	0.04	0.50	2.00–2.75	—	—	—

(*Continues*)

TABLE 10.2. (*Continued*)

Chemical Composition, Percent[b]

AWS Classification[a]	C	Mn	P	S	Si	Ni	Cr	Mo	V
E7016-C1L	0.05	1.25	0.03	0.04	0.50	2.00–2.75	—	—	—
E7018-C1L	0.05	1.25	0.03	0.04	0.50	2.00–2.75	—	—	—
E8016-C2	0.12	1.25	0.03	0.04	0.60	3.00–3.75	—	—	—
E8018-C2	0.12	1.25	0.03	0.04	0.80	3.00–3.75	—	—	—
E7015-C2L	0.05	1.25	0.03	0.04	0.50	3.00–3.75	—	—	—
E7016-C2L	0.05	1.25	0.03	0.04	0.50	3.00–3.75	—	—	—
E7018-C2L	0.05	1.25	0.03	0.04	0.50	3.00–3.75	—	—	—
E8016-C3[c]	0.12	0.40–1.25	0.03	0.03	0.80	0.80–1.10	0.15	0.35	0.05
E8018-C3	0.12	0.40–1.25	0.03	0.03	0.80	0.80–1.10	0.15	0.35	0.05
Nickel-molybdenum steel electrodes									
E8018-NM[d]	0.10	0.80–1.25	0.02	0.03	0.60	0.80–1.10	0.05	0.40–0.65	0.02
Manganese-molybdenum steel electrodes									
E9015-D1	0.12	1.25–1.75	0.03	0.04	0.60	—	—	0.25–0.45	—
E9018-DI	0.12	1.25–1.75	0.03	0.04	0.80	—	—	0.25–0.45	—
E8016-D3	0.12	1.00–1.75	0.03	0.04	0.60	—	—	0.40–0.65	—
E8018-D3	0.12	1.00–1.75	0.03	0.04	0.80	—	—	0.40–0.65	—
E10015-D2	0.15	1.65–2.00	0.03	0.04	0.60	—	—	0.25–0.45	—
E10016-D2	0.15	1.65–2.00	0.03	0.04	0.60	—	—	0.25–0.45	—
E10018-D2	0.15	1.65–2.00	0.03	0.04	0.80	—	—	0.25–0.45	—
All other low-alloy steel electrodes[e]									
EXXIO-G[e]	—	1.00 min[f]	—	—	0.80 min[f]	0.50 min[f]	0.30 min[f]	0.20 min[f]	0.10 min[f]
EXX11-G									
EXX13-G									
EXX15-G									
EXX16-G									
EXX18-G									
E7020-G									
E9018-M[c]		0.100			0.80	1.40–1.80	0.15	0.35	0.05
E10018-M[c]	0.10	0.75–1.70	0.030	0.030	0.60	1.40–2.10	0.35	0.25–0.50	0.05
E11018-M[c]	0.10		1.3		0.60	1.25–2.50	0.40	0.25–0.50	0.05
E12018-M[c]	0.10	1.30–2.25	0.030	0.030	0.60	1.75–2.50	0.30–1.50	0.30–0.55	0.05
E12018-M1[c]	0.10	0.80–1.60	0.015	0.012	0.65	3.00–3.80	0.65	0.20–0.30	0.05
E7018-W[g]	0.12	0.40–0.70	0.025	0.025	0.40–0.70	0.20–0.40	0.15–0.30	—	0.08
E8018-W1[g]	0.12	0.50–1.30	0.03	0.04	0.35–0.80	0.40–0.80	0.45–0.70	—	—

Note: Single values shown are *maximum* percentages, except where otherwise specified.

[a]The suffixes A1, B3, C2, etc. designate the chemical composition of the electrode classification.

[b]For determining the chemical composition, DECN (electrode negative) may be used where DC, both polarities, is specified.

[c]These classifications are intended to conform to classifications covered by the military specifications for similar compositions.

[d]Copper shall be 0.10% max and aluminum shall be 0.05% max for E8018-NM electrodes.

[e]The letters "XX" used in the classification designations in this table stand for the various strength levels (70, 80, 90, 100, 110, and 120) of electrodes.

[f]In order to meet the alloy requirements of the G group, the weld deposit need have the minimum, as specified in the table, of only one of the elements listed. Additional chemical requirements may be agreed between supplier and purchaser.

[g]Copper shall be 0.30 to 0.60% for E7018-W electrodes.

Only the fabricated flux-cored electrodes are suited to this method of welding, since coiling extruded flux-coated electrodes would damage the coating. In addition, metal-to-metal contact at the electrode's surface is necessary to transfer the welding current from the welding gun into the electrode. This is impossible if the electrode is covered.

A typical application of semiautomatic and fully automatic equipment for FCAW is shown in Figure 10.11. For a given cross section of electrode wire, much higher welding amperage can be applied with semiautomatic and fully automatic processes. This is because the current travels only a very short distance along the bare metal electrode, since contact between the current-carrying gun and the bare metal electrode occurs close to the arc. In manual welding, the welding current must travel the entire length of the electrode, and the amount of current is limited to the current-carrying capacity of the wire. The higher currents used with automatic welding result in a high weld metal deposition. This increases travel speed and reduces welding time, thereby lowering costs.

10.3. GAS-SHIELDED METAL ARC WELDING (GMAW)

The GMAW process, sometimes called metal inert gas (MIG) welding, incorporates the automatic feeding of a continuous consumable electrode that is shielded by an externally supplied gas. Since the equipment provides for automatic control of the arc, only the travel speed, gun positioning, and guidance are controlled manually. Process control and function are achieved through the basic elements of equipment shown in Figure 10.12. The gun guides the consumable electrode and conducts the electric current and shielding gas to the workpiece. The electrode feed unit and power source are used in a system that provides automatic regulation of the arc length. The basic combination used to produce this regulation consists of a constant-voltage power source (characteristically providing an essentially flat volt-ampere curve) in conjunction with a constant-speed electrode feed unit.

10.3.1. GMAW for Maintenance Welding

In terms of maintenance welding applications, GMAW has the following advantages over SMAW:

1. Can be used in all positions with the low-energy modes.
2. Produces virtually no slag to remove or be trapped in weld.
3. Requires less operator training time than SMAW.
4. Adaptable to semiautomatic or machine welding.
5. Low-hydrogen process.
6. Faster welding speeds than SMAW.
7. Suitable for welding carbon steels, alloy steels, stainless steels, aluminum, and other nonferrous metals. Table 10.3 lists recommended filler metals for GMAW.

10.3.2. Gas Selection for GMAW

There are many different gases and combinations of gases that can be used with the GMAW process. These choices vary with the base metal, whether a spray arc or short-circuiting arc is desired, or sometimes just according to operator preference. Recommended gas choices are given in Tables 10.4 and 10.5.

FIGURE 10.11.

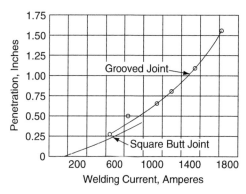

FIGURE 10.12.

TABLE 10.3.

Base Metal Type	Recommended Electrode		AWS Filler Metal Specification (use latest edition)	Electrode Diameter		Current Range, Amperes
	Material Type	Electrode Classification		in.	mm	
Aluminum and aluminum alloys	1100	ER1100 or ER4043		0.030	0.8	50–175
	3003, 3004	ER1100 or ER5356		3/64	1.2	90–250
	5052, 5454	ER5554, ER5356, or ER5183	A5.10	1/16	1.6	160–350
				3/32	2.4	225–400
	5083, 5086, 5456	ER5556 or ER5356		1/8	3.2	350–475
	6061, 6063	ER4043 or ER5356				
Magnesium alloys	AZ10A	ERAZ61A, ERAZ92A		0.040	1.0	150–300[2]
	AZ3IB, AZ61A, AZ80A	ERAZ61A, ERAZ92A		3/64	1.2	160–320[2]
	ZE10A	ERAZ61A, ERAZ92A		1/16	1.6	210–400[2]
	ZK21A	ERAZ61A, ERAZ92A		3/32	2.4	320–510[2]
	AZ63A, AZ81A, AZ91C	ERAZ92A	A5.19	1/8	3.2	400–600[2]
	AZ92A, AM100A	ERAZ92A				
	HK31A, HM21A, HM31A	EREZ33A				
	LA141A	EREZ33A				
Copper and copper alloys	Silicon bronze	ERCuSi-A				
	Deoxidized copper	ERCu		0.035	0.9	150–300
	Cu-Ni alloys	ERCuNi	A5.7	0.045	1.2	200–400
	Aluminum bronze	ERCuA1-A1,A2orA3		1/16	1.6	250–450
	Phosphor bronze	ERCuSn-A		3/32	2.4	350–550
Nickel and nickel alloys				0.020	0.5	—
	Monel[a] Alloy 400	ERNiCu-7		0.030	0.8	—
	Inconel[a] Alloy 600	ERNiCrFe-5	A5.14	0.035	0.9	100–160
				0.045	1.2	150–260
				1/16	1.6	100–400
Titanium and titanium alloys	Commercially pure	Use a filler metal one or two grades lower		0.030	0.8	—
				0.035	0.9	—
	Ti-0.15Pd	ERTi-0.2Pd	A5.16	0.045	1.2	—
	Ti-5A1–2.5Sn	ERTi-5A1–2.5Sn or commercially pure				
Austenitic stainless steels	Type 201	ER308		0.020	0.5	—
	Types 301, 302, 304 and 308	ER308		0.025	0.6	—
				0.030	0.8	75–150
	Type 304L	ER308L		0.035	0.9	100–160
	Type 310	ER310	A5.9	0.045	1.2	140–310
	Type 316	ER316		1/16	1.6	280–450
	Type 321	ER321		5/64	2.0	—
	Type 347	ER347		3/32	2.4	—
				7/64	2.8	—
				8	3.2	—

TABLE 10.3. (*Continued*)

Base Metal Type	Recommended Electrode		AWS Filler Metal Specification (use latest edition)	Electrode Diameter		Current Range, Amperes
	Material Type	Electrode Classification		in.	mm	
Steel	Hot rolled or cold-drawn plain carbon steels	ER70S-3 or ER70S-1 ER70S-2, ER70S-4 ER70S-5. ER70S-6		0.020	0.5	—
				0.025	0.6	—
				0.030	0.8	40–220
				0.035	0.9	60–280
			A5.18	0.045	1.2	125–380
				0.052	1.3	260–460
				1/16	1.6	275–450
				5/64	2.0	—
				3/32	2.4	—
				1/8	3.2	—
Steel	Higher strength carbon steels and some low alloy steels	ER80S-D2 ER80S-Ni1 ER100S-G		0.035	0.9	60–280
				0.045	1.2	125–380
				1/16	1.6	275–50
			A5.28	5/64	2.0	—
				3/32	2.4	—
				1/8	3.2	—
				5/32	4.0	—

[a]Trademark—International Nickel Co.

TABLE 10.4.

Metal	Shielding Gas	Advantages
Aluminum	Argon	0 to 1 in. (0 to 25 mm) thick: best metal transfer and arc stability; least spatter.
	35% argon + 65% helium	1 to 3 in. (25 to 76 mm) thick: higher heat input than straight argon; improved fusion characteristics with 5XXX series Al-Mg alloys.
	25% argon + 75% helium	Over 3 in. (76 mm) thick: highest heat input; minimizes porosity.
Magnesium	Argon	Excellent cleaning action.
Carbon steel	Argon + 1–5% oxygen	Improves arc stability; produces a more fluid and controllable weld puddle; good coalescence and bead contour, minimizes undercutting; permits higher speeds than pure argon.
	Argon + 3–10% CO_2	Good bead shape; minimizes spatter; reduces chance of cold lapping; cut—not weld—out of position.
Low-alloy steel	Argon + 2% oxygen	Minimizes undercutting; provides good toughness.
Stainless steel	Argon + 1% oxygen	Improves arc stability; produces a more fluid and controllable weld puddle, good coalescence and bead contour, minimizes undercutting on heavier stainless steels.
	Argon + 2% oxygen	Provides better arc stability, coalescence, and welding speed than 1% oxygen mixture for thinner stainless steel materials.
Copper, nickel, and their alloys	Argon	Provides good wetting; decreases fluidity of weld metal for thickness up to 1/8 in. (3.2 mm).
	Argon + helium	Higher heat inputs of 50 and 75% thelium mixtures offset high heat dissipation of heavier gauges.
Titanium	Argon	Good arc stability; minimum weld contamination; inert gas backing is required to prevent air contamination on back of weld area.

TABLE 10.5.

Metal	Shielding Gas	Advantages
Carbon steel	75% argon + 25% CO_2	Less than 1/8 in. (3.2 mm) thick: high welding speeds without burn-thru; minimum distortion and spatter.
	75% argon + 25% CO_2	More than 1/8 in. (3.2 mm) thick: minimum spatter; clean weld appearance; good puddle control in vertical and overhead positions.
	CO_2	Deeper penetration; faster welding speeds.
Stainless steel	90% helium + 7.5% argon + 2.5% CO_2	No effect on corrosion resistance; small heat-affected zone; no undercutting; minimum distortion.
Low alloy steel	60–70% helium + 25–35% argon + 4–5% CO_2	Minimum reactivity; excellent toughness; excellent arc stability, wetting characteristics, and bead contour, little spatter.
	75% argon + 25% CO_2	Fair toughness; excellent arc stability, wetting characteristics, and bead contour; little spatter.
Aluminum, copper, magnesium, nickel, and their alloys	Argon and argon + helium	Argon satisfactory on sheet metal; argon-helium preferred on thicker sheet material [over 1/8 in. (3.2 mm)].

10.4. GAS TUNGSTEN ARC WELDING (GTAW)

The gas tungsten arc welding (GTAW) process, also referred to as the tungsten inert gas (TIG) process, derives the heat for welding from an electric arc established between a tungsten electrode and the part to be welded (Figure 10.13). The arc zone must be filled with an inert gas to protect the tungsten electrode and molten metal from oxidation and to provide a conducting path for the arc current. The process was developed in 1941 primarily to provide a suitable means for welding magnesium and aluminum, where it was necessary to have a process superior to the shielded metal arc (stick electrode) process. Since that time, GTAW has been refined and has been used to weld almost all metals and alloys.

The GTAW process requires a gas- or water-cooled torch to hold the tungsten electrode; the torch is connected to the weld power supply by a power cable. In the lower-current gas-cooled torches (Figure 10.14), the power cable is inside the gas hose, which also provides insulation for the conductor. Water-cooled torches (Figure 10.15) require three hoses: one for the water supply, one for the water return, and one for the gas supply. The power cable is usually located in the water-return hose. Water cooling of the power cable allows use of a smaller conductor than that used in a gas-cooled torch of the same current rating.

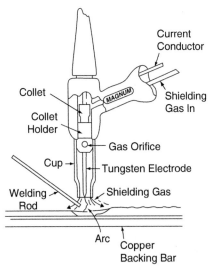

FIGURE 10.13.

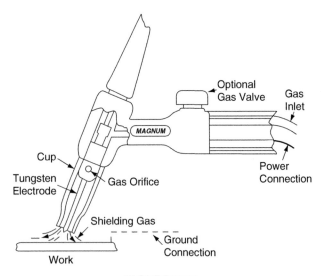

FIGURE 10.14.

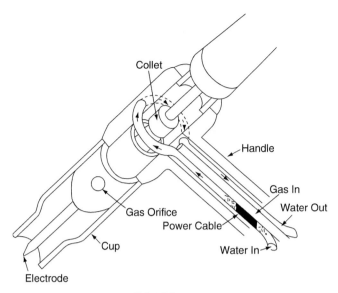

FIGURE 10.15.

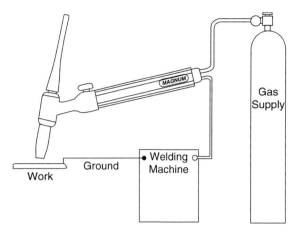

FIGURE 10.16.   Complete gas tungsten arc welding arrangement.

10.4.1. Applicability of GTAW

The GTAW process is capable of producing very high-quality welds in almost all metals and alloys. However, it produces the lowest metal deposition rate of all the arc welding processes. Therefore, it normally would not be used on steel, where a high deposition rate is required and very high quality usually is not necessary. The GTAW process can be used for making root passes on carbon and low-alloy steel piping with consumable insert rings or with added filler metal. The remainder of the groove would be filled using the coated-electrode process or one of the semiautomatic processes such as GMAW (with solid wire) or FCAW (with flux-cored wire).

A constant-current or drooping-characteristic power supply is required for GTAW, either DC or AC and with or without pulsing capabilities. For water-cooled torches, a water cooler circulator is preferred over the use of tap water.

For automatic or machine welding, additional equipment is required to provide a means of moving the part in relation to the torch and feeding the wire into the weld pool. A fully automatic system may require a programmer consisting of a microprocessor to control weld current, travel speed, and filler wire feed rate. An inert-gas supply (argon, helium, or a mixture of these), including pressure regulators, flowmeters, and hoses, is required for this process. The gases may be supplied from cylinders or liquid containers. A schematic diagram of a complete gas tungsten arc welding arrangement is shown in Figure 10.16.

GTAW would be used for those alloys for which high-quality welds and freedom from atmospheric contamination are critical. Examples of these are the reac-

tive and refractory metals such as titanium, zirconium, and columbium, where very small amounts of oxygen, nitrogen, and hydrogen can cause loss of ductility and corrosion resistance. It can be used on stainless steels and nickel-base superalloys, where welds exhibiting high quality with respect to porosity and fissuring are required. The GTAW process is well suited for welding thin sheet and foil of all weldable metals because it can be controlled at the very low amperages (2 to 5 amperes) required for these thicknesses. GTAW would not be used for welding the very low-melting metals, such as tin-lead solders and zinc-base alloys, because the high temperature of the arc would be difficult to control.

10.4.2. Advantages and Disadvantages of GTAW

The main advantage of GTAW is that high-quality welds can be made in all weldable metals and alloys except the very low-melting alloys. This is because the inert gas surrounding the arc and weld zone protects the hot metal from contamination. Another major advantage is that filler metal can be added to the weld pool independently of the arc current. With other arc welding processes, the rate of filler metal addition controls the arc current. Additional advantages are very low spatter, portability in the manual mode, and adaptability to a variety of automatic and semiautomatic applications.

The main disadvantage of GTAW is the low filler metal deposition rate. Further disadvantages are that it requires greater operator skill and is generally more costly than other arc welding processes.

10.4.3. Principles of Operating GTAW

In the GTAW process, an electric arc is established in an inert-gas atmosphere between a tungsten electrode and the metal to be welded. The arc is surrounded by the inert gas, which may be argon, helium, or a mixture of these two. The heat developed in the arc is the product of the arc current times the arc voltage, where approximately 70% of the heat is generated at the positive terminal of the arc. Arc current is carried primarily by electrons (Figure 10.17), which are emitted by the heated negative terminal (cathode) and obtained by ionization of the gas atoms. These electrons are attracted to the positive terminal (anode), where they generate approximately 70% of the arc heat. A smaller portion of the arc current is carried by positive gas ions which are attracted to the negative terminal (cathode), where they generate approximately 30% of the arc heat. The cathode loses heat by the emission of elec-

trons, and this energy is transferred as heat when the electrons deposit or condense on the anode. This is one reason why a significantly greater amount of heat is developed at the anode than at the cathode.

The voltage across an arc is made up of three components: the cathode voltage, the arc column voltage, and the anode voltage. In general, the total voltage of the gas tungsten arc will increase with arc length (Figure 10.18), although current and shielding gas have effects on voltage, which will be discussed later. The total arc voltage can be measured readily, but attempts to measure the cathode and anode voltages accurately have been unsuccessful. However, if the total arc voltage is plotted against arc length and extrapolated to zero arc length, a voltage that approximates the sum

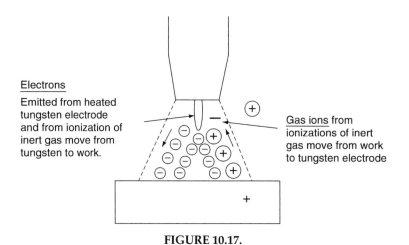

Electrons

Emitted from heated tungsten electrode and from ionization of inert gas move from tungsten to work.

Gas ions from ionizations of inert gas move from work to tungsten electrode

FIGURE 10.17.

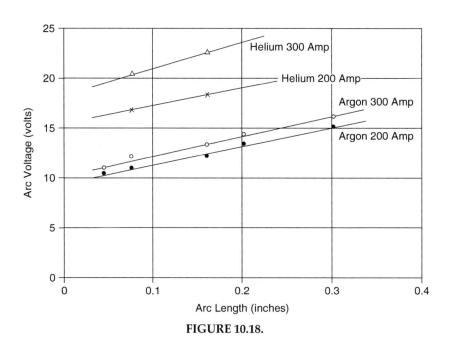

FIGURE 10.18.

of cathode voltage plus anode voltage is obtained. The total cathode plus anode voltage determined in this manner is between 7 and 10 volts for a tungsten cathode in argon. Since the greater amount of heat is generated at the anode, the GTAW process is normally operated with the tungsten electrode or cathode negative (negative polarity) and the work or anode positive. This puts the heat where it is needed, at the work.

10.4.4. Polarity and GTAW

The GTAW process can be operated in three different modes: electrode-negative (straight) polarity, electrode-positive (reverse) polarity, or AC (Figure 10.19). In the electrode-negative mode, the greatest amount of heat is developed at the work. For this reason, electrode-negative (straight) polarity is used with GTAW for welding most metals. Electrode-negative (straight) polarity has one disadvantage—it does not provide cleaning action on the work surface. This is of little consequence for most metals, because their oxides decompose or melt under the heat of the arc so that molten metal deposits will wet the joint surfaces. However, the oxides of aluminum and magnesium are very stable and have melting points well above that of the metal. They would not be removed by the arc heat and would remain on the metal surface and restrict wetting.

In the electrode-positive (reverse) polarity mode, cleaning action takes place on the work surface by the impact of gas ions. This removes a thin oxide layer while the surface is under the cover of an inert gas, allowing molten metal to wet the surface before more oxide can form.

When using AC gas tungsten arc welding aluminum, rectification occurs, and more current will flow when the electrode is negative (Fig. 10.20). This condition exists because the clean aluminum surface does not emit electrons as readily as the hot tungsten electrode. It will occur with standard AC welding power supplies. More advanced GTA welders incorporate circuits that can balance the negative- and positive-polarity half-cycles. Generally, this balanced condition is desirable for welding aluminum. The newest GTA power supplies include solid-state control boards, which allow adjustment of the AC current so as to favor either the positive- or negative-polarity half-cycle. These power supplies also chop the tip of the positive and negative half-cycles to produce a squarewave AC rather than a sinusoidal AC. When maximum cleaning is desired, the electrode-positive mode is favored; when maximum heat is desired, the electrode-negative mode is favored.

10.4.5. GTAW Shielding Gases and Flow Rates

Any of the inert gases could be used for GTAW. However, only helium (atomic weight 4) and argon (atomic weight 40) are used commercially, because they are much more plentiful and much less costly than the other inert gases. Typical flow rates are 15 to 40 cubic feet per hour (cfh).

Argon is used more extensively than helium for GTAW because:

1. It produces a smoother, quieter arc action;
2. It operates at a lower arc voltage for any given current and arc length;
3. There is greater cleaning action in the welding of materials such as aluminum and magnesium in the AC mode;

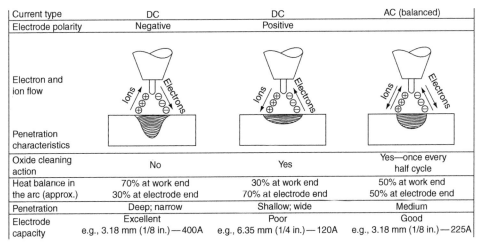

Current type	DC	DC	AC (balanced)
Electrode polarity	Negative	Positive	
Electron and ion flow			
Penetration characteristics			
Oxide cleaning action	No	Yes	Yes—once every half cycle
Heat balance in the arc (approx.)	70% at work end 30% at electrode end	30% at work end 70% at electrode end	50% at work end 50% at electrode end
Penetration	Deep; narrow	Shallow; wide	Medium
Electrode capacity	Excellent e.g., 3.18 mm (1/8 in.)—400A	Poor e.g., 6.35 mm (1/4 in.)—120A	Good e.g., 3.18 mm (1/8 in.)—225A

FIGURE 10.19.

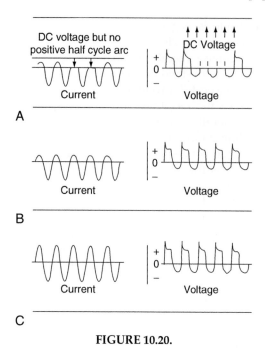

FIGURE 10.20.

4. Argon is more available and lower in cost than helium;
5. Good shielding can be obtained with lower flow rates;
6. Argon is more resistant to arc zone contamination by cross drafts;
7. The arc is easier to start in argon.

The density of argon is approximately 1.3 times that of air and 10 times that of helium. For this reason, argon will blanket a weld area and be more resistant than helium to cross drafts. Helium, being much lighter than air, tends to rise rapidly and cause turbulence, which will bring air into the arc atmosphere. Since helium costs about three times as much as argon, and its required flow rate is two to three times that for argon, the cost of helium used as a shielding gas can be as much as nine times that of argon.

Although either helium or argon can be used successfully for most GTAW applications, argon is selected most frequently because of the smoother arc operation and lower overall cost. Argon is preferred for welding thin sheets to prevent melt-through. Helium is preferred for welding thick materials and materials of high thermal conductivity such as copper and aluminum.

10.4.6. Electrode Material for GTAW

In selecting electrodes for GTAW, five factors must be considered: material, size, tip shape, electrode holder, and nozzle. Electrodes for GTAW are classified as pure tungsten, tungsten containing 1 or 2% thoria, tungsten containing 0.15 to 0.4% zirconia, and tungsten that contains an internal lateral segment of thoriated tungsten. The internal segment runs the full length of the electrode and contains 1 or 2% thoria. Overall, these electrodes contain 0.35 to 0.55% thoria. All tungsten electrodes are normally available in diameters from 0.010 to 0.250" and lengths from 3" to 24". Chemical composition requirements for these electrodes are given in AWS A5.12, "Specification for Tungsten Arc Welding Electrodes."

Pure tungsten electrodes, which are 99.5% pure, are the least expensive but also have the lowest current-carrying capacity on AC power and a low resistance to contamination. Tungsten electrodes containing 1 or 2% thoria have greater electron emissivity than pure tungsten and, therefore, greater current-carrying capacity and longer life. Arc starting is easier, and the arc is more stable, which helps make the electrodes more resistant to contamination from the base metal. These electrodes maintain a well sharpened point for welding steel.

Tungsten electrodes containing zirconia have properties in between those of pure tungsten and thoriated tungsten electrodes with regard to arc starting and current-carrying capacity. These electrodes are recommended for AC welding of aluminum over pure tungsten or thoriated tungsten electrodes because they retain a balled end during welding and have a high resistance to contamination. Another advantage of the tungsten-zirconia electrodes is their freedom from the radioactive element thorium, which, although not harmful in the levels used in electrodes, is of concern to some welders.

10.4.7. GTAW Electrode Size and Tip Shape

The electrode material, size, and tip shape (Figure 10.21) will depend on the welding application, material, thickness, type of joint, and quantity. Electrodes used for AC or electrode-positive polarity will be of larger diameter than those used for electrode-negative polarity.

The total length of an electrode will be limited by the length that can be accommodated by the GTAW torch. Longer lengths allow for more redressing of the tip than short lengths and are therefore more economical. The extension of the electrode from the collet or holder determines the heating and voltage drop in the electrode. Since this heat is of no value to the weld, the electrode extension should be kept as short as necessary to provide access to the joint.

It is recommended that electrodes to be used for DC negative-polarity welding be of the 2% thoria type and

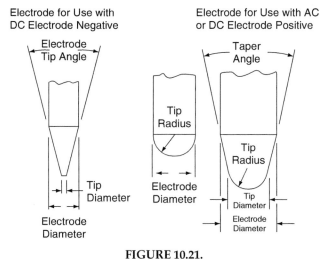

Electrode for Use with
DC Electrode Negative

Electrode
Tip Angle

Tip
Diameter

Electrode
Diameter

Electrode for Use with AC
or DC Electrode Positive

Taper
Angle

Tip
Radius

Tip
Radius

Tip
Diameter

Electrode
Diameter

Electrode
Diameter

FIGURE 10.21.

be ground to a truncated conical tip. Excessive current will cause the electrode to overheat and melt. Too low a current will permit cathode bombardment and erosion caused by the low operating temperature and resulting arc instability. Although a sharp point on the tip promotes easy arc starting, it is not recommended because it will melt and form a small ball on the end.

For AC and DC electrode-positive welding, the desirable electrode tip shape is a hemisphere of the same diameter as the electrode. This tip shape on the larger electrodes required for AC and DC electrode-positive welding provides a stable surface within the operating current range. Zirconia-type electrodes are preferred for AC and DC electrode-positive operation because they have a higher current-carrying capacity than the pure tungsten electrodes, yet they will readily form a molten ball under standard operating conditions. Thoriated electrodes do not ball readily and, therefore, are not recommended for AC or DC electrode-positive welding.

The degree of taper on the electrode tip affects weld penetration, where the smaller taper angles tend to reduce the width of the weld bead and thus increase penetration. When preparing the tip angle on an electrode, grinding should be done parallel to the length of the electrode. Special machines are available for grinding electrodes. These can be set to accurately grind any angle required.

10.4.8. GTAW Electrode Holders and Gas Nozzles

Electrode holders usually consist of a two-piece collet made to fit each standard-sized tungsten electrode. These holders and the part of the GTAW torch into which they fit must be capable of handling the required welding current without overheating. These holders are made of a hardenable copper alloy.

The function of the gas nozzle is to direct the flow of inert gas around the holder and electrode and then to the weld area. The nozzles are made of a hard, heat-resistant material such as ceramic and are available in various sizes and shapes. Large sizes give a more complete inert gas coverage of the weld area but may be too big to fit into restricted areas. Small nozzles can provide adequate gas coverage in restricted areas where features of the component help keep the inert gas at the joint. Most nozzles have internal threads that screw over threads on the electrode holder. Some nozzles are fitted with a washer-like device that consists of several layers of fine-wire screen or porous powder metal. These units provide a nonturbulent or lamellar gas flow from the torch, which results in improved inert gas coverage at a greater distance from the nozzle. In machine or automatic welding, more complete gas coverage may be provided by backup gas shielding from the fixture and a trailer shield attached to the torch.

10.4.9. Characteristics of GTAW Power Supplies

Power supplies for use with GTAW should be of the constant-current, drooping-voltage type (Figure 10.22). They may have other optional features such as up slope, down slope, pulsing, and current programming capabilities. Constant-voltage power supplies should not be used for GTAW.

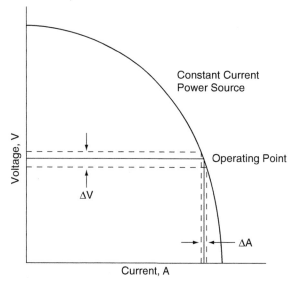

Constant Current
Power Source

Operating Point

ΔV

ΔA

Voltage, V

Current, A

FIGURE 10.22.

The power supply may be a single-phase transformer-rectifier, which also can supply AC for welding aluminum. Engine generator-type power supplies are usually driven by a gasoline or diesel engine and will produce DC with either a constant-current, drooping-voltage, or constant-voltage characteristics. Engine alternator power supplies will produce AC for GTAW. A power supply capable of operating on either constant current or constant voltage should be set for the constant-current mode for GTAW.

Power supplies made specifically for GTAW normally will include a high-frequency source for arc starting and valves that control the flow of inert gas and cooling water for the torch. Timers allow the valves to be opened a short time before the arc is initiated and closed a short time after the arc is extinguished. The high frequency is necessary for arc starting instead of torch starting, where tungsten contamination of the weld is likely. It should be possible to set the high frequency for arc starting only, or for continuous operation in the AC mode.

Power supplies should include a secondary contactor and a means of controlling arc current remotely. For manual welding, a foot pedal would perform these functions of operating the contactor and controlling weld current. A power supply with a single current range is desirable because it allows the welder to vary the arc current between minimum and maximum without changing a range switch.

The more advanced power supplies incorporate features that permit pulsing the current in the DC mode with essentially square pulses. Both background and pulse peak current can be adjusted, as well as pulse duration and pulsing frequency (Figure 10.23). In the AC mode, the basic 60-Hz sine wave can be modified to produce a rectangular wave. Other controls permit the AC wave to be balanced or varied to favor the positive or negative half-cycles. This feature is particularly useful when welding aluminum and magnesium, where the control can be set to favor the positive half-cycle for maximum cleaning. In the DC mode, the pulsing capability allows welds to be made in thin material, root passes, and overhead with less chance of melt-through or droop.

10.4.10. GTAW Torches

A torch for GTAW must perform the following functions:

1. Hold the tungsten electrode so that it can be manipulated along the weld path.
2. Provide an electrical connection to the electrode.
3. Provide inert-gas coverage of the electrode tip, arc, and hot weld zone.

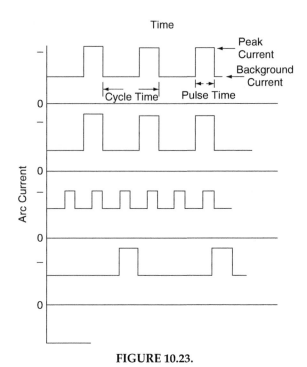

FIGURE 10.23.

4. Insulate the electrode and electrical connections from the operator or mounting bracket.

Typical GTAW torches consist of a metallic body, a collet holder, a collet, and a tightening cap to hold the tungsten electrode. The electrical cable is connected to the torch body, which is enclosed in a plastic insulating outer sheath. For manual torches, a handle is connected to the sheath. Power, gas, and water connections pass through the handle or, in the case of automatic operation, through the top of the torch. In the smaller, low-current torches, the electrode, collet, and internal components are cooled by the inert-gas flow. Larger, high-current torches are water-cooled and require connections to tap water and a drain or to a water cooler circulator. A cooler circulator with distilled or deionized water is preferred to prevent buildup of mineral deposits from tap water inside the torch.

Inert gas flows through the torch body and through holes in the collet holder to the arc end of the torch. A cup or nozzle is fitted over the arc end of the torch to direct inert gas over the electrode and the weld pool. The nozzles normally screw onto the torch and are made of a hard, heat-resistant ceramic. Some are made of high-temperature glass such as Vicor and are pressed on over a compressible plastic taper. Some nozzles can be fitted with an insert washer made up of several layers of fine-wire screen sometimes called a gas lens. This produces a lamellar rather than turbulent flow of inert gas to increase the efficiency of shielding.

On most manual GTAW torches, the handle is fixed at an angle of approximately 70 degrees to the torch body. Some makes of torches have a flexible neck between the handle and torch body which allows the angle between the handle and the torch body to be adjusted over a range from about 50 degrees to 90 degrees.

10.4.11. Manual GTAW Techniques

To become proficient in manual gas tungsten arc welding, the welder must develop skills in manipulating the torch with one hand, while controlling weld current with a foot pedal or thumb control and feeding filler metal with the other hand (see Figure 10.24). Before welding is started on any job, a rough idea of the welding conditions, such as filler material, current, shielding gas, etc., is needed.

10.4.12. Establishing Welding Parameters for GTAW

The material, thickness, joint design, and service requirements will determine the weld current, inert gas, voltage, and travel speed. This information may be available in a "welding procedure specification" (WPS) or from handbook data on the material and thickness. If welding parameters are not provided in a WPS, the information given in Tables 10.6, 10.7, and 10.8 can be used as starting-point parameters for carbon and low-alloy steels, stainless steels, and aluminum. These should be considered starting values; final values should be established by running a number of test parts.

FIGURE 10.24.

10.4.13. Gas Tungsten Arc Starting Methods

The gas tungsten arc may be started by touching the work with the electrode, by a superimposed high-frequency pulse, or by a high-voltage pulse.

The touch method is not recommended for critical work because there is a strong possibility of tungsten contamination with this technique. Most weld power supplies intended for GTAW contain a high-frequency generator (usually a spark-gap oscillator), which superimposes the high-frequency pulse on the main weld power circuit. When welding with DC electrode-negative or -positive, the high-frequency switch should be set in the HT start position. When the welder presses the foot pedal to start welding, a timer is activated, which starts the high-frequency pulse and stops it when the arc initiates. Once started, the arc will continue after the high-frequency pulse stops as long as the power and proper arc gap are maintained. When welding with AC, the switch should be set in the HF continuous position to ensure that the arc restarts after voltage reversal on each half-cycle. High-frequency generators on welders produce frequencies in the radio communications range. Therefore, manufacturers of power supplies must certify that the radio frequency radiation from the power supply does not exceed limitations established by the Federal Communications Commission (FCC). The allowable radiation may be harmful to some computer and microprocessor systems and to communications systems. These possibilities for interference should be investigated before high-frequency starting is used. Installation instructions provided with the power supply should be studied and followed carefully.

10.5. OXYACETYLENE CUTTING

Steel can be cut with great accuracy using an oxyacetylene torch (see Figure 10.25). However, not all metals cut as readily as steel. Cast iron, stainless steel, manganese steels, and nonferrous materials cannot be cut and shaped satisfactorily with the oxyacetylene process because of their reluctance to oxidize. In these cases, plasma arc cutting is recommended.

The cutting of steel is a chemical action. The oxygen combines readily with the iron to form iron oxide. In cast iron, this action is hindered by the presence of carbon in graphite form, so cast iron cannot be cut as readily as steel. Higher temperatures are necessary, and cutting is slower. In steel, the action starts at bright-red heat, whereas in cast iron, the temperature must be nearer the melting point in order to obtain a sufficient reaction.

TABLE 10.6.

Welder Size	60-Hz Input Voltage	Ampere Rating	3 Wires in Conduit or 3-Conductor Cable, Type R	Grounding Conductor
200	230	44	8	8
	460	22	12	10
	575	18	12	14
300	230	62	6	8
	460	31	10	10
	575	25	10	12
400	230	78	6	6
	460	39	8	8
	575	31	10	10
600	230	124	2	6
	460	62	6	8
	575	50	8	8
900	230	158	1	3
	460	79	6	6
	575	63	4	8

TABLE 10.7.

Welder Size	Volts Input	Amp Input With Condenser	Amp Input Without Condenser	Wire Size (3 in conduit) With Condenser	Wire Size (3 in conduit) Without Condenser	Wire Size (3 in conduit) Ground Conduction	Wire Size (3 in free air) With Condenser	Wire Size (3 in free air) Without Condenser	Wire Size (3 in free air) Ground Conduction
300	200	84	104	2	1	1	4	4	4
	440	42	52	6	6	6	8	8	8
	550	38	42	8	6	6	10	8	8
400	220	115	143	0	00	00	3	1	1
	440	57.5	71.5	4	3	3	6	6	6
	550	46	57.2	8	4	4	8	6	6
500	220	148	180	000	0000	0000	1	0	0
	440	74	90	3	2	2	6	4	4
	550	61	72	4	3	3	6	6	6

Because of the very high temperature, the speed of cutting is usually fairly high. However, since the process is essentially one of melting without any great action, tending to force the molten metal out of the cut, some provision must be made for permitting the metal to flow readily away from the cut. This is usually done by starting at a point from which the molten metal can flow readily. This method is followed until the desired amount of metal has been melted away.

10.6. AIR-CARBON ARC CUTTING AND GOUGING

Air-carbon arc cutting (CAC-A) is a physical means of removing base metal or weld metal using a carbon electrode, an electric arc, and compressed air (see Figure 10.26). In the air-carbon arc process, the intense heat of the arc between the carbon electrode and the workpiece melts a portion of the base metal or weld.

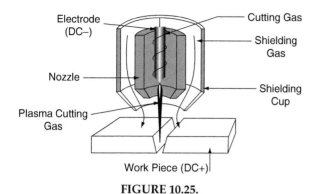

Electrode
(DC–)

Nozzle

Plasma Cutting
Gas

Cutting Gas

Shielding
Gas

Shielding
Cup

Work Piece (DC+)

FIGURE 10.25.

TABLE 10.8.

| Machine Size′ | Cable Sizes for Lengths (electrode plus ground) | | |
Amp	Up to 50 ft	50–100 ft	100–250 ft
200	2	2	1/0
300	1/0	1/0	3/0
400	2/0	2/0	4/0~
600	3/0	3/0	4/0~
900	Automatic application only		

Recommended longest length of 4/0 cable for 400-amp welder, 150 ft; for 600-amp welder, 100 ft. For greater distances, cable size should be increased; this may be a question of cost—consider ease of handling versus moving of welder closer to work.

TABLE 10.9.

Material	Electrode	Power
Steel	DC	DCEP
	AC	AC
Stainless steel	DC	DCEP
	AC	AC
Iron (cast iron, ductile iron, malleable iron)	AC	AC or DCEN
	DC	DCEP (high-amperage)
Copper alloys	AC	AC or DCEN
	DC	DCEP
Nickel alloys	AC	AC or DCEN

Note: AC is not the preferred method.
Source: *AWS Handbook*, 6th ed., Section 3A.

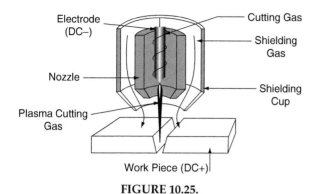

FIGURE 10.26.

Simultaneously, a jet of air is passed through the arc of sufficient volume and velocity to blow away the molten material. This sequence can be repeated until the required groove or cut has been obtained. Since CAC-A does not depend on oxidation to maintain the cut, it is capable of cutting metals that oxyacetylene cutting will not cut. It is used to cut carbon steel, stainless steel, many copper alloys, and cast iron.

Arc gouging can be used to remove material approximately five times as fast as chipping. Depth of cut can be closely controlled, and welding slag does not deflect or hamper the cutting action, as it would with cutting tools. Gouging equipment generally costs less to operate than chipping hammers or gas cutting torches. An arc-gouged surface is clean and smooth and usually can be welded without further preparation. Drawbacks of the process include the fact that it requires *large* volumes of compressed air, and it is not as good as other processes for through-cutting.

Electrode type and air supply specifications for arc gouging are outlined in Tables 10.9 and 10.10.

10.6.1. Applications

The CAC-A process may be used to provide a suitable bevel or groove when preparing plates for welding (see Figure 10.27). It also may be used to back-gouge a seam prior to welding the side. CAC-A provides an excellent means of removing defective welds or misplaced welds and has many applications in

TABLE 10.10

Maximum Electrode Size (in.)	Application	Pressure (psi)	Consumption (cfm)
1/4	Intermittent-duty, manual torch	40	3
1/4	Intermittent-duty, manual torch	80	9
3/8	General-purpose	80	16
3/4	Heavy-duty	80	29
5/8	Semiautomatic mechanized torch	80	25

metal fabrication, casting finishing, construction, mining, and general repair and maintenance. When using CAC-A, normal safety precautions must be taken and, in addition, ear plugs must be worn.

10.6.2. Power Sources

While it is possible to arc air gouge with AC, this is not the preferred method. A DC power source of sufficient capacity and a minimum of 60 open circuit volts, either rectifier or motor generator, will give the best results. With DC, it is operated with DCEP (electrode-positive). Arc voltages normally range from about 35 to 56 volts. Table 10.11 lists recommended power sources, and Table 10.12 lists suggested current ranges. It is recommended that the power source

Conditions	SMAW (Stick)	GMAW (mig)	FCAW Gas	FCAW Self shielded	GTAW (TIG)	SAW (Sub arc)
Carbon steel						
a. Sheet metal	★★	★★★★		★★	★★★★	
b. Plate	★★★	★★★	★★★★	★★★★		★★★★
Stainless						
a. Sheet metal	★	★★★★			★★★★	
b. Plate	★★	★★★			★★	★★★
Aluminum						
a. Sheet metal					★★★★	
b. Plate		★★★			★★	
Nonferrous Copper, Bronze, etc.					★★★★	
High dep. rate	★	★★	★★★	★★★		★★★★
Portability of equipment	★★★	★★★	★★★	★★★	★★	
User friendly	★	★★	★★	★	★★★★	★
Min. weld cleaning	★	★★★	★★	★	★★★★	★
All position welding	★★★★	★★★★	★★	★★	★★★★	

Rating   ★★★★   Excellent
        ★★★   Very good
        ★★   Good
        ★   Fair
     (none)   Not recommended

A

FIGURE 10.27.

Weldment:_____   Quality Category:

Drawing No._____   Critical [ ]

    Functional [ ]

Process	Electrode Size/Type	Polarity	Volts	WFS/AMP (either)	ESO	Shield. Gas/Flux
SMAW						
GMAW						
SAW						
TIG						
FCAW						

Sketch of the Joint

Special instructions:_____
_____
_____
_____
_____
_____

Welder: _____   Proced. Apprvd by: _____

Welder: _____   Date: _____

B

FIGURE 10.27. (*Continued*)

have overload protection in the output circuit. High current surges of short duration occur with arc gouging, and these surges can overload the power source.

10.7. PLASMA ARC CUTTING

Plasma arc cutting has become an essential requirement for any properly equipped maintenance department. It provides the best, fastest, and cheapest method of cutting carbon or alloy steel, stainless steel, aluminum, nonferrous metals, and cast iron. In fact, it will cut any conductive material. The cuts are clean and

TABLE 10.11.

Type of Current	Type of Power Source	Remarks
DC	Variable-voltage motor-generator, rectifier or resistor-grid equipment	Recommended for all electrode sizes
AC	Constant-voltage motor-generator, or rectifier	Recommended only for electrodes above ¼-in. diameter
AC–DC	Transformer	Should be used only with AC electrodes

TABLE 10.12.

Type of Electrode and Power	Maximum and Minimum Current (amp)					
	Electrode Size (in.)					
	$^5/_{32}$	$^3/_{16}$	$^1/_4$	$^5/_{16}$	$^3/_8$	$^1/_2$
DC electrodes, DECP power	90–150	150–200	200–400	250–450	350–600	600–1000
AC electrodes, AC power	—	150–200	200–300	—	300–500	400–600
AC electrodes, DCEN power	—	150–180	200–250	—	300–400	400–500

precise, with very little dross or slag to remove. The heat is so concentrated within the immediate area of the cut that very little distortion takes place (see Figure 10.28). On gauge thickness material, the speed is limited only by the skill of the operator.

Plasma arc cutting operates on the principle of passing an electric arc through a quantity of gas through a restricted outlet. The electric arc heats the gas as it travels through the arc. This turns the gas into a plasma that is extremely hot. It is the heat in the plasma that heats the metal. A typical power source for plasma arc cutting is about the size of a small transformer welder and comes equipped with a special torch, as shown in Figure 10.29.

Plasma arc torch consumables include the electrode and the orifice. The electrode is copper, with a small hafnium insert at the center tip. The arc emanates from the hafnium, which gradually erodes with use, and this requires electrode replacement periodically. The torch tip contains the orifice which constricts the plasma arc. Various sized orifices are available, and the smaller diameters operate at lower amperages and produce a more constricted narrow arc than the larger diameters. The orifices also gradually wear with use and must be replaced when the arc becomes too wide.

The plasma arc cutting torch can be used for arc gouging. The main changes are that the tip orifice is

FIGURE 10.28.

FIGURE 10.29.

larger than for cutting, and the torch is held at an angle of about 30 degrees from horizontal rather than at 90 degrees, as in cutting. Plasma gouging can be used on all metals and is particularly suitable for aluminum and stainless steels, where oxyacetylene cutting is ineffective and carbon arc gouging tends to cause carbon contamination.

10.8. WELDING PROCEDURES

Much of the welding done by maintenance welders does not normally require detailed written welding procedures. The judgment and skill of the welders generally are sufficient to get the job done properly. However, there are some maintenance applications that demand the attention of a welding engineer or a supervisor, someone who knows more than the typical welder about the service conditions of the weldment or perhaps the weldability of it. Or there may be metallurgical factors or high-stress service requirements that must be given special consideration. A distinction should be made between "casual" and "critical" welding. When an application is considered critical, a welding engineer or a properly qualified supervisor should provide the welding operator with a detailed written procedure specification (WPS) providing all the information needed to make the weld properly.

10.9. QUALIFICATION OF WELDERS

If the nature of the welding is critical and requires a written procedure, then equal consideration should be given to making sure the operator is qualified to do the kind of welding called for in the procedure. This can be done by having the operator make test welds that simulate the real thing. These welds can then be examined either destructively or nondestructively to see if the welder has demonstrated the required skills.

Note: Some welding may fall under local, state, or federal code requirements that do not permit welding to be done by in-house maintenance welders unless they have been certified.

10.10. PLASMA ARC WELDING

Plasma arc welding exists in several forms. The basic principle is that of an arc or jet created by the electrical heating of a plasma-forming gas (such as argon with additions of helium or hydrogen) to such a high temperature that its molecules become ionized atoms possessing extremely high energy. When properly controlled, this process results in very high melting temperatures. Plasma arc holds the potential solution to the easier joining of many hard-to-weld materials. Another application is the depositing of materials having high melting temperatures to produce surfaces of high resistance to extreme wear, corrosion, or temperature. As discussed earlier, when plasma arc technology is applied to metal cutting, it achieves unusually high speeds and has become an essential tool for a variety of maintenance applications.

10.11. BASE METALS

10.11.1. The Carbon Steels

Carbon steels are widely used in all types of manufacturing. The weldability of the different types (low, medium, and high) varies considerably. The preferred analysis range of the common elements found in carbon steels is shown in Table 10.13. Welding metals whose elements vary above or below the range usually calls for special welding procedures.

Low-Carbon Steels (0.10 to 0.30% Carbon)

Steels of low-carbon content represent the bulk of the carbon steel tonnage used by industry. These steels are usually more ductile and easier to form than higher-carbon steels. For this reason, low-carbon steels are used in most applications requiring considerable cold forming, such as stampings and rolled or bent shapes in bar stock, structural shapes, or sheet. Steels with less than 0.13% carbon and 0.30% manganese have a slightly greater tendency to internal porosity than steels of higher carbon and manganese content.

Medium-Carbon Steels (0.31 to 0.45%)

The increased carbon content in medium-carbon steel usually raises the tensile strength, hardness, and

TABLE 10.13

	Low, %	Preferred, %	High, %
Carbon	0.06	0.10 to 0.25	0.35
Manganese	0.30	0.35 to 0.80	1.40
Silicon		0.10 or under	0.30 max
Sulfur		0.035 or under	0.05 max
Phosphorus		0.03 or under	0.04 max

wear resistance of the material. These steels are selectively used by manufacturers of railroad equipment, farm machinery, construction machinery, material-handling equipment, and other similar products. The medium-carbon steels can be welded successfully with the E60XX electrode if certain simple precautions are taken and the cooling rate is controlled to prevent excessive hardness.

High-Carbon Steels (0.46% and Higher)

The high-carbon steels are generally used in a hardened condition. This group includes most of the steels used in tools for forming, shaping, and cutting. Tools used in metalworking, woodworking, mining, and farming, such as lathe tools, drills, dies, knives, scraper blades, and plowshares, are typical examples. The high-carbon steels are often described as being "difficult to weld" and are not suited to mild steel welding procedures. Usually, low-hydrogen electrodes or processes are required, and controlled welding procedures, including preheating and postheating, are needed to produce crack-free welds.

The higher the carbon content of the steel, the harder the material becomes when it is quenched from above the critical temperature. Welding raises steel above the critical temperature, and the cold mass of metal surrounding the weld area creates a quench effect. Hardness and the absence of ductility result in cracking as the weld cools and contracts. Preheating from 300°F to 600°F and slow cooling will usually prevent cracking. Figure 10.30 shows a calculator for determining preheat and interpass temperatures.

FIGURE 10.30.

For steels in the higher carbon ranges (over 0.30%), special electrodes are recommended. The lime ferritic low-hydrogen electrodes (E7016 or E7018) can be used to good advantage in overcoming the cracking tendencies of high-carbon steels. A 308 stainless steel electrode also can be used to give good physical properties to a weld in high-carbon steel.

Cast Iron

Cast iron is a complex alloy with a very high carbon content. Quickly cooled cast iron is harder and more brittle than slowly cooled cast iron. The metal also naturally exhibits low ductility, which results in considerable strain on parts of a casting when one local area is heated. The brittleness and the uneven contraction and expansion of cast iron are the principal concerns when welding it.

Each job must be analyzed to predetermine the effect of welding heat so that corresponding procedures can be adopted. Welds can be deposited in short lengths, allowing each to cool. Peening of the weld metal while it is red hot may be used to stretch the weld deposit. Steel, cast iron, carbon, or nonferrous electrodes may be used. All oil, dirt, and foreign matter must be removed from the joint before welding. With steel electrodes, intermittent welds no longer than 3" should be used with light peening. To reduce contraction, the work should never be allowed to get too hot in one spot. Preheating will help to reduce hardening of the deposit to make it more machinable.

For the most machinable welds, a nonferrous alloy rod should be used. A two-layer deposit will have a softer fusion zone than a single-layer deposit. When it is practical, heating of the entire casting to a dull red heat is recommended in order to further soften the fusion practice to use a steel electrode for welding cast iron to fill up the joint to within approximately 1/8" of the surface and then finish the weld with the more machinable nonferrous deposit, usually a 95 to 98% nickel electrode.

10.11.2. The Alloy Steels

High-Tensile, Low-Alloy Steels

These steels are finding increasing use in metal fabricating because their higher strength levels permit the use of thinner sections, thereby saving metal and reducing weight. They are made with a number of different alloys and can be readily welded with specially designed electrodes that produce excellent welds of the same mechanical properties as the base metal. However, it is not necessary to have a core wire of exactly the same composition as the steel.

Stainless Steels

Electrodes are made to match various types of stainless steels so that corrosion-resistance properties are not destroyed in welding. The most commonly used types of stainless steels for welded structures are the 304, 308, 309, and 310 groups. Group 304 stainless, with a maximum carbon content of 0.8%, is commonly specified for weldments.

Welding procedures are much the same as for welding mild steel, except that one must take into account the higher electrical resistance of stainless steels and reduce the current accordingly. It is important to work carefully, cleaning all edges of foreign material. Light-gauge work must be clamped firmly to prevent distortion and buckling. Small-diameter and short electrodes should be used to prevent loss of chromium and undue overheating of the electrode. The weld deposit should be approximately the same analysis as the plate (see Tables 10.14 and 10.15).

Stainless Clad Steel

The significant precautions in welding this material are in joint design, including edge preparation, procedure, and choice of electrode. The electrode should be of the correct analysis for the cladding being welded. The joint must be prepared and welded to prevent dilution of the clad surface by the steel backing material. The backing material is welded with a mild steel electrode but in multiple passes to prevent excessive penetration into the cladding. The clad side is also welded in small passes to prevent penetration into the backing material and resulting dilution of the stainless joint. When one is welding thin-gauge material and it is necessary to make the weld in one pass, a 309 stainless electrode should be used for the steel side as well as for the stainless side. The design and preparation of the joint can do much to prevent iron pickup, as well as reduce the labor costs of making the joint.

TABLE 10.14.

AISI Type	Composition* (%)			
	Carbon	Chromium	Nickel	Other†
201	0.15	16.0–18.0	3.5–5.5	0.25 N, 5.5–7.5 Mn, 0.060 P
202	0.15	17.0–19.0	4.0–6.0	0.25 N, 7.5–10.0 Mn, 0.060 P
301	0.15	16.0–18.0	6.0–8.0	—
302	0.15	17.0–19.0	8.0–10.0	—
302B	0.15	17.0–19.0	8.0–10.0	2.0–3.0 Si
303	0.15	17.0–19.0	8.0–10.0	0.20 P, 0.15 S (min), 0.60 Mo (opt)
303Se	0.15	17.0–19.0	8.0–10.0	0.20 P, 0.06 S, 0.15 Se (min)
304	0.08	18.0–20.0	8.0–12.0	—
304L	0.03	18.0–20.0	8.0–12.0	—
305	0.12	17.0–19.0	10.0–13.0	—
308	0.08	19.0–21.0	10.0–12.0	—
309	0.20	22.0–24.0	12.0–15.0	—
309S	0.08	22.0–24.0	12.0–15.0	—
310	0.25	24.0–26.0	19.0–22.0	1.5 Si
310S	0.08	24.0–26.0	19.0–22.0	1.5 Si
314	0.25	23.0–26.0	19.0–22.0	1.5–3.0 Si
316	0.06	16.0–18.0	10.0–14.0	2.0–3.0 Mo
316L	0.04	16.0–18.0	10.0–14.0	2.0–3.0 Mo
317	0.06	18.0–20.0	11.0–15.0	3.0–4.0 Mo
321	0.06	17.0–19.0	9.0–12.0	Ti (5 × %C min)
347	0.08	17.0–19.0	9.0–13.0	Cb + Ta (10 × % C min)
348	0.08	17.0–19.0	9.0–13.0	Cb + Ta (10 × % C min but 0.10 Ta max), 0.20 Co

*Single values denote maximum percentage unless otherwise noted.
†Unless otherwise noted, other elements of all alloys listed include maximum contents of 2.0% Mn, 1.0% Si, 0.045% P, and 0.030% S. Balance is Fe.

TABLE 10.15.

AISI Type	Composition* (%)			
	Carbon	Chromium	Manganese	Other†
405	0.08	11.5–14.5	1.0	0.1–0.3 Al
430	0.12	14.0–18.0	1.0	—
430 F	0.12	14.0–18.0	1.25	0.060 P, 0.15 S (min), 0.60 Mo (opt)
430 F Se	0.12	14.0–18.0	1.25	0.060 P, 0.060 S, 0.15 Se (min)
442	0.20	18.0–23.0	1.0	—
446	0.20	23.0–27.0	1.5	0.25 N

*Single values denote maximum percentage unless otherwise noted.

†Unless otherwise noted, other elements of all alloys listed include maximum contents of 1.0% Si, 0.040% P, and 0.030% S. Balance is Fe.

Straight Chromium Steels

The intense air-hardening property of these steels, which is proportional to the carbon and chromium content, is the chief consideration in establishing welding procedures. Considerable care must be taken to keep the work warm during welding, and it must be annealed afterward; otherwise, the welds and the areas adjacent to the welds will be brittle. It is a good idea to consult steel suppliers for specific details of proper heat treatment.

High-Manganese Steels

High-manganese steels (11 to 14% Mn) are very tough and are work-hardening, which makes them ideally suited for surfaces that must resist abrasion or wear as well as shock. When building up parts made of high-manganese steel, an electrode of similar analysis should be used.

10.11.3. The Nonferrous Metals

Aluminum

Most fusion welding of aluminum alloys is done with either the gas metal arc (GMAW) process or the gas tungsten arc (GTAW) process. In either case, inert-gas shielding is used.

With GMAW, the electrode is aluminum filler fed continuously from a reel into the weld pool. This action propels the filler metal across the arc to the workpiece in line with the axis of the electrode, regardless of the orientation of the electrode.

Because of this, and because of aluminum's qualities of density, surface tension, and cooling rate, horizontal, vertical, and overhead welds can be made with relative ease. High deposition rates are practical, producing less distortion, greater weld strength, and lower welding costs than can be attained with other fusion welding processes.

GTAW uses a nonconsumable tungsten electrode, with aluminum alloy filler material added separately, either from a handheld rod or from a reel. Alternating current (AC) is preferred by many users for both manual and automatic gas tungsten arc welding of aluminum because AC GTAW achieves an efficient balance between penetration and cleaning.

Copper and Copper Alloys

Copper and its alloys can be welded with shielded metal arc, gas-shielded carbon arc, or gas tungsten arc welding. Of all of these, gas-shielded arc welding with an inert gas is preferred. Decrease in tensile strength as temperature rises and a high coefficient of contraction may make welding of copper complicated. Preheating usually is necessary on thicker sections because of the high heat conductivity of the metal. Keeping the work hot and pointing the electrode at an angle so that the flame is directed back over the work will aid in permitting gases to escape. It is also advisable to put as much metal down per bead as is practical.

10.12. CONTROL OF DISTORTION

The heat of welding can distort the base metal; this sometimes becomes a problem in welding sheet metal or unrestrained large sections. The following suggestions will help in overcoming problems of distortion:

1. Reduce the effective shrinkage force.
   a. Avoid overwelding. Use as little weld metal as possible by taking advantage of the penetrating effect of the arc force.
   b. Use correct edge preparation and fit-up to obtain required fusion at the root of the weld.
   c. Use fewer passes.
   d. Place welds near a neutral axis.
   e. Use intermittent welds.
   f. Use back-step welding method.
2. Make shrinkage forces work to minimize distortion.
   a. Preset the parts so that when the weld shrinks, they will be in the correct position after cooling.
   b. Space parts to allow for shrinkage.
   c. Prebend parts so that contraction will pull the parts into alignment.

3. Balance shrinkage forces with other forces (where natural rigidity of parts is insufficient to resist contraction).

   a. Balance one force with another by correct welding sequence so that contraction caused by the weld counteracts the forces of welds previously made.

   b. Peen beads to stretch weld metal. Care must be taken so that weld metal is not damaged.

   c. Use jigs and fixtures to hold the work in a rigid position with sufficient strength to prevent parts from distorting. Fixtures can actually cause weld metal to stretch, preventing distortion.

10.13. SPECIAL APPLICATIONS

10.13.1. Sheet Metal Welding

Plant maintenance frequently calls for sheet metal welding. The principles of good welding practice apply in welding sheet metal as elsewhere, but welding thin-gauge metals poses the specific challenges of potential distortion and/or burn-through. Special attention should therefore be given to all the factors involved in controlling distortion: the speed of welding, the choice of proper joints, good fit-up, position, proper current selection, use of clamping devices and fixtures, number of passes, and sequence of beads.

Good welding practice normally calls for the highest arc speeds and the highest currents within the limits of good weld appearance. In sheet metal work, however, there is always the limitation imposed by the threat of burn-through. As the gap in the work increases in size, the current must be decreased to prevent burn-through; this, of course, will reduce welding speeds. A clamping fixture will improve the fit-up of joints, making higher speeds possible. If equipped with a copper backing strip, the clamping fixture will make welding easier by decreasing the tendency to burn-through and also removing some of the heat that can cause warpage. Where possible, sheet metal joints should be welded downhill at about a 45-degree angle with the same currents that are used in the flat position or slightly higher. Tables 10.16 and 10.17 offer guides to the selection of proper current, voltage, and electrodes for the various types of joints used with 20- to 8-gauge sheet metal.

10.13.2. Hard Surfacing

The building up of a layer of metal or a metal surface by electric arc welding, commonly known as hard surfacing, has important and useful applications in equipment maintenance. These may include restoring worn cutting edges and teeth on excavators, building up worn shafts with low- or medium-carbon deposits, lining a carbon-steel bin or chute with a stainless steel corrosion-resistant alloy deposit, putting a tool-steel cutting edge on a medium-carbon steel base, and applying wear-resistant surfaces to metal machine parts of all kinds. The dragline bucket shown in Figure 10.31 is being returned to "new" condition by rebuilding and hard surfacing. Arc weld surfacing techniques include, but are not limited to, hard surfacing. There are many buildup applications that do not require hard surfacing. Excluding the effects of corrosion, wear of machinery parts results from various combinations of abrasion and impact. Abrasive wear results from one material scratching another, and impact wear results from one material hitting another.

10.13.3. Resisting Abrasive Wear

Abrasive wear is resisted by materials with a high scratch hardness. Sand quickly wears metals with a low scratch hardness, but under the same conditions, it will wear a metal of high scratch hardness very slowly. Scratch hardness, however, is not necessarily measured by standard hardness tests. Brinell and Rockwell hardness tests are not reliable measures for determining the abrasive wear resistance of a metal. A hard-surfacing material of the chromium carbide type may have a hardness of 50 Rockwell C. Sand will wear this material at a slower rate than it will a steel hardened to 60 Rockwell C. The sand will scratch all the way across the surface of the steel. On the surfacing alloy, the scratch will progress through the matrix material and then stop when the sand grain comes up against one of the microscopic crystals of chromium carbide, which has a higher scratch hardness than sand. If two metals of the same type have the same kind of microscopic constituents, however, the metal having the higher Rockwell hardness will be more resistant to abrasive wear.

10.13.4. Resisting Impact Wear

Whereas abrasive wear is resisted by the surface properties of a metal, impact wear is resisted by the properties of the metal beneath the surface. To resist impact, a tough material is used, one that does not readily bend, break, chip, or crack. It yields so as to distribute or absorb the load created by impact, and the ultimate strength of the metal is not exceeded. Included in impact wear is that caused by bending or compression at low velocity without impact, resulting in loss of metal by cracking, chipping, upsetting, flowing, or crushing.

TABLE 10.16.

Type of Welded Joint	20ga			18ga			16ga			14ga			12ga			10ga			8ga		
	F*	V*	O*	F	V	O	F	V	O	F	V	O	F	V	O	F	V	O	F	V	O
Plain butt	30†	30†	30†	40†	40†	40†	70†	70†	70†	85†	80	85†	115	110	110	135	120	115	190	130	120
Lap	40†	40†	40†	60†	60†	60†	100	100	100	130	130	130	135	120	120	155	130	120	165	140	120
Fillet	40†			40†	40†	40†	70†	70†	70†	100	90	85	150	140	120	160	150	130	160	160	130
Corner	40†	40†	40†	60†	60†	60†	90†	90†	90†	90	80	75	125	110	110	140	130	125	175	130	125
Edge	40†	40†	40†	60†	60†	60†	80†	80†	80†	110	80	80	145	110	110	150	120	120	160	120	120

*F—fat position; V—vertical; O—overhead.
†Electrode negative, work positive.

TABLE 10.17.

Type of Welded Joint	20ga			18ga			16ga			14ga			12ga			10ga			8ga		
	F*	V*	O*	F	V	O	F	V	O	F	V	O	F	V	O	F	V	O	F	V	O
Plain butt	3/32	3/32	3/32	8/32	8/32	8/32	1/8	1/8	1/8	1/8	1/8	1/8	5/32	5/32	5/32	5/32	5/32	5/32	8/16	5/32	5/32
Lap	3/32	3/32	3/32	8/32	8/32	8/32	1/8	1/8	1/8	5/32	5/32	5/32	5/32	5/32	5/32	3/16	3/16	5/32	3/16	3/16	5/32
Fillet	3/32	3/32		8/32	8/32	8/32	1/8	1/8	1/8	1/8	1/8	1/8	5/32	5/32	5/32	3/16	3/16	5/32	3/16	3/16	5/32
Corner	3/32	3/32	3/32	8/32	8/32	8/32	1/8	1/8	1/8	1/8	1/8	1/8	3/16	5/32	5/32	3/16	5/32	5/32	3/16	5/32	5/32
Edge	3/32	3/32	3/32	8/32	8/32	8/32	1/8	1/8	1/8	1/8	1/8	1/8	3/16	5/32	5/32	3/16	5/32	5/32	3/16	5/32	5/32

*F—fat position; V—vertical; O—overhead.

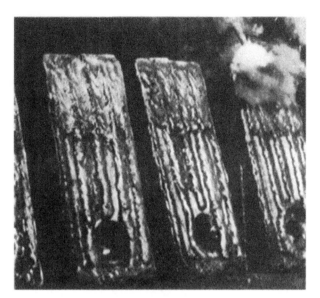

FIGURE 10.31.

10.13.5. Types of Surfacing Electrodes

Many different kinds of surfacing electrodes are available. The problem is to find the best one to do a given job. Yet because service conditions vary so widely, no universal standard can be established for determining the ability of the surfacing to resist impact or abrasion. Furthermore, there is no ideal surfacing material that resists impact and abrasion equally well. In manufacturing the surfacing electrodes, it is necessary to sacrifice one quality somewhat to gain the other. High impact resistance is gained by sacrificing abrasion resistance, and vice versa.

Price is no index of electrode quality. An expensive electrode ingredient does not necessarily impart wear resistance. Therefore, the user of surfacing materials must rely on a combination of the manufacturer's recommendations and the user's own tests to select the best surfacing material for a particular purpose.

10.13.6. Choosing Hard-Facing Material

The chart shown in Figure 10.32 lists the relative characteristics of manual hard-facing materials. This chart is a guide to selecting the two items in the following list:

1. The hard-facing electrode best suited for a job not hard-faced before.
2. A more suitable hard-facing electrode for a job where the present material has not produced the desired results.

Example 1

Application: Dragline bucket tooth, as shown in Figure 10.32.

Service: Sandy gravel with some good-sized rocks.

Maximum wear that can be economically obtained is the goal of most hard-facing applications. The material chosen should rate as highly as possible in the resistance-to-abrasion column, unless some other characteristics shown in the other columns make it unsuited for this particular application.

First, consider the tungsten carbide types. Notice that they are composed of very hard particles in a softer and less abrasion-resistant matrix. Although such material is the best for resisting sliding abrasion on hard material, in sand the matrix is apt to scour out slightly, and then the brittle particles are exposed. These particles are rated poor in impact resistance, and they may break and spall off when they encounter the rocks.

Next best in terms of abrasion, as listed in the chart, is the high-chromium carbide type shown in the electrode size column to be a powder. It can be applied only in a thin layer and also is not rated high in impact resistance. This makes it of dubious use in this rocky soil.

The rod-type high-chromium carbides also rate very high in abrasion resistance but do not rate high in impact resistance. However, the second does show sufficient impact rating to be considered if two or three different materials are to be tested in a field test. Given the possibility that it has enough impact resistance to do this job, there may be reluctance to pass up its very good wearing properties.

Nevertheless, the semiaustenitic type is balanced in both abrasion and impact resistance. It is much better in resistance to impact than the materials that rate higher in abrasion resistance. Therefore, the semiaustenitic is the first choice for this job, considering that the added impact resistance of the austenitic type is not necessary, since the impact in this application is not extreme.

Example 2

Application: Same dragline tooth used in Example 1.
Service: Soil changed to clay and shale.

The semiaustenitic type selected in the first example stands up well, but the teeth wear only half as long as the bucket lip. With double the wear on the teeth, only half the downtime periods would be needed for resurfacing, and both teeth and bucket could be done together. Since impact wear is now negligible with the new soil conditions, a material higher in the abrasion column should be considered. A good selection would

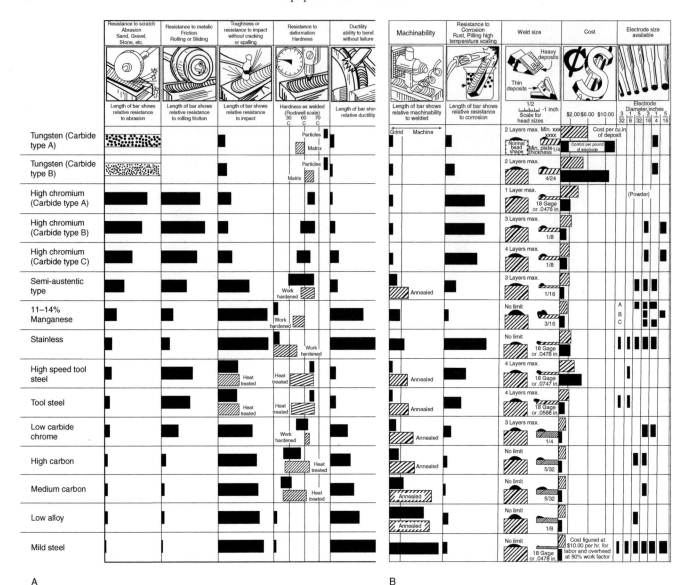

FIGURE 10.32.

be the first high-chromium carbide rod, which could give twice the wear by controlling the size of bead applied while still staying within a reasonable cost range.

Example 3

Application: Same dragline tooth used as in Examples 1 and 2.

Service: Soil changed to obtain large rocks.

If the earth contains many hard and large rocks, and the teeth are failing because of spalling under impact, one should move down the abrasion-resisting column

to a more impact-resistant material, such as the semi-austenitic type.

These examples demonstrate that where a dragline operates in all kinds of soils, a material that is resistant to both impact and abrasion, such as a semiaustenitic type, is the best choice. When the same type of reasoning is used to check the important characteristics, an appropriate material can be chosen for any application. If, for any reason, the first choice does not prove satisfactory, it is usually easy to improve the next application by choosing a material that is rated higher in the characteristic that was lacking. Where failures occur because of cracking or spalling, it usually indicates that a material higher in impact or ductility rating should be used. Where normal wear alone seems too rapid, a material with a higher abrasion rating is indicated.

10.13.7. Check Welding Procedure

Often, hard-facing failures due to cracking or spalling may be caused by improper welding procedures. Before changing the hard-surfacing material, consider whether or not the material has been properly applied. For almost any hard-facing application, very good results can be obtained by following these precautions:

1. Do not apply hard-surfacing material over cracked or porous areas. Remove any defective areas down to sound base metal.
2. Preheat. Preheating to 400°F to 500°F improves the resistance to cracking and spalling. This minimum temperature should be maintained until welding is completed. The exception to the rule is 11 to 14% manganese steel, which should be kept cool.
3. Cool slowly. If possible, allow the finished weldment to cool under an insulating material such as lime or sand.
4. Do not apply more than the recommended number of layers.

When more than normal buildup is required, apply intermediate layers of either medium carbon or stainless steel. This will provide a good bond to the base metal and will eliminate excessively thick layers of hard-surfacing material that might otherwise spall off. Stainless steel is also an excellent choice for intermediate layers on manganese steels or for hard-to-weld steels where preheating is not practical.

10.13.8. Check Before the Part Is Completely Worn

Whenever possible, examine a surfaced part when it is only partly worn. Examination of a part after it is completely worn is unsatisfactory. Did the surface crumble off, or was it scratched off? Is a tougher surface needed, or is additional abrasion resistance required? Should a heavier layer of surfacing be used? Should surfacing be reduced? All these questions can be answered by examination of a partially worn part and with a knowledge of the surfacing costs and service requirements.

When it is impossible to analyze the service conditions thoroughly in advance, it is always on the safe side to choose a material tougher than is thought to be required. A tough material will not spall or chip off and will offer some resistance to abrasion. A hard, abrasion-resistant

material is more susceptible to chipping, and surfacing material does no good if it falls off.

After some experience with surfacing materials, various combinations of materials can be tried to improve product performance. For example, on a part which is normally surfaced with a tough, semiaustenitic electrode, it may be possible to get additional abrasion resistance without sacrificing resistance to cracking. A little of the powdered chromium carbide material can be fused to critical areas where additional protection is needed.

Many badly worn parts are first built up to almost finished size with a high-carbon electrode, then surfaced with an austenitic rod, and finally a few beads of chromium carbide deposit are placed in spots requiring maximum protection against abrasion. Regardless of the circumstances, a careful analysis of the surfacing problem will be well worthwhile. Examples of jobs are shown in Figures 10.33 to 10.35.

10.13.9. Hard Surfacing with SAW

The submerged arc process offers several advantages for hard surfacing. The greater uniformity of the surface makes for better wearing qualities. The speed of SAW creates major economies in hard-surfacing areas that require the deposition of large amounts of metal. These areas may be either flat or curved surfaces. Mixer bottom plates, scraper blades, fan blades, chutes, and refinery vessels are examples of the flat plate to be surfaced. Shafts, blooming mill spindles, skelp rolls, crane wheels, tractor idlers and rollers, and rams are examples of cylindrical surfaces (Figures 10.36 to 10.39).

FIGURE 10.33.

FIGURE 10.34.

FIGURE 10.37.

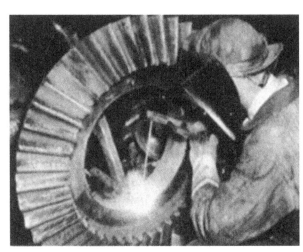

FIGURE 10.35.

FIGURE 10.38.

FIGURE 10.36.

FIGURE 10.39.

The process can be used with either fully automatic or semiautomatic equipment depending on the economics of the application. Fully automatic equipment can be quickly fitted with auxiliary accessories, resulting in more economical metal deposition. An oscillating device can be added to an automatic head to create a bead up to 3″ wide in a single pass. Another attachment permits the feeding of two electrode wires through a single head and a single contact jaw. Both of these attachments are useful in hard surfacing.

Hard surfacing with a submerged arc can be done with several different types of materials. The hard-surfacing deposit can be created by using solid alloy wires and a neutral granular flux. It also can be created by using a solid mild-steel wire and an agglomerated-alloy flux, the alloys being added to the deposit through the flux rather than through the wire. Also available are tubular wires that contain alloying material in the hollow portion of the mild-steel tube. All the methods have specific advantages. With SAW, considerable variation in the hard-surfacing deposit can be made by changing the welding procedure to control admixture and the heat-treatment effect of the welding cycle. Methods and procedures should be established with the help of qualified engineers.

10.14. SELECTION AND MAINTENANCE OF EQUIPMENT

10.14.1. Machines

Satisfactory welding can be accomplished with either alternating or direct welding current. Each type of current, however, has particular advantages that make it best suited for certain types of welding and welding conditions. The chief advantage of alternating current is its elimination of arc blow, which may be encountered when welding on heavy plate or into a corner. The magnetic fields set up in the plate deflect the path of the arc. Alternating current tends to minimize this deflection and also will increase the speed of welding with larger electrodes, over ³⁄₁₆″ diameter, and with the iron powder type of electrodes.

The chief advantages of direct current are the stability of the arc and the fact that the current output of the motor-generator type of welder will remain constant in spite of variations in the input voltage that affect a transformer-type welder. Direct current, therefore, is a more versatile welding current. Certain electrodes, such as stainless, require a very stable arc; these electrodes operate much better with direct current. Direct current, because of its stability, is also better for sheet metal welding, where the danger of burn-through is present. The DC arc also can be more readily varied to meet different welding conditions. A wider range of control over both voltage and current permits closer adjustment of the arc for difficult welding conditions, such as might be encountered in vertical or overhead welding. Because of its versatility, direct current should be available for maintenance welding.

Direct current welders (Figures 10.40 and 10.41) are made either as motor-generator sets or as transformer-rectifier sets. Motor-generator sets are powered by AC or DC motors. Generators are also powered by small air-cooled gasoline engines (Figure 10.42) The advantage of this type of set is that for on-the-spot maintenance welding, it is not necessary to string electric power lines to the job site. Engine-driven welders powered by gasoline engines are also available and come in larger sizes than the air-cooled engine sets (Figure 10.43). These are suitable where the size of the plant maintenance operation warrants a larger welder.

For most general maintenance welding, a 250-amp output capacity is ample. Several manufacturers make compact, portable machines especially for this type of welding. Higher amperages may be required in particular applications; for these, heavy-duty machines should be used.

Another type of welding machine is one that produces both alternating and direct welding current, either of which is available at the flip of a switch (Figure 10.44). This is ideal for maintenance welding, since it makes any kind of welding arc available, offering complete flexibility.

FIGURE 10.40.

FIGURE 10.41.

FIGURE 10.43.

FIGURE 10.42.

FIGURE 10.44.

10.14.2. Accessory Equipment

The varied and severe service demands made on maintenance welding equipment require that the best in accessories be used. Most maintenance welders make racks or other storage conveniences, which they attach directly to the welding machine to facilitate storing and transporting electrodes and accessories. While these arrangements will vary to suit individual tastes and needs, the end result is to have everything immediately available for use.

A fire extinguisher is an essential accessory. Many electrode holders are available, but only a few combine all the desirable features. The operator holds the electrode clamped in a holder, and the current from the welding set passes through the holder to the electrode. The clamping device should be designed to hold the electrode securely in position yet permit the quick and easy exchange of electrodes. It should be light in weight, properly balanced and easy to handle, yet sturdy enough to withstand rough use. It should be designed to remain cool enough to be handled comfortably (see Figure 10.45).

Face or head shields are generally constructed of some kind of pressed fiber insulating material, usually black to reduce glare. The shield should be light in weight and comfortable to molten wear. The glass windows in the shield should be of a material that absorbs infrared rays, ultraviolet rays, and most visible rays emanating from the arc. The welding lens should be protected from metal spatter and breakage by a chemically treated clear "nonspatter" glass covering the exposed side of the lens. The operator should always wear a protective shield when welding and should never look at the arc with the naked eye. When a new lens is put into the shield, care should be taken to make sure no light leaks in around the glass. If practical, the welding room should be painted a dead black or some other dark color to prevent the reflection of light and glare. Others working around the welding area can be easily shielded from light and sparks by the use of portable screens.

FIGURE 10.45.

Special goggles are used by welders' helpers, foremen, supervisors, inspectors, and others working close to a welding arc to protect their eyes from occasional flashes. A good set of goggles has an adjustable elastic head band and is lightweight, cool, well ventilated, and comfortable. Clear cover glasses and tinted lenses in various shades are available for this type of goggle.

During the arc welding process, some sparks and globules of molten metal are thrown out from the arc. For protection from possible burns, the operator is advised to wear an apron of leather or other protective material. Some operators also wear spats or leggings and sleevelets of leather or other fire-resistant material. Some sort of protection should be provided for the operator's ankles and feet, since a globule of molten metal can cause a painful burn before it can be extracted from the shoe. A gauntlet type of glove, preferably made of leather, is generally used by operators to protect their hands from the arc rays, spatters of molten metal, sparks, etc. Gloves also provide protection when the operator is handling the work.

Other tools of value in any shop where welding is done include wire brushes for cleaning the welds, cold chisels for chipping, clamps for holding work in position for welding, wedges, and, where work is large or heavy, a crane or chain block. A drill, air hammer, and grinder are also valuable accessories.

10.15. INSTALLATION OF EQUIPMENT

Good welding begins with the proper installation of equipment. Installations should be made in locations that are as clean as possible, and there should be provisions for a continuous supply of clean air for ventilation. It is important to provide separate enclosures if the atmosphere is excessively moist or contains corrosive vapors. If welding must be done where the ambient temperature is high, place the equipment in a different location. Sets operated outdoors should be equipped with protection against inclement weather.

When installing welding equipment, consider the following:

1. Contact the local power company to ensure an adequate supply of electric power.
2. Provide an adequate and level support for the equipment.
3. Protect adequately against mechanical abuse and atmospheric conditions.
4. Provide fresh air for ventilation and cooling.

5. Electrically ground the frame of the welder.
6. Check electrical connections to make sure they are clean and mechanically tight.
7. The fuses for a motor-generator welder should be of the "high lag" type and be rated two or three times the input-current rating of the welder.
8. Provide welding leads of sufficient capacity to handle the required current.
9. Check the set before operating it to make sure that no parts are visibly loose or in poor condition.

10.16. EQUIPMENT OPERATION AND MAINTENANCE

The following precautions will do much to ensure maximum service and performance from arc welding equipment.

10.16.1. Keep the Machine Clean and Cool

Because of the large volume of air pulled through welders by the fans in order to keep the machines cool, the greatest enemies of continuous efficient performance are airborne dust and abrasive materials. Machines that are exposed to ordinary dust should be blown out at least once a week with dry, clean compressed air at a pressure not exceeding 30 psi. Higher pressures may damage windings.

In foundries or machine shops, where cast iron or steel dust is present, vacuum cleaning should be substituted for compressed air. Compressed air under high pressure tends to drive the abrasive dust into the windings.

Abrasive material in the atmosphere grooves and pits the commutator and wears out brushes. Greasy dirt or lint-laden dust quickly clogs air passages between coils and causes them to overheat. Since resistance of the coils is raised and the conductivity lowered by heat, it reduces efficiency and can result in burned-out coils if the machine is not protected against overload. Overheating makes the insulation between coils dry and brittle. Neither the air intake nor the exhaust vents should be blocked, because this will interrupt the flow of air through the machine. The welder covers should be kept on; removing them destroys the proper path of ventilation.

10.16.2. Do Not Abuse the Machine

Never leave the electrode grounded to the work. This can create a "dead" short circuit. The machine is forced to generate much higher current than it was designed for, which can result in a burned-out machine.

10.16.3. Do Not Work the Machine Over Its Rated Capacity

A 200-amp machine will not do the work of a 400-amp machine. Operating above capacity causes overheating, which can destroy the insulation or melt the solder in the commutator connections.

Use extreme care in operating a machine on a steady load other than arc welding, such as thawing water pipes, supplying current for lighting, running motors, charging batteries, or operating heating equipment. For example, a DC machine, NEMA-rated 300 amp to 40 volts or 12 kW, should not be used for any continuous load greater than 9.6 kW and not more than 240 amp. This precaution applies to machines with a duty cycle of at least 60%. Machines with lower load-factor ratings must be operated at still lower percentages of the rated load.

10.16.4. Do Not Handle Roughly

A welder is a precisely aligned and balanced machine. Mechanical abuse, rough handling, or severe shock may disturb the alignment and balance of the machine, resulting in serious trouble. Misalignment can cause bearing failure, bracket failure, unbalanced air gap, or unbalance in the armature.

Never pry on the ventilating fan or commutator to try to move the armature. To do so will damage the fan or commutator. If the armature is jammed, inspect the unit for the cause of the trouble. Check for dirt or foreign particles between the armature and frames. Inspect the banding wire on the armature. Look for a frozen bearing.

Do not neglect the engine if the welder is an engine-driven unit. It deteriorates rapidly if not properly cared for. Follow the engine manufacturer's recommendations. Change the oil regularly. Keep air filters and oil strainers clean. Do not allow grease and oil from the engine to leak back into the generator. Grease quickly accumulates dirt and dust, clogging the air passages between the coils.

10.16.5. Maintain the Machine Regularly

Bearings

The ball bearings in modern welders have sufficient grease to last the life of the machine under normal conditions. Under severe conditions—heavy use or a dirty location—the bearings should be greased about once a year. An ounce of grease a year is sufficient for each bearing. A pad of grease approximately one cubic inch in volume weighs close to 1 ounce. Dirt is responsible for more bearing failures than any other cause. This dirt may get into the grease cup when it is removed to refill, or it may get into the grease in its original container. Before the grease cup or pipe plug is removed, it is important to wipe it absolutely clean. A piece of dirt no larger than the period at the end of this sentence may cause a bearing to fail in a short time. Even small particles of grit that float around in the factory atmosphere are dangerous.

If too little grease is applied, bearings fail. If the grease is too light, it will run out. Grease containing solid materials may ruin antifriction bearings. Rancid grease will not lubricate. Dirty grease or dirty fittings or pipes can cause bearing failures.

Generally, bearings do not need inspection. They are sealed against dirt and should not be opened. If bearings must be pulled, it should be done using a special puller designed to act against the inner race.

Never clean new bearings before installing them. Handle them with care. Put them in place by driving against the inner race. Make sure that they fit squarely against the shoulders.

Brackets or End Bolts

If it becomes necessary to remove a bracket, to replace a bearing, or to disassemble the machine, do so by removing the bolts and tapping lightly and evenly with a babbitt hammer all around the outside diameter of the bracket ring. Do not drive off with a heavy steel hammer. The bearing may become worn over size, caused by the pounding of the bearing when the armature is out of balance. The bearing should slide into the housing with a light drive fit. Replace the bracket if the housing is oversize.

Brushes and Brush Holders

Set brush holders approximately $1/32''$ to $3/32''$ above the surface of the commutator. If brush holders have been removed, be certain that they are set squarely in the rocker slot when replaced. Do not force the brush holder into the slot by driving on the insulation. Check to ensure that the brush holder insulation is squarely set. Tighten brush holders firmly. When properly set, they are parallel to the mica segments between commutator bars. Use the grade of brushes recommended by the manufacturer of the welding set. Brushes that are too hard or too soft may damage the commutator. Brushes will be damaged by excessive clearance in the brush holder or uneven brush spring pressure. High

commutator bars, high mica segments, excessive brush spring pressure, and abrasive dust will also wear out brushes rapidly.

Inspect brushes and holders regularly. A brush may wear down and lose spring tension. It will then start to arc, with damage to the commutator and other brushes. Keep the brush contact surface of the holder clean and free from pit marks. Brushes must be able to move freely in the holder. Replace them when the pigtails are within $1/8''$ of the commutator or when the limit of spring travel is reached.

New brushes must be sanded in to conform to the shape of the commutator. This may be done by stoning the commutator with a stone or by using fine sandpaper (not emery cloth or paper). Place the sandpaper under the brush, and move it back and forth while holding the brush down in the normal position under slight pressure with the fingers. See that the brush holders and springs seat squarely and firmly against the brushes and that the pigtails are fastened securely.

Commutators

Commutators normally need little care. They will build up a surface film of brown copper oxide, which is highly conductive, hard, and smooth. This surface helps to protect the commutator. Do not try to keep a commutator bright and shiny by constant stoning. The brown copper oxide film prevents the buildup of a black abrasive oxide film that has high resistance and causes excessive brush and commutator wear. Wipe clean occasionally with a rag or canvas to remove grease discoloration from fumes or other unnatural film. If brushes are chattering because of high bars, high mica, or grooves, stone by hand or remove and turn in a lathe, if necessary.

Most commutator trouble starts because the wrong grade of brushes is used. Brushes that contain too much abrasive material or have too high a copper content usually scratch the commutator and prevent the desired surface film from building up. A brush that is too soft may smudge the surface with the same result as far as surface film is concerned. In general, brushes that have a low voltage drop will give poor commutation. Conversely, a brush with high voltage drop commutates better but may cause overheating of the commutator surface.

If the commutator is burned, it may be dressed down by pressing a commutator stone against the surface with the brushes raised. If the surface is badly pitted or out of round, the armature must be removed from the machine and the commutator turned in a lathe. It is good practice for the commutator to run within a radial tolerance of 0.003". The mica separating the bars of the commutator is undercut to a depth of $1/32"$ to $1/16"$. Mica exposed at the commutator surface causes brush and commutator wear and poor commutation. If the mica is even with the surface, undercut it. When the commutator is operating properly, there is very little visible sparking. The brush surface is shiny and smooth, with no evidence of scratches.

Generator Frame

The generator frame and coils need no attention other than inspection to ensure tight connections and cleanliness. Blow out dust and dirt with compressed air. Grease may be cleaned off with naphtha. Keep air gaps between armature and pole pieces clean and even.

Armature

The armature must be kept clean to ensure proper balance. Unbalance in the set will pound out the bearings and wear the bearing housing oversize. Blow out the armature regularly with clean, dry compressed air. Clean out the inside of the armature thoroughly by attaching a long pipe to the compressed air line and reaching into the armature coils.

Motor Stator

Keep the stator clean and free from grease. When reconnecting it for use on another voltage, solder all connections. If the set is to be used frequently on different voltages, it may save time to place lugs on the ends of all the stator leads. This eliminates the necessity for loosening and resoldering to make connections, since the lugs may be safely joined with a screw, nut, and lock washer.

Exciter Generator

If the machine has a separate exciter generator, its armature, coils, brushes, and brush holders will need the same general care recommended for the welder set. Keep the covers over the exciter armature, since the commutator can be damaged easily.

Controls

Inspect the controls frequently to ensure that the ground and electrode cables are connected tightly to the output terminals. Loose connections cause arcing that destroys the insulation around the terminals and burns them. Do not bump or hit the control handles—it damages the controls, resulting in poor electrical contacts. If the handles are tight or jammed, inspect them for the

cause. Check the contact fingers of the magnetic starting switch regularly. Keep the fingers free from deep pits or other defects that will interfere with a smooth, sliding contact. Copper fingers may be filed lightly. All fingers should make contact simultaneously. Keep the switch clean and free from dust. Blow out the entire control box with low-pressure compressed air.

Connections of the leads from the motor stator to the switch must be tight. Keep the lugs in a vertical position. The line voltage is high enough to jump between the lugs on the stator leads if they are allowed to become loose and cocked to one side or the other. Keep the cover on the control box at all times.

Condensers

Condensers may be placed in an AC welder to raise the power factor. When condensers fail, it is not readily apparent from the appearance of the condenser. Consequently, to check a condenser, one should see if the input current reading corresponds to the nameplate amperes at the rated input voltage and with the welder drawing the rated output load current. If the reading is 10 to 20% more, at least one condenser has failed. Caution: Never touch the condenser terminals without first disconnecting the welder from the input power source; then discharge the condenser by touching the two terminals with an insulated screwdriver.

Delay Relays

The delay relay contacts may be cleaned by passing a cloth soaked in naphtha between them. Do not force the contact arms or use any abrasives to clean the points. Do not file the silver contacts. The pilot relay is enclosed in a dust-proof box and should need no attention. Relays are usually adjusted at the factory and should not be tampered with unless faulty operation is obvious. Table 10.18, a troubleshooting chart, may prove to be a great timesaver.

10.17. SAFETY

Arc welding can be done safely, provided that sufficient measures are taken to protect the operator from the potential hazards. If the proper measures are ignored or overlooked, welding operators can be exposed to such dangers as electrical shock and overexposure to radiation, fumes and gases, and fire and explosion, any of which could cause severe injury or even death. With the diversification of the welding that may be done by maintenance departments, it is vitally important that the appropriate safety measures be evaluated on a job-by-job basis and that they be rigidly enforced.

A quick guide to welding safety is provided in Figure 10.46. All the potential hazards, as well as the proper safety measures, may be found in *ANSI Z-49.1,* published by the American National Standards Institute and the American Welding Society. A similar publication, "Arc Welding Safety," is available from the Lincoln Electric Company.

TABLE 10.18.

Trouble	Cause	Remedy
Welder will not start (starter not operating)	Power circuit dead	Check voltage
	Broken power lead	Repair
	Wrong supply voltage	Check nameplate against supply
	Open power switches	Close
	Blown fuses	Replace
	Overload relay tripped	Let set cool. Remove cause of overloading
	Open circuit to starter button	Repair
	Defective operating coil	Replace
	Mechanical obstruction in contactor	Remove
Welder will not start (starter operating)	Wrong motor connections	Check connection diagram
	Wrong supply voltage	Check nameplate against supply
	Rotor stuck	Try turning by hand
	Power circuit single-phased	Replace fuse; repair open line
	Starter single-phased	Check contact of starter tips
	Poor motor connection	Tighten
	Open circuit in windings	Repair
Starter operates and blows fuse	Fuse too small	Should be two to three times rated motor current
	Short circuit in motor connections	Check starter and motor leads for insulation from ground and from each other
Welder starts but will not deliver welding current	Wrong direction of rotation	Check connection diagram
	Brushes worn or missing current	Check that all brushes bear on commutator with sufficient tension
	Brush connections loose	Tighten
	Open field circuit	Check connection to rheostat, resistor, and auxiliary brush studs
	Series field and armature circuit open	Check with test lamp or bell ringer

Hazard	Factors to consider	Precaution summary
Electric shock can kill	• Wetness • Welder in or workpiece • Confined space • Electrode holder and cable insulation	• Insulate welder from workpiece and ground using dry insulation. Rubber mat or dry wood. • Wear dry, *hole-free* gloves. (Change as necessary to keep dry.) • Do not touch electrically "hot" parts or electrode with bare skin or wet clothing. • If wet area and welder cannot be insulated from workpiece with dry insulation, use a semiautomatic, constant-voltage welder or stick welder with voltage reducing device. • Keep electrode holder and cable insulation in good condition. Do not use if insulation damaged or missing.
Fumes and gases can be dangerous	• Confined area • Positioning of welder's head • Lack of general ventilation • Electrode types, i.e., manganese, chromium, etc., see MSDS • Base metal coatings, galvanize, paint	• Use ventillation or exhaust to keep air breathing zone clear, comfortable. • Use helmet and positioning of head to minimize fume in breathing zone. • Read warnings on electrode container and material safety data sheet (MSDS) for electrode. • Provide additional ventilation/exhaust where special ventillation requirements exist. • Use special care when welding in a confined area. • Do not weld unless ventillation is adequate.
Welding sparks can cause fire or explosion	• Containers which have held combustibiles • Flammable materials	• Do not weld on containers which have held combustible materials (unless strict AWS F4.1 procedures are followed). Check before welding. • Remove flammable materiels from welding area or shield from sparks, heat. • Keep a fire watch in area during and after welding. • Keep a fire extinguisher in the welding area. • Wear fire retardant clothing and hat. Use earplugs when welding overhead
Arc rays can burn eyes and skin	• Process: gas-shielded arc most severe	• Select a filter lens which is comfortable for you while welding. • Always use helmet when welding. • Provide nonflammable shielding to protect others. • Wear clothing which protects skin while welding.
Confined space	• Metal enclosure • Wetness • Restricted entry • Heavier than air gas • Welder inside or on workpiece	• Carefully evaluate adequacy of ventilation especially where electrode requires special ventilation or where gas may displace breathing air. • If basic electric shock precautions cannot be followed to insulate welder from work and electrode, use semiautomatic, constant-voltage equipment with cold electrode or stick welder with voltage reducing device. • Provide welder helper and method of welder retrieval from outside enclosure.

FIGURE 10.46.

CHAPTER

11

Bearings*

A bearing is a machine element that supports a part, such as a shaft, that rotates, slides, or oscillates in or on it. There are two broad classifications of bearings: plain and rolling element (also called antifriction). Plain bearings are based on sliding motion made possible through the use of a lubricant. Antifriction bearings are based on rolling motion, which is made possible by balls or other types of rollers. In modern rotor systems operating at relatively high speeds and loads, the proper selection and design of the bearings and bearing-support structure are key factors affecting system life.

11.1. TYPES OF MOVEMENT

The type of bearing used in a particular application is determined by the nature of the relative movement and other application constraints. Movement can be grouped into the following categories: rotation about a point, rotation about a line, translation along a line, rotation in a plane, and translation in a plane. These movements can be either continuous or oscillating.

Although many bearings perform more than one function, they can generally be classified based on types of movement. There are three major classifications of both plain and rolling element bearings: radial, thrust, and guide. *Radial* bearings support loads that act radially and at right angles to the shaft center line. These loads may be visualized as radiating into or away from a center point like the spokes on a bicycle wheel. *Thrust* bearings support or resist loads that act axially. These may be described as endwise loads that act

*Source: Ricky Smith and Keith Mobley, *Industrial Machinery Repair: Best Maintenance Practices Pocket Guide* (Boston: Butterworth–Heinemann, 2003), pp. 71–119.

parallel to the center line toward the ends of the shaft. This type of bearing prevents lengthwise or axial motion of a rotating shaft. *Guide* bearings support and align members having sliding or reciprocating motion. This type of bearing guides a machine element in its lengthwise motion, usually without rotation of the element.

Table 11.1 gives examples of bearings that are suitable for continuous movement; Table 11.2 shows bearings that are appropriate for oscillatory movement only. For the bearings that allow movements in addition to the one listed, the effect on machine design is described in the column "Effect of the other degrees of freedom." Table 11.3 compares the characteristics, advantages, and disadvantages of plain and rolling element bearings.

11.1.1. About a Point (Rotational)

Continuous movement about a point is rotation, a motion that requires repeated use of accurate surfaces. If the motion is oscillatory rather than continuous, some additional arrangements must be made in which the geometric layout prevents continuous rotation.

11.1.2. About a Line (Rotational)

Continuous movement about a line is also referred to as rotation, and the same comments apply as for movement about a point.

11.1.3. Along a Line (Translational)

Movement along a line is referred to as translation. One surface is generally long and continuous, and the moving component is usually supported on a fluid

TABLE 11.1.   Bearing Selection Guide (Continuous Movement)

Constraint Applied to the Movement	Examples of Arrangements Which Allow Movement Only Within This Constraint	Examples of Arrangements Which Allow Movement But Also Have Other Degrees of Freedom	Effect of the Other Degrees of Freedom
About a point	Gimbals	Ball on a recessed plate	Ball must be forced into contact with the plate
About a line	Journal bearing with double thrust location	Journal bearing	Simple journal bearing allows free axial movement as well
	Double conical bearing	Screw and nut	Gives some related axial movement as well
		Ball joint or spherical roller bearing	Allows some angular freedom to the line of rotation
	Crane wheels restrained between two rails	Railway or crane wheel on a track	These arrangements need to be loaded into contact. This is usually done by gravity. Wheels on a single rail or cable need restraint to prevent rotation about the track member
		Pulley wheel on a cable	
		Hovercraft or hoverpad on a track	
In a plane (rotation)	Double thrust bearing	Single thrust bearing	Single thrust bearing must be loaded into contact
In a plane (translation)		Hovercraft or hoverpad	Needs to be loaded into contact usually by gravity

Source: M. J. Neale, Society of Automotive Engineers Inc., *Bearings—A Tribology Handbook* (Oxford: Butterworth–Heinemann, 1993).

TABLE 11.2.   Bearing Selection Guide (Oscillatory Movement)

Constraint Applied to the Movement	Examples of Arrangements Which Allow Movement Only Within This Constraint	Examples of Arrangements Which Allow This Movement But Also Have Other Degrees of Freedom	Effect of the Other Degrees of Freedom
About a point	Hookes joint	Cable connection between components	Cable needs to be kept in tension
About a line	Crossed strip flexure pivot	Torsion suspension	A single torsion suspension gives no lateral location
		Knife-edge pivot	Must be loaded into contact
		Rubber bush	Gives some axial and lateral flexibility as well
		Rocker pad	Gives some related translation as well. Must be loaded into contact
	Crosshead and guide bars	Piston and cylinder	Piston can rotate as well unless it is located by connecting rod
In a plane (rotation)		Rubber ring or disc	Gives some axial and lateral flexibility as well
In a plane (translation)	Plate between upper and lower guide blocks	Block sliding on a plate	Must be loaded into contact

Source: M. J. Neale, Society of Automotive Engineers Inc., *Bearings—A Tribology Handbook* (Oxford: Butterworth–Heinemann, 1993).

TABLE 11.3.   Comparison of Plain and Rolling Element Bearings

Rolling Element	Plain
Assembly on crankshaft is virtually impossible, except with very short or built-up crankshafts	Assembly on crankshaft is no problem as split bearings can be used
Cost relatively high	Cost relatively low
Hardness of shaft unimportant	Hardness of shaft important with harder bearings
Heavier than plain bearings	Lighter than rolling element bearings
Housing requirement not critical	Rigidity and clamping most important housing requirement
Less rigid than plain bearings	More rigid than rolling element bearings
Life limited by material fatigue	Life not generally limited by material fatigue
Lower friction results in lower power consumption	Higher friction causes more power consumption
Lubrication easy to accomplish; the required flow is low except at high speed	Lubrication pressure feed critically important; required flow is large, susceptible to damage by contaminants and interrupted lubricant flow
Noisy operation	Quiet operation
Poor tolerance of shaft deflection	Moderate tolerance of shaft deflection
Poor tolerance of hard dirt particles	Moderate tolerance of dirt particles, depending on hardness of bearing
Requires more overall space	Requires less overall space
Length: Smaller than plain	Length: Larger than rolling element
Diameter: Larger than plain	Diameter: Smaller than rolling element
Running friction	Running friction
Very low at low speeds	Higher at low speeds
May be high at high speeds	Moderate at usual crank speeds
Smaller radial clearance than plain	Larger radial clearance than rolling element

Source: Integrated Systems Inc.

film or rolling contact in order to achieve an acceptable wear rate. If the translational movement is reciprocation, the application makes repeated use of accurate surfaces, and a variety of economical bearing mechanisms are available.

11.1.4. In a Plane (Rotational/ Translational)

If the movement in a plane is rotational or both rotational and oscillatory, the same comments apply as for movement about a point. If the movement in a plane is translational or both translational and oscillatory, the same comments apply as for movement along a line.

11.2. COMMONLY USED BEARING TYPES

As mentioned before, the major bearing classifications are plain and rolling element. These types of bearings are discussed in the sections to follow. Table 11.4 is a bearings characteristics summary. Table 11.5

is a selection guide for bearings operating with continuous rotation and special environmental conditions. Table 11.6 is a selection guide for bearings operating with continuous rotation and special performance requirements. Table 11.7 is a selection guide for oscillating movement and special environment or performance requirements.

11.2.1. Plain Bearings

All plain bearings also are referred to as fluid-film bearings. In addition, radial plain bearings also are commonly referred to as journal bearings. Plain bearings are available in a wide variety of types or styles and may be self-contained units or built into a machine assembly. Table 11.8 is a selection guide for radial and thrust plain bearings.

Plain bearings are dependent on maintaining an adequate lubricant film to prevent the bearing and shaft surfaces from coming into contact, which is necessary to prevent premature bearing failure. However, this is difficult to achieve, and some contact usually occurs

TABLE 11.4. Bearings Characteristic Summary

Bearing Type	Description
Plain	See Table 11.3.
Lobed	See Radial, Elliptical.
Radial or journal	
Cylindrical	Gas lubricated, low-speed applications.
Elliptical	Oil lubricated, gear and turbine applications, stiffer and somewhat more stable bearing.
Four-axial grooved	Oil lubricated, higher-speed applications than cylindrical.
Partial arc	Not a bearing type, but a theoretical component of grooved and lobed bearing configurations.
Tilting pad	High-speed applications where hydrodynamic instability and misalignment are common problems.
Thrust	Semifluid lubrication state, relatively high friction, lower service pressures with multicollar version, used at low speeds.
Rolling element	See Table 11.3. Radial and axial loads, moderate- to high-speed applications
Ball	Higher speed and lighter load applications than roller bearings.
Single row	
Radial nonfilling slot	Also referred to as Conrad or deep-groove bearing. Sustains combined radial and thrust loads, or thrust loads alone, in either direction, even at high speeds. Not self-aligning.
Radial filling slot	Handles heavier loads than nonfilling slot.
Angular contact radial thrust	Radial loads combined with thrust loads, or heavy thrust loads alone. Axial deflection must be limited.
Ball-thrust	Very high thrust loads in one direction only, no radial loading, cannot be operated at high speeds.
Double row	Heavy radial with minimal bearing deflection and light thrust loads.
Double roll, self aligning	Moderate radial and limited thrust loads.
Roller	Handles heavier loads and shock better than ball bearings, but are more limited in speed than ball bearings.
Cylindrical	Heavy radial loads, fairly high speeds, can allow free axial shaft movement.
Needle-type cylindrical	Does not normally support thrust loads, used in barrel space-limited applications, angular mounting of rolls in double-row version tolerates combined axial and thrust loads.
Spherical	High radial and moderate-to-heavy thrust loads, usually comes in double-row mounting that is inherently self-aligning.
Tapered	Heavy radial and thrust loads. Can be preloaded for maximum system rigidity.

Source: Integrated Systems Inc.

TABLE 11.5. Bearings Selection Guide for Special Environmental Conditions (Continuous Rotation)

Bearing Type	High Temperature	Low Temperature	Vacuum	Wet/Humid	Dirt/Dust	External Vibration
Plain, externally pressurized	1 (with gas lubrication)	2	No (affected by lubricant feed)	2	2 (1 when gas lubricated)	1
Plain, porous metal (oil impregnated)	4 (lubricant oxidizes)	3 (may have high starting torque)	Possible with special lubricant	2	Seals essential	2
Plain, rubbing (nonmetalic)	2 (up to temp. limit of material)	2	1	2 (shaft must not corrode)	2 (seals help)	2
Plain, fluid film	2 (up to temp. limit of lubricant)	2 (may have high starting torque)	Possible with special lubricant	2	2 (with seals and filtration)	2
Rolling	Consult manufacturer above 150°C	2	3 (with special lubricant)	3 (with seals)	Sealing essential	3 (consult manufacturers)
Things to watch with all bearings	Effect of thermal expansion on fits	Effect of thermal expansion on fits		Corrosion		Fretting

Rating: 1—Excellent, 2—Good, 3—Fair, 4—Poor.
Source: Adapted by Integrated Systems Inc. from M. J. Neale, Society of Automotive Engineers Inc., *Bearings—A Tribology Handbook* (Oxford: Butterworth–Heinemann, 1993).

TABLE 11.6. Bearings Selection Guide for Particular Performance Requirements (Continuous Rotation)

Bearing Type	Accurate Radial Location	Axial Load Capacity as Well	Low Starting Torque	Silent Running	Standard Parts Available	Simple Lubrication
Plain, externally pressurized	1	No (needs separate thrust bearing)	1	1	No	4 (needs special system)
Plain, fluid film	3	No (needs separate thrust bearing)	2	1	Some	2 (usually requires circulation system)
Plain, porous metal (oil impregnated)	2	Some	2	1	Yes	1
Plain, rubbing (nonmetalic)	4	Some in most instances	4	3	Some	1
Rolling	2	Yes in most instances	1	Usually satisfactory	Yes	2 (when grease lubricated)

Rating: 1—Excellent, 2—Good, 3—Fair, 4—Poor.
Source: Adapted by Integrated Systems Inc. from M. J. Neale, Society of Automotive Engineers Inc., *Bearings—A Tribology Handbook* (Oxford: Butterworth–Heinemann, 1993).

TABLE 11.7. Bearings Selection Guide for Special Environments or Performance (Oscillating Movement)

Bearing Type	High Temperature	Low Temperature	Low Friction	Wet/Humid	Dirt/Dust	External Vibration
Knife edge pivots	2	2	1	2 (watch corrosion)	2	4
Plain, porous metal (oil impregnated)	4 (lubricant oxidizes)	3 (friction can be high)	2	2	Sealing essential	2
Plain, rubbing	2 (up to temp. limit of material)	1	2 (with PTFE)	3 (shaft must not corrode)	2 (sealing helps)	1
Rolling	Consult manufacturer above 150°C	2	1	2 (with seals)	Sealing essential	4
Rubber brushes	4	4	Elastically stiff	1	1	1
Strip flexures	2	1	1	2 (watch corrosion)	1	1

Rating: 1—Excellent, 2—Good, 3—Fair, 4—Poor.
Source: Adapted by Integrated Systems Inc. from M. J. Neale, Society of Automotive Engineers Inc., *Bearings—A Tribology Handbook* (Oxford: Butterworth–Heinemann, 1993).

during operation. Material selection plays a critical role in the amount of friction and the resulting seizure and wear that occurs with surface contact. Note that fluid-film bearings do not have the ability to carry the full load of the rotor assembly at any speed and must have turning gear to support the rotor's weight at low speeds.

Thrust or Fixed

Thrust plain bearings consist of fixed shaft shoulders or collars that rest against flat bearing rings.

The lubrication state may be semifluid, and friction is relatively high. In multicollar thrust bearings, allowable service pressures are considerably lower because of the difficulty in distributing the load evenly between several collars. However, thrust ring performance can be improved by introducing tapered grooves. Figure 11.1 shows a mounting half-section for a vertical thrust bearing.

Radial or Journal

Plain radial, or journal, bearings also are referred to as sleeve or babbitt bearings. The most common type is

TABLE 11.8.   Plain Bearing Selection Guide

Journal Bearings Characteristics	Direct Lined	Insert Liners
Accuracy	Dependent upon facilities and skills available	Precision components
Quality (consistency)	Doubtful	Consistent
Cost	Initial cost may be lower	Initial cost may be higher
Ease of repair	Difficult and costly	Easily done by replacement
Condition upon extensive use	Likely to be in fatigue	Ability to sustain higher peak loads
Materials used	Limited to white metals	Extensive range available

Thrust Bearings Characteristics	Flanged Journal Bearings	Separate Thrust Washer
Cost	Costly to manufacture	Much lower initial cost
Replacement	Involves whole journal/thrust component	Easily replaced without moving journal bearing
Materials used	Thrust face materials limited in larger sizes	Extensive range available
Benefits	Aids assembly on a production line	Aligns itself with the housing

Source: Adapted by Integrated Systems Inc. from M. J. Neale, Society of Automotive Engineers Inc., *Bearings—A Tribology Handbook* (Oxford: Butterworth–Heinemann, 1993).

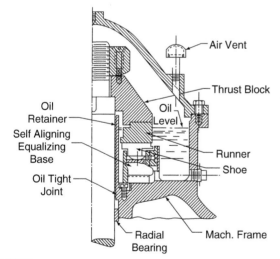

FIGURE 11.1.   Half section of mounting for vertical thrust bearing.

the full journal bearing, which has 360-degree contact with its mating journal. The partial journal bearing has less than 180-degree contact and is used when the load direction is constant. The sections to follow describe the major types of fluid-film journal bearings: plain cylindrical, four-axial groove, elliptical, partial-arc, and tilting-pad.

Plain Cylindrical

The plain cylindrical journal bearing (Figure 11.2) is the simplest of all journal bearing types. The performance characteristics of cylindrical bearings are well established, and extensive design information is available. Practically, use of the unmodified cylindrical bearing is generally limited to gas-lubricated bearings and low-speed machinery.

Four-Axial Groove Bearing

To make the plain cylindrical bearing practical for oil or other liquid lubricants, it is necessary to modify it by the addition of grooves or holes through which the lubricant can be introduced. Sometimes, a single circumferential groove in the middle of the bearing is used. In other cases, one or more axial grooves are provided.

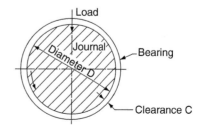

FIGURE 11.2.   Plain cylindrical bearing.

The four-axial groove bearing is the most commonly used oil-lubricated sleeve bearing. The oil is supplied at a nominal gauge pressure that ensures an adequate oil flow and some cooling capability. Figure 11.3 illustrates this type of bearing.

Elliptical Bearing

The elliptical bearing is oil-lubricated and typically is used in gear and turbine applications. It is classified as a lobed bearing in contrast to a grooved bearing. Where the grooved bearing consists of a number of partial arcs with a common center, the lobed bearing is made up of partial arcs whose centers do not coincide. The elliptical bearing consists of two partial arcs where the bottom arc has its center a distance above the bearing center. This arrangement has the effect of preloading the bearing, where the journal center eccentricity with respect to the loaded arc is increased and never becomes zero. This results in the bearing being stiffened, somewhat improving its stability. An elliptical bearing is shown in Figure 11.4.

Partial-Arc Bearings

A partial-arc bearing is not a separate type of bearing. Instead, it refers to a variation of previously discussed bearings (e.g., grooved and lobed bearings) that incorporates partial arcs. It is necessary to use partial-arc bearing data to incorporate partial arcs

in a variety of grooved and lobed bearing configurations. In all cases, the lubricant is a liquid and the bearing film is laminar. Figure 11.5 illustrates a typical partial-arc bearing.

Tilting-Pad Bearings

Tilting-pad bearings are widely used in high-speed applications where hydrodynamic instability and misalignment are common problems. This bearing consists of a number of shoes mounted on pivots, with each shoe being a partial-arc bearing. The shoes adjust and follow the motions of the journal, ensuring inherent stability if the inertia of the shoes does not interfere with the adjustment ability of the bearing. The load direction may either pass between the two bottom shoes or it may pass through the pivot of the bottom shoe. The lubricant is incompressible (i.e., liquid), and the lubricant film is laminar. Figure 11.6 illustrates a tilting-pad bearing.

11.2.2. Rolling Element or Antifriction

Rolling element antifriction bearings are one of the most common types used in machinery. Antifriction bearings are based on rolling motion as opposed to the sliding motion of plain bearings. The use of rolling elements between rotating and stationary surfaces

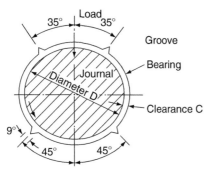

FIGURE 11.3. Four-axial groove bearing.

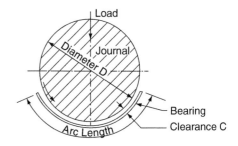

FIGURE 11.5. Partial arc.

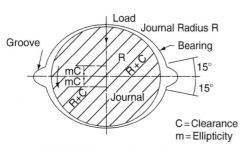

FIGURE 11.4. Elliptical bearing.

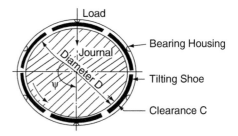

FIGURE 11.6. Tilting-pad bearing.

reduces the friction to a fraction of that resulting with the use of plain bearings. Use of rolling element bearings is determined by many factors, including load, speed, misalignment sensitivity, space limitations, and desire for precise shaft positioning. They support both radial and axial loads and are generally used in moderate- to high-speed applications.

Unlike fluid-film plain bearings, rolling element bearings have the added ability to carry the full load of the rotor assembly at any speed. Where fluid-film bearings must have turning gear to support the rotor's weight at low speeds, rolling element bearings can maintain the proper shaft centerline through the entire speed range of the machine.

Grade Classifications

Rolling element bearings are available in either commercial- or precision-grade classifications. Most *commercial-grade bearings* are made to nonspecific standards and are not manufactured to the same precise standards as precision-grade bearings. This limits the speeds at which they can operate efficiently, and given the brand of bearings may or may not be interchangeable.

Precision bearings are used extensively on many machines such as pumps, air compressors, gear drives, electric motors, and gas turbines. The shape of the rolling elements determines the use of the bearing in machinery. Because of standardization in bearing envelope dimensions, precision bearings were once considered to be interchangeable, even if manufactured by different companies. It has been discovered, however, that interchanging bearings is a major cause of machinery failure and should be done with extreme caution.

Rolling Element Types

There are two major classifications of rolling elements: ball and roller. Ball bearings function on point contact and are suited for higher speeds and lighter loads than roller bearings. Roller element bearings function on line contact and generally are more expensive than ball bearings, except for the larger sizes. Roller bearings carry heavy loads and handle shock more satisfactorily than ball bearings, but are more limited in speed. Figure 11.7 provides general guidelines to determine if a ball or roller bearing should be selected. This figure is based on a rated life of 30,000 hours.

Although there are many types of rolling elements, each bearing design is based on a series of hardened rolling elements sandwiched between hardened inner and outer rings. The rings provide continuous tracks or races for the rollers or balls to roll in. Each ball or

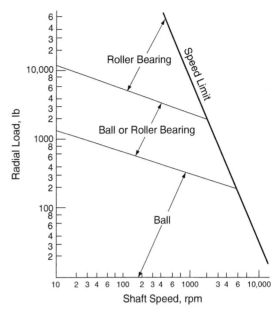

FIGURE 11.7. Guide to selecting ball or roller bearings.

roller is separated from its neighbor by a separator cage or retainer, which properly spaces the rolling elements around the track and guides them through the load zone. Bearing size is usually given in terms of boundary dimensions: outside diameter, bore, and width.

Ball Bearings

Common functional groupings of ball bearings are radial, thrust, and angular-contact bearings. Radial bearings carry a load in a direction perpendicular to the axis of rotation. Thrust bearings carry only thrust loads, a force parallel to the axis of rotation tending to cause endwise motion of the shaft. Angular-contact bearings support combined radial and thrust loads. These loads are illustrated in Figure 11.8. Another common classification of ball bearings is single-row (also referred to as Conrad or deep-groove bearing) and double-row.

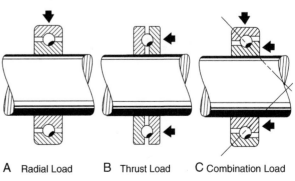

A Radial Load   B Thrust Load   C Combination Load

FIGURE 11.8. Three principal types of ball-bearing loads.

Single-Row

Types of single-row ball bearings are radial nonfilling slot bearings, radial filling slot bearings, angular contact bearings, and ball thrust bearings.

Radial, Nonfilling Slot Bearings

This ball bearing is often referred to as the Conrad-type or deep-groove bearing and is the most widely used of all ball bearings (and probably of all antifriction bearings). It is available in many variations, with single or double shields or seals. They sustain combined radial and thrust loads, or thrust loads alone, in either direction—even at extremely high speeds. This bearing is not designed to be self-aligning; therefore, it is imperative that the shaft and the housing bore be accurately aligned (Figure 11.9).

Figure 11.10 labels the parts of the Conrad antifriction ball bearing. This design is widely used and is versatile because the deep-grooved raceways permit the rotating balls to rapidly adjust to radial and thrust loadings, or a combination of these loadings.

Radial, Filling Slot Bearing

The geometry of this ball bearing is similar to the Conrad bearing, except for the filling slot. This slot allows more balls in the complement and thus can carry

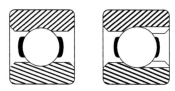

FIGURE 11.9. Single-row radial, nonfilling slot bearing.

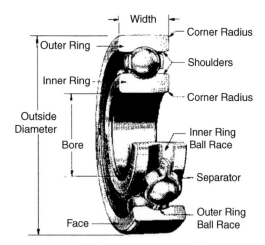

FIGURE 11.10. Conrad antifriction ball-bearing parts.

heavier radial loads. The bearing is assembled with as many balls that fit in the gap created by eccentrically displacing the inner ring. The balls are evenly spaced by a slight spreading of the rings and heat expansion of the outer ring. However, because of the filling slot, the thrust capacity in both directions is reduced. In combination with radial loads, this bearing design accommodates thrust of less than 60% of the radial load.

Angular Contact Radial Thrust

This ball bearing is designed to support radial loads combined with thrust loads, or heavy thrust loads (depending on the contact-angle magnitude). The outer ring is designed with one shoulder higher than the other, which allows it to accommodate thrust loads. The shoulder on the other side of the ring is just high enough to prevent the bearing from separating. This type of bearing is used for pure thrust load in one direction and is applied either in opposed pairs (duplex), or one at each end of the shaft. They can be mounted either face to face or back to back and in tandem for constant thrust in one direction. This bearing is designed for combination loads where the thrust component is greater than the capacity of single-row, deep-groove bearings. Axial deflection must be confined to very close tolerances.

Ball-Thrust Bearing

The ball-thrust bearing supports very high thrust loads in one direction only, but supports *no* radial loading. To operate successfully, this type of bearing must be at least moderately thrust-loaded at all times. It should not be operated at high speeds, since centrifugal force causes excessive loading of the outer edges of the races.

Double-Row

Double-row ball bearings accommodate heavy radial and light thrust loads without increasing the outer diameter of the bearing. However, the double-row bearing is approximately 60 to 80% wider than a comparable single-row bearing. The double-row bearing incorporates a filling slot, which requires the thrust load to be light. Figure 11.11 shows a double-row-type ball bearing.

This unit is, in effect, two single-row angular contact bearings built as a unit with the internal fit between balls and raceway fixed during assembly. As a result, fit and internal stiffness are not dependent upon mounting methods. These bearings usually have a known amount of internal preload, or compression, built in for maximum resistance to deflection under combined loads with thrust from either direction. As a result of

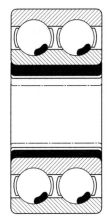

FIGURE 11.11.   Double-row type ball bearing.

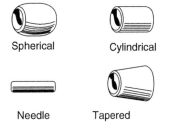

FIGURE 11.13.   Types of roller elements.

this compression prior to external loading, the bearings are very effective for radial loads where bearing deflection must be minimized.

Another double-row ball bearing is the i nternal self-aligning type, which is shown in Figure 11.12. It compensates for angular misalignment, which can be caused by errors in mounting, shaft deflection, misalignment, etc. This bearing supports moderate radial loads and limited thrust loads.

11.2.3. Roller

As with plain and ball bearings, roller bearings also may be classified by their ability to support radial, thrust, and combination loads. Note that combination load–supporting roller bearings are *not* called angular-contact bearings as they are with ball bearings. For example, the taper-roller bearing is a combination load–carrying bearing by virtue of the shape of its rollers.

Figure 11.13 shows the different types of roller elements used in these bearings. Roller elements are

classified as cylindrical, barrel, spherical, and tapered. Note that barrel rollers are called needle rollers when they are less than ¼" in diameter and have a relatively high ratio of length to diameter.

Cylindrical

Cylindrical bearings have solid or helically wound hollow cylindrically shaped rollers, which have an approximate length-diameter ratio ranging from 1:1 to 1:3. They normally are used for heavy radial loads beyond the capacities of comparably sized radial ball bearings.

Cylindrical bearings are especially useful for free axial movement of the shaft. The free ring may have a restraining flange to provide some restraint to endwise movement in one direction. Another configuration comes without a flange, which allows the bearing rings to be displaced axially.

Either the rolls or the roller path on the races may be slightly crowned to prevent edge loading under slight shaft misalignment. Low friction makes this bearing type suitable for fairly high speeds. Figure 11.14 shows a typical cylindrical roller bearing.

Figure 11.15 shows separable inner-ring cylindrical roller bearings. Figure 11.16 shows separable inner-ring cylindrical roller bearings with a different inner ring.

The roller assembly in Figure 11.15 is located in the outer ring with retaining rings. The inner ring can be omitted and the roller operated on hardened ground shaft surfaces.

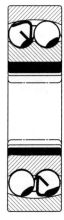

FIGURE 11.12.   Double-row internal self-aligning bearing.

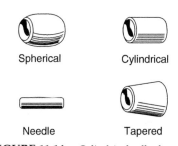

FIGURE 11.14.   Cylindrical roller bearing.

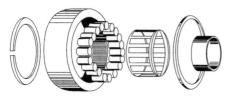

FIGURE 11.15. Separable inner-ring type cylindrical roller bearings.

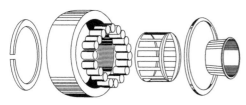

FIGURE 11.16. Separable inner-ring type roller bearings with different inner ring.

The style in Figure 11.16 is similar to the one in Figure 11.15, except the rib on the inner ring is different. This prohibits the outer ring from moving in a direction toward the rib.

Figure 11.17 shows separable inner-ring type cylindrical roller bearings with elimination of a retainer ring on one side.

The style shown in Figure 11.17 is similar to the two previous styles except for the elimination of a retainer ring on one side. It can carry small thrust loads in only *one* direction.

Needle-Type Cylindrical or Barrel

Needle-type cylindrical bearings (Figure 11.18) incorporate rollers that are symmetrical, with a length at least four times their diameter. They are sometimes referred to as barrel rollers. These bearings are most useful where space is limited and thrust-load support is not required. They are available with or without an inner race. If a shaft takes the place of an inner race, it must be hardened and ground. The full-complement type is used for high loads and oscillating or slow speeds. The cage type should be used for rotational motion.

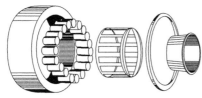

FIGURE 11.17. Separable inner-ring type cylindrical roller bearings with elimination of a retainer ring on one side.

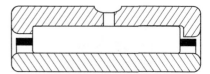

FIGURE 11.18. Needle bearings.

They come in both single-row and double-row mountings. As with all cylindrical roller bearings, the single-row mounting type has a low thrust capacity, but angular mounting of rolls in the double-row type permits its use for combined axial and thrust loads.

Spherical

Spherical bearings are usually furnished in a double-row mounting that is inherently self-aligning. Both rows of rollers have a common spherical outer raceway. The rollers are barrel-shaped with one end smaller to provide a small thrust to keep the rollers in contact with the center guide flange.

This type of roller bearing has a high radial and moderate-to-heavy thrust load-carrying capacity. It maintains this capability with some degree of shaft and bearing housing misalignment. While their internal self-aligning feature is useful, care should be taken in specifying this type of bearing to compensate for misalignment. Figure 11.19 shows a typical spherical roller bearing assembly. Figure 11.20 shows a series of spherical roller bearings for a given shaft size.

Tapered

Tapered bearings are used for heavy radial and thrust loads. They have straight tapered rollers, which are held in accurate alignment by means of a guide flange on the inner ring. Figure 11.21 shows a typical tapered-roller bearing. Figure 11.22

FIGURE 11.19. Spherical roller bearing assembly.

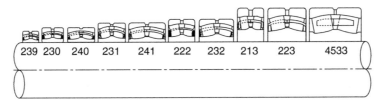

FIGURE 11.20. Series of spherical roller bearings for a given shaft size (available in several series).

shows necessary information to identify a taper-roller bearing. Figure 11.23 shows various types of tapered roller bearings.

True rolling occurs because they are designed so all elements in the rolling surface and the raceways intersect at a common point on the axis. The basic

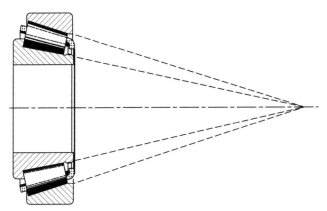

FIGURE 11.21. Tapered-roller bearing.

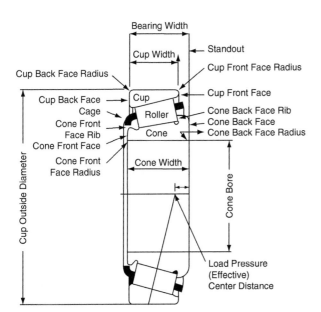

FIGURE 11.22. Information needed to identify a taper-roller bearing.

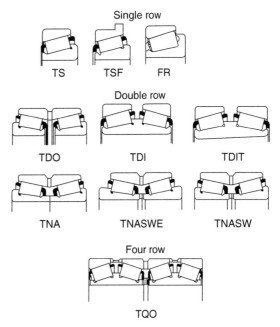

FIGURE 11.23. Various types of tapered roller bearings.

characteristic of these bearings is that if the apexes of the tapered working surfaces of both rollers and races were extended, they would coincide on the bearing axis. Where maximum system rigidity is required, they can be adjusted for a preload. These bearings are separable.

11.3. BEARING MATERIALS

Because two contacting metal surfaces are in motion in bearing applications, material selection plays a crucial role in their life. Properties of the materials used in bearing construction determine the amount of sliding friction that occurs, a key factor affecting bearing life. When two similar metals are in contact without the presence of adequate lubrication, friction is generally high, and the surfaces will seize (i.e., weld) at relatively low pressures or surface loads. However, certain combinations of materials support substantial loads without seizing or welding as a result of their low frictional qualities.

In most machinery, shafts are made of steel. Bearings are generally made of softer materials that have low frictional as well as sacrificial qualities when in contact with steel. A softer, sacrificial material is used for bearings because it is easier and cheaper to replace a worn bearing as opposed to a worn shaft. Common bearing materials are cast iron, bronze, and babbitt. Other less commonly used materials include wood, plastics, and other synthetics.

There are several important characteristics to consider when specifying bearing materials, including: (1) strength or the ability to withstand loads without plastic deformation; (2) ability to permit embedding of grit or dirt particles that are present in the lubricant; (3) ability to elastically deform in order to permit load distribution over the full bearing surface; (4) ability to dissipate heat and prevent hot spots that might seize; and (5) corrosion resistance.

11.3.1. Plain

As indicated above, dissimilar metals with low frictional characteristics are most suitable for plain bearing applications. With steel shafts, plain bearings made of bronze or babbitt are commonly used. Bronze is one of the harder bearing materials and is generally used for low speeds and heavy loads.

A plain bearing may sometimes be made of a combination of materials. The outer portion may be constructed of bronze, steel, or iron to provide the strength needed to provide a load-carrying capability. The bearing may be lined with a softer material such as babbitt to provide the sacrificial capability needed to protect the shaft.

11.3.2. Rolling Element

A specially developed steel alloy is used for an estimated 98% of all rolling element bearing uses. In certain special applications, however, materials such as glass, plastic, and other substances are sometimes used in rolling element construction.

Bearing steel is a high-carbon chrome alloy with high hardenability and good toughness characteristics in the hardened and drawn state. All load-carrying members of most rolling contact bearings are made with this steel.

Controlled procedures and practices are necessary to ensure specification of the proper alloy, maintain material cleanliness, and ensure freedom from defects—all of which affect bearing reliability. Alloying practices that conform to rigid specifications are required to reduce anomalies and inclusions that adversely affect a bearing's useful life. Magnaflux inspections ensure

that rolling elements are free from material defects and cracks. Light etching is used between rough and finish grinding processes to stop burning during heavy machining operations.

11.4. LUBRICATION

It is critical to consider lubrication requirements when specifying bearings. Factors affecting lubricants include relatively high speeds, difficulty in performing relubrication, nonhorizontal shafts, and applications where leakage cannot be tolerated. This section briefly discusses lubrication mechanisms and techniques for bearings.

11.4.1. Plain Bearings

In plain bearings, the lubricating fluid must be replenished to compensate for end leakage in order to maintain the bearings' load-carrying capacity. Pressure lubrication from a pump- or gravity-fed tank, or automatic lubricating devices such as oil rings or oil disks, are provided in self-contained bearings. Another means of lubrication is to submerge the bearing (in particular, thrust bearings for vertical shafts) in an oil bath.

Lubricating Fluids

Almost any process fluid may be used to lubricate plain bearings if parameters such as viscosity, corrosive action, toxicity, change in state (where a liquid is close to its boiling point), and in the case of a gaseous fluid, its compressibility, are appropriate for the application. Fluid-film journal and thrust bearings have run successfully, for example, on water, kerosene, gasoline, acid, liquid refrigerants, mercury, molten metals, and a wide variety of gases.

Gases, however, lack the cooling and boundary-lubrication capabilities of most liquid lubricants. Therefore, the operation of self-acting gas bearings is restricted by start/stop friction and wear. If start/stop is performed under load, then the design is limited to about seven pounds per square inch (lb/in^2) or 48 kiloNewtons per square meter (kN/m^2) on the projected bearing area, depending upon the choice of materials. In general, the materials used for these bearings are those of dry rubbing bearings (e.g., either a hard/hard combination such as ceramics with or without a molecular layer of boundary lubricant, or a hard/soft combination with a plastic surface).

Externally pressurized gas journal bearings have the same principle of operation as hydrostatic liquid-lubricated bearings. Any clear gas can be used, but many of the design charts are based on air. There are three forms of external flow restrictors in use with these bearings: pocketed (simple) orifice, unpocketed (annular) orifice, and slot.

State of Lubrication

Fluid or complete lubrication, the condition where the surfaces are completely separated by a fluid film, provides the lowest friction losses and prevents wear.

The semifluid lubrication state exists between the journal and bearing when a load-carrying fluid film does not form to separate the surfaces. This occurs at comparatively low speed with intermittent or oscillating motion, heavy load, and insufficient oil supply to the bearing. Semifluid lubrication also may exist in thrust bearings with fixed parallel-thrust collars; guide bearings of machine tools; bearings with plenty of lubrication that have a bent or misaligned shaft; or where the bearing surface has improperly arranged oil grooves. The coefficient of friction in such bearings may range from 0.02 to 0.08.

In situations where the bearing is well lubricated, but the speed of rotation is very slow or the bearing is barely greasy, boundary lubrication takes place. In this situation, which occurs in bearings when the shaft is starting from rest, the coefficient of friction may vary from 0.08 to 0.14.

A bearing may run completely dry in exceptional cases of design or with a complete failure of lubrication. Depending on the contacting surface materials, the coefficient of friction will be between 0.25 and 0.40.

11.4.2. Rolling Element Bearings

Rolling element bearings also need a lubricant to meet or exceed their rated life. In the absence of high temperatures, however, excellent performance can be obtained with a very small quantity of lubricant. Excess lubricant causes excessive heating, which accelerates lubricant deterioration.

The most popular type of lubrication is the sealed grease ball-bearing cartridge. Grease is commonly used for lubrication because of its convenience and minimum maintenance requirements. A high-quality lithium-based NLGI 2 grease is commonly used for temperatures up to 180°F (82°C). Grease must be replenished and relubrication intervals in hours of operation are dependent on temperature, speed, and bearing size. Table 11.9 is a general guide to the time after which it is advisable to add a small amount of grease.

Some applications, however, cannot use the cartridge design, for example, when the operating environment

TABLE 11.9. Ball-Bearing Grease Relubrication Intervals (Hours of Operation)

Bearing Bore, mm	Bearing Speed, rpm				
	5,000	3,600	1,750	1,000	200
10	8,700	12,000	25,000	44,000	220,000
20	5,500	8,000	17,000	30,000	150,000
30	4,000	6,000	13,000	24,000	127,000
40	2,800	4,500	11,000	20,000	111,000
50		3,500	9,300	18,000	97,000
60		2,600	8,000	16,000	88,000
70			6,700	14,000	81,000
80			5,700	12,000	75,000
90			4,800	11,000	70,000
100			4,000	10,000	66,000

Source: Theodore Baumeister, ed., *Marks' Standard Handbook for Mechanical Engineers,* 8th ed. (New York: McGraw-Hill, 1978).

is too hot for the seals. Another example is when minute leaks or the accumulation of traces of dirt at the lip seals cannot be tolerated (e.g., food processing machines). In these cases, bearings with specialized sealing and lubrication systems must be used.

In applications involving high speed, oil lubrication is typically required. Table 11.10 is a general guide in selecting oil of the proper viscosity for these bearings. For applications involving high-speed shafts, bearing selection must take into account the inherent speed limitations of certain bearing designs, cooling needs, and lubrication issues such as churning and aeration suppression. A typical case is the effect of cage design and roller-end thrust-flange contact on the lubrication requirements in taper roller

TABLE 11.10. Oil Lubrication Viscosity (ISO Identification Numbers)

Bearing Bore, mm	Bearing Speed, rpm				
	10,000	3,600	1,800	600	50
4–7	68	150	220		
10–20	32	68	150	220	460
25–45	10	32	68	150	320
50–70	7	22	68	150	320
75–90	3	10	22	68	220
100	3	7	22	68	220

Source: Theodore Baumeister, ed. *Marks' Standard Handbook for Mechanical Engineers,* 8th ed. (New York: McGraw-Hill, 1978).

bearings. These design elements limit the speed and the thrust load that these bearings can endure. As a result, it is important always to refer to the bearing manufacturer's instructions on load-carrying design and lubrication specifications.

11.5. INSTALLATION AND GENERAL HANDLING PRECAUTIONS

Proper handling and installation practices are crucial to optimal bearing performance and life. In addition to standard handling and installation practices, the issue of emergency bearing substitutions is an area of critical importance. If substitute bearings are used as an emergency means of getting a machine back into production quickly, the substitution should be entered into the historical records for that machine. This documents the temporary change and avoids the possibility of the substitute bearing becoming a *permanent* replacement. This error can be extremely costly, particularly if the incorrectly specified bearing continually fails prematurely. It is important that an inferior substitute be removed as soon as possible and replaced with the originally specified bearing.

11.5.1. Plain Bearing Installation

It is important to keep plain bearings from shifting sideways during installation and to ensure an axial position that does not interfere with shaft fillets. Both of these can be accomplished with a locating lug at the parting line.

Less frequently used is a dowel in the housing, which protrudes partially into a mating hole in the bearing.

The distance across the outside parting edges of a plain bearing are manufactured slightly greater than the housing bore diameter. During installation, a light force is necessary to snap it into place and, once installed, the bearing stays in place because of the pressure against the housing bore.

It is necessary to prevent a bearing from spinning during operation, which can cause a catastrophic failure. Spinning is prevented by what is referred to as "crush." Bearings are slightly longer circumferentially than their mating housings and upon installation this excess length is elastically deformed or "crushed." This sets up a high radial contact pressure between the bearing and housing, which ensures good back contact for heat conduction and, in combination with the bore-to-bearing friction, prevents spinning. It is important that *under no circumstances* should the

bearing parting lines be filed or otherwise altered to remove the crush.

11.5.2. Roller Bearing Installation

A basic rule of rolling element bearing installation is that one ring must be mounted on its mating shaft or in its housing with an interference fit to prevent rotation. This is necessary because it is virtually impossible to prevent rotation by clamping the ring axially.

Mounting Hardware

Bearings come as separate parts that require mounting hardware or as premounted units that are supplied with their own housings, adapters, and seals.

Bearing Mountings

Typical bearing mountings, which are shown in Figure 11.24, locate and hold the shaft axially and allow for thermal expansion and/or contraction of the shaft. Locating and holding the shaft axially is generally accomplished by clamping one of the bearings on the shaft so that all machine parts remain in proper relationship dimensionally. The inner ring is locked axially relative to the shaft by locating it between a shaft shoulder and some type of removable locking device once the inner ring has a tight fit. Typical removable locking devices are specially designed nuts, which are used for a through shaft, and clamp plates, which are commonly used when the bearing is

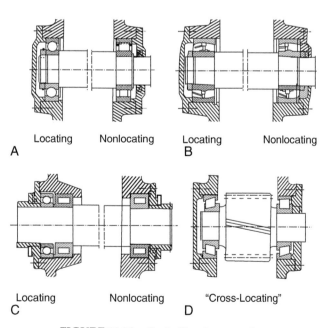

Locating   Nonlocating   Locating   Nonlocating
A   B

Locating   Nonlocating   "Cross-Locating"
C   D

FIGURE 11.24.   Typical bearing mounting.

mounted on the end of the shaft. For the locating or held bearing, the outer ring is clamped axially, usually between housing shoulders or end-cap pilots.

With general types of cylindrical roller bearings, shaft expansion is absorbed internally simply by allowing one ring to move relative to the other (Figure 11.24a and 11.24c, nonlocating positions). The advantage of this type of mounting is that both inner and outer rings may have a tight fit, which is desirable or even mandatory if significant vibration and/or imbalance exists in addition to the applied load.

Premounted Bearing

Premounted bearings, referred to as pillow-block and flanged-housing mountings, are of considerable importance to millwrights. They are particularly adaptable to "line-shafting" applications, which are a series of ball and roller bearings supplied with their own housings, adapters, and seals. Premounted bearings come with a wide variety of flange mountings, which permit them to be located on faces parallel or perpendicular to the shaft axis. Figure 11.25 shows a typical pillow block. Figure 11.26 shows a flanged bearing unit.

Inner races can be mounted directly on ground shafts or can be adapter-mounted to "drill-rod" or to commercial-shafting. For installations sensitive to imbalance and vibration, the use of accurately ground shaft seats is recommended.

FIGURE 11.25.   Typical pillow block.

FIGURE 11.26.   Flanged bearing unit.

Most pillow-block designs incorporate self-aligning bearing types and do not require the precision mountings utilized with other bearing installations.

Mounting Techniques

When mounting or dismounting a roller bearing, the most important thing to remember is to apply the mounting or dismounting force to the side face of the ring with the interference fit. This force should not pass from one ring to the other through the ball or roller set, as internal damage can easily occur.

Mounting tapered-bore bearings can be accomplished simply by tightening the locknut or clamping plate. This locates it on the shaft until the bearing is forced the proper distance up the taper. This technique requires a significant amount of force, particularly for large bearings.

Cold Mounting

Cold mounting, or force fitting a bearing onto a shaft or into a housing, is appropriate for all small bearings (i.e., 4" bore and smaller). The force, however, must be applied as uniformly as possible around the side face of the bearing and to the ring to be press-fitted. Mounting fixtures, such as a simple piece of tubing of appropriate size and a flat plate, should be used. It is not appropriate to use a drift and hammer to force the bearing on, which will cause the bearing to cock. It is possible to apply force by striking the plate with a hammer or by an arbor press. However, before forcing the bearing on the shaft, a coat of light oil should be applied to the bearing seat on the shaft and the bearing bores. All sealed and shielded ball bearings should be cold mounted in this manner.

Temperature Mounting

The simplest way to mount any open straight-bore bearing regardless of its size is temperature mounting, which entails heating the entire bearing, pushing it on its seat, and holding it in place until it cools enough to grip the shaft. The housing may be heated if practical for tight outside-diameter fits; however, temperatures should not exceed 250°F. If heating of the housing is not practical, the bearing may be cooled with dry ice. The risk of cooling is that if the ambient conditions are humid, moisture is introduced and there is a potential for corrosion in the future. Acceptable ways of heating bearings are by hot plate, temperature-controlled oven, induction heaters, and hot-oil bath.

With the hot plate method, the bearing is simply laid on the plate until it reaches the approved temperature, using a pyrometer or Tempilstik to make certain it is not overheated. Difficulty in controlling the temperature is the major disadvantage of this method.

When using a temperature-controlled oven, the bearings should be left in the oven long enough to heat thoroughly, but they should never be left overnight.

The use of induction heaters is a quick method of heating bearings. However, some method of measuring the ring temperature (e.g., pyrometer or a Tempilstik) must be used or damage to the bearing may occur. Note that bearings must be demagnetized after the use of this method.

The use of a hot-oil bath is the most practical means of heating larger bearings. Disadvantages are that the temperature of the oil is hard to control and may ignite or overheat the bearing. The use of a soluble oil and water mixture (10 to 15% oil) can eliminate these problems and still attain a boiling temperature of 210°F. The bearing should be kept off the bottom of the container by a grate or screen located several inches off the bottom. This is important to allow contaminants to sink to the bottom of the container and away from the bearing.

Dismounting

Commercially available bearing pullers allow rolling element bearings to be dismounted from their seats without damage. When removing a bearing, force should be applied to the ring with the tight fit, although sometimes it is necessary to use supplementary plates or fixtures. An arbor press is equally effective at removing smaller bearings as well as mounting them.

Ball Installation

Figure 11.27 shows the ball installation procedure for roller bearings. The designed load carrying capacity of Conrad-type bearings is determined by the number of balls that can be installed between the rings. Ball installation is accomplished by the following procedure:

- Slip the inner ring slightly to one side;
- Insert balls into the gap, which centers the inner ring as the balls are positioned between the rings;

1. The inner ring is moved to one side

2. Balls are installed in the gap

3. The inner ring is centered to the balls and equality positioned in place

4. A retainer is installed

FIGURE 11.27.   Ball installation procedures.

- Place stamped retainer rings on either side of the balls before riveting together. This positions the balls equidistant around the bearing.

11.5.3. General Roller-Element Bearing Handling Precautions

In order for rolling element bearings to achieve their design life and perform with no abnormal noise, temperature rise, or shaft excursions, the following precautions should be taken:

- Always select the best bearing design for the application and not the cheapest. The cost of the original bearing is usually small by comparison to the costs of replacement components and the downtime in production when premature bearing failure occurs because an inappropriate bearing was used.
- If in doubt about bearings and their uses, consult the manufacturer's representative and the product literature.
- Bearings should always be handled with great care. Never ignore the handling and installation instructions from the manufacturer.
- Always work with clean hands, clean tools, and the cleanest environment available.
- Never wash or wipe bearings prior to installation unless the instructions specifically state that this should be done. Exceptions to this rule are when oil-mist lubrication is to be used and the slushing compound has hardened in storage or is blocking lubrication holes in the bearing rings. In this situation, it is best to clean the bearing with kerosene or other appropriate petroleum-based solvent. The other exception is if the slushing compound has been contaminated with dirt or foreign matter before mounting.
- Keep new bearings in their greased paper wrappings until they are ready to install. Place unwrapped bearings on clean paper or lint-free cloth if they cannot be kept in their original containers. Wrap bearings in clean, oil-proof paper when not in use.
- Never use wooden mallets, brittle or chipped tools, or dirty fixtures and tools when bearings are being installed.
- Do not spin bearings (particularly dirty ones) with compressed service air.
- Avoid scratching or nicking bearing surfaces. Care must be taken when polishing bearings with emery cloth to avoid scratching.
- Never strike or press on race flanges.
- Always use adapters for mounting that ensure uniform steady pressure rather than hammering on a drift or sleeve. *Never* use brass or bronze drifts to

install bearings as these materials chip very easily into minute particles that will quickly damage a bearing.

- Avoid cocking bearings onto shafts during installation.
- Always inspect the mounting surface on the shaft and housing to insure that there are no burrs or defects.
- When bearings are being removed, clean housings and shafts before exposing the bearings.
- Dirt is abrasive and detrimental to the designed life span of bearings.
- Always treat used bearings as if they are new, especially if they are to be reused.
- Protect dismantled bearings from moisture and dirt.
- Use clean filtered, water-free Stoddard's solvent or flushing oil to clean bearings.
- When heating is used to mount bearings onto shafts, follow the manufacturer's instructions.
- When assembling and mounting bearings onto shafts, *never* strike the outer race or press on it to force the inner race. Apply the pressure on the inner race only. When dismantling, follow the same procedure.
- Never press, strike, or otherwise force the seal or shield on factory-sealed bearings.

11.6. BEARING FAILURES, DEFICIENCIES, AND THEIR CAUSES

The general classifications of failures and deficiencies requiring bearing removal are overheating, vibration, turning on the shaft, binding of the shaft, noise during operation, and lubricant leakage. Table 11.11 is a troubleshooting guide that lists the common causes for each of these failures and deficiencies. As indicated by the causes of failure listed, bearing failures are rarely caused by the bearing itself.

Many abnormal vibrations generated by actual bearing problems are the result of improper sizing of the bearing liner or improper lubrication. However, numerous machine and process-related problems generate abnormal vibration spectra in bearing data. The primary contributors to abnormal bearing signatures are: (1) imbalance, (2) misalignment, (3) rotor instability, (4) excessive or abnormal loads, and (5) mechanical looseness.

Defective bearings that leave the manufacturer are very rare, and it is estimated that defective bearings contribute to only 2% of total failures. The failure is invariably linked to symptoms of misalignment, imbalance, resonance, and lubrication—or the lack of it. Most of the problems that occur result from

the following reasons: dirt, shipping damage, storage and handling, poor fit resulting in installation damage, wrong type of bearing design, overloading, improper lubrication practices, misalignment, bent shaft, imbalance, resonance, and soft foot. Anyone of these conditions will eventually destroy a bearing—two or more of these problems can result in disaster!

Although most industrial machine designers provide adequate bearings for their equipment, there are some cases in which bearings are improperly designed, manufactured, or installed at the factory. Usually, however, the trouble is caused by one or more of the following reasons: (1) improper on-site bearing selection and/or installation, (2) incorrect grooving, (3) unsuitable surface finish, (4) insufficient clearance, (5) faulty relining practices, (6) operating conditions, (7) excessive operating temperature, (8) contaminated oil supply, and (9) oil-film instability.

11.6.1. Improper Bearing Selection and/or Installation

There are several things to consider when selecting and installing bearings, including the issue of interchangeability, materials of construction, and damage that might have occurred during shipping, storage, and handling.

Interchangeability

Because of the standardization in envelope dimensions, precision bearings were once regarded as interchangeable among manufacturers. This interchangeability has since been considered a major cause of failures in machinery, and the practice should be used with extreme caution.

Most of the problems with interchangeability stem from selecting and replacing bearings based only on bore size and outside diameters. Often, very little consideration is paid to the number of rolling elements contained in the bearings. This can seriously affect the operational frequency vibrations of the bearing and may generate destructive resonance in the host machine or adjacent machines.

More bearings are destroyed during their installation than fail in operation. Installation with a heavy hammer is the usual method in many plants. Heating the bearing with an oxyacetylene burner is another classical method. However, the bearing does not stand a chance of reaching its life expectancy when either of these installation practices are used. The bearing manufacturer's installation instructions should always be followed.

TABLE 11.11   Troubleshooting Guide

Overheating	Vibration	Turning on the Shaft	Binding of the Shaft	Noisy Bearing	Lubricant Leakage
Inadequate or insufficient lubrication	Dirt or chips in bearing	Growth of race due to overheating	Lubrication breakdown	Lubrication breakdown	Overfilling of lubricant
Excessive lubrication	Fatigued race or rolling elements	Fretting wear	Contamination by abrasive or corrosive materials	Inadequate lubrication	Grease churning due to too soft a consistency
Grease liquefaction or aeration	Rotor imbalance	Improper initial fit	Housing distortion or out-of-round pinching bearing	Pinched bearing	Grease deterioration due to excessive operating temperature
Oil foaming	Out-of-round shaft	Excessive shaft deflection	Uneven shimming of housing with loss of clearance	Contamination	Operating beyond grease life
Abrasion or corrosion due to contaminants	Race misalignment	Initial coarse finish on shaft	Tight rubbing seals	Seal rubbing	Seal wear
Housing distortion due to warping or out-of-round	Housing resonance	Seal rub on inner race	Preloaded bearings	Bearing slipping on shaft or in housing	Wrong shaft attitude (bearing seals designed for horizontal mounting only)
Seal rubbing or failure	Cage wear		Cocked races	Flatted roller or ball	Seal failure
Inadequate or blocked scavenge oil passages	Flats on races or rolling elements		Loss of clearance due to excessive adapter tightening	Brinelling due to assembly abuse, handling, or shock loads	Clogged breather
Inadequate bearing clearance or bearing preload	Race turning		Thermal shaft expansion	Variation in size of rolling elements	Oil foaming due to churning or air flow through housing
Race turning	Excessive clearance			Out-of-round or lobular shaft	Gasket (O-ring) failure or misapplication
Cage wear	Corrosion	False brinelling or indentation of races		Housing bore waviness	Porous housing or closure
	Electrical arcing			Chips or scores under bearing seat	Lubricator set at the wrong flow rate
	Mixed rolling element diameters				
	Out-of-square rolling paths in races				

Source: Integrated Systems Inc.

Shipping Damage

Bearings and the machinery containing them should be properly packaged to avoid damage during shipping. However, many installed bearings are exposed to vibrations, bending, and massive shock loadings through bad handling practices during shipping. It has been estimated that approximately 40% of newly received machines have "bad" bearings.

Because of this, all new machinery should be thoroughly inspected for defects before installation. Acceptance criteria should include guidelines that clearly define acceptable design/operational specifications. This practice pays big dividends by increasing productivity and decreasing unscheduled downtime.

Storage and Handling

Storeroom and other appropriate personnel must be made aware of the potential havoc they can cause by their mishandling of bearings. Bearing failure often starts in the storeroom rather than the machinery. Premature opening of packages containing bearings should be avoided whenever possible. If packages must be opened for inspection, they should be protected from exposure to harmful dirt sources and then resealed in the original wrappings. The bearing should never be dropped or bumped as this can cause shock loading on the bearing surface.

Incorrect Placement of Oil Grooves

Incorrectly placed oil grooves can cause bearing failure. Locating the grooves in high-pressure areas causes them to act as pressure-relief passages. This interferes with the formation of the hydrodynamic film, resulting in reduced load-carrying capability.

Unsuitable Surface Finish

Smooth surface finishes on both the shaft and the bearing are important to prevent surface variations from penetrating the oil film. Rough surfaces can cause scoring, overheating, and bearing failure. The smoother the finishes, the closer the shaft may approach the bearing without danger of surface contact. Although important in all bearing applications, surface finish is critical with the use of harder bearing materials such as bronze.

Insufficient Clearance

There must be sufficient clearance between the journal and bearing in order to allow an oil film to form. An average diametral clearance of 0.001 inches per inch of shaft diameter is often used. This value may be adjusted depending on the type of bearing material, the load, speed, and the accuracy of the shaft position desired.

Faulty Relining

Faulty relining occurs primarily with babbitted bearings rather than precision machine-made inserts. Babbitted bearings are fabricated by a pouring process that should be performed under carefully controlled conditions. Some reasons for faulty relining are: (1) improper preparation of the bonding surface, (2) poor pouring technique, (3) contamination of babbitt, and (4) pouring bearing to size with journal in place.

Operating Conditions

Abnormal operating conditions or neglect of necessary maintenance precautions cause most bearing failures. Bearings may experience premature and/or catastrophic failure on machines that are operated heavily loaded, speeded up, or being used for a purpose not appropriate for the system design. Improper use of lubricants can also result in bearing failure. Some typical causes of premature failure include: (1) excessive operating temperatures, (2) foreign material in the lubricant supply, (3) corrosion, (4) material fatigue, and (5) use of unsuitable lubricants.

Excessive Temperatures

Excessive temperatures affect the strength, hardness, and life of bearing materials. Lower temperatures are required for thick babbitt liners than for thin precision babbitt inserts. Not only do high temperatures affect bearing materials, they also reduce the viscosity of the lubricant and affect the thickness of the film, which affects the bearing's load-carrying capacity. In addition, high temperatures result in more rapid oxidation of the lubricating oil, which can result in unsatisfactory performance.

Dirt and Contamination in Oil Supply

Dirt is one of the biggest culprits in the demise of bearings. Dirt makes its appearance in bearings in many subtle ways, and it can be introduced by bad work habits. It also can be introduced through lubricants that have been exposed to dirt, a problem that is responsible for approximately half of bearing failures throughout the industry.

To combat this problem, soft materials such as babbitt are used when it is known that a bearing will be exposed to abrasive materials. Babbitt metal embeds hard particles, which protects the shaft against abrasion. When harder materials are used in the presence of abrasives, scoring and galling occurs as a result of abrasives caught between the journal and bearing.

In addition to the use of softer bearing materials for applications where abrasives may potentially be present, it is important to properly maintain filters and breathers, which should regularly be examined. In order to avoid oil supply contamination, foreign material that collects at the bottom of the bearing sump should be removed on a regular basis.

Oil-Film Instability

The primary vibration frequency components associated with fluid-film bearings problems are in fact displays of turbulent or nonuniform oil film. Such instability problems are classified as either *oil whirl* or *oil whip* depending on the severity of the instability.

Machine-trains that use sleeve bearings are designed based on the assumption that rotating elements and shafts operate in a balanced and, therefore, centered position. Under this assumption, the machine-train shaft will operate with an even, concentric oil film between the shaft and sleeve bearing.

For a normal machine, this assumption is valid after the rotating element has achieved equilibrium. When the forces associated with rotation are in balance, the rotating element will center the shaft within the bearing. However, several problems directly affect

this self-centering operation. First, the machine-train must be at designed operating speed and load to achieve equilibrium. Second, any imbalance or abnormal operation limits the machine-train's ability to center itself within the bearing.

A typical example is a steam turbine. A turbine must be supported by auxiliary running gear during startup or shutdown to prevent damage to the sleeve bearings. The lower speeds during the startup and shutdown phase of operation prevent the self-centering ability of the rotating element. Once the turbine has achieved full speed and load, the rotating element and shaft should operate without assistance in the center of the sleeve bearings.

Oil Whirl

In an abnormal mode of operation, the rotating shaft may not hold the centerline of the sleeve bearing. When this happens, an instability called oil whirl occurs. Oil whirl is an imbalance in the hydraulic forces within a sleeve bearing. Under normal operation, the hydraulic forces such as velocity and pressure are balanced. If the rotating shaft is offset from the true centerline of the bearing, instability occurs.

As Figure 11.28 illustrates, a restriction is created by the offset. This restriction creates a high pressure

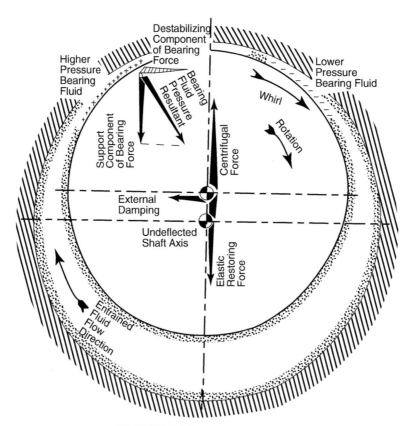

FIGURE 11.28.   Oil whirl, oil whip.

and another force vector in the direction of rotation. Oil whirl accelerates the wear and failure of the bearing and bearing support structure.

Oil Whip

The most severe damage results if the oil whirl is allowed to degrade into oil whip. Oil whip occurs when the clearance between the rotating shaft and sleeve bearing is allowed to close to a point approaching actual metal-to-metal contact. When the clearance between the shaft and bearing approaches contact, the oil film is no longer free to flow between the shaft and bearing. As a result, the oil film is forced to change directions. When this occurs, the high-pressure area created in the region behind the shaft is greatly increased. This vortex of oil increases the abnormal force vector created by the off-set and rotational force to the point that metal-to-metal contact between the shaft and bearing occurs. In almost all instances where oil whip is allowed, severe damage to the sleeve bearing occurs.

CHAPTER

12

Compressors[*]

A compressor is a machine that is used to increase the pressure of a gas or vapor. They can be grouped into two major classifications: centrifugal and positive displacement. This section provides a general discussion of these types of compressors.

12.1. CENTRIFUGAL

In general, the centrifugal designation is used when the gas flow is radial and the energy transfer is predominantly due to a change in the centrifugal forces acting on the gas. The force utilized by the centrifugal compressor is the same as that utilized by centrifugal pumps.

In a centrifugal compressor, air or gas at atmospheric pressure enters the eye of the impeller. As the impeller rotates, the gas is accelerated by the rotating element within the confined space that is created by the volute of the compressor's casing. The gas is compressed as more gas is forced into the volute by the impeller blades. The pressure of the gas increases as it is pushed through the reduced free space within the volute.

As in centrifugal pumps, there may be several stages to a centrifugal air compressor. In these multistage units, a progressively higher pressure is produced by each stage of compression.

12.1.1. Configuration

The actual dynamics of centrifugal compressors are determined by their design. Common designs are: overhung or cantilever, centerline, and bullgear.

*Source: Ricky Smith and Keith Mobley, *Industrial Machinery Repair: Best Maintenance Practices Pocket Guide* (Boston: Butterworth–Heinemann, 2003), pp. 133–179.

Overhung or Cantilever

The cantilever design is more susceptible to process instability than centerline centrifugal compressors. Figure 12.1 illustrates a typical cantilever design.

The overhung design of the rotor (i.e., no outboard bearing) increases the potential for radical shaft deflection. Any variation in laminar flow, volume, or load of the inlet or discharge gas forces the shaft to bend or deflect from its true centerline. As a result, the mode shape of the shaft must be monitored closely.

Centerline

Centerline designs, such as horizontal and vertical split-case, are more stable over a wider operating range, but should not be operated in a variable-demand system. Figure 12.2 illustrates the normal airflow pattern through a horizontal split-case compressor. Inlet air enters the first stage of the compressor, where pressure and velocity increases occur. The partially compressed air is routed to the second stage where the velocity and pressure are increased further. Adding additional stages until the desired final discharge pressure is achieved can continue this process.

Two factors are critical to the operation of these compressors: impeller configuration and laminar flow, which must be maintained through all of the stages.

The impeller configuration has a major impact on stability and operating envelope. There are two impeller configurations: in-line and back-to-back, or opposed. With the in-line design, all impellers face in the same direction. With the opposed design, impeller direction is reversed in adjacent stages.

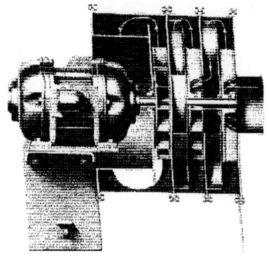

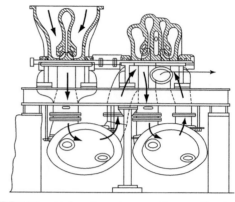

FIGURE 12.2. Airflow through a centerline centrifugal compressor.

FIGURE 12.1. Cantilever centrifugal compressor is susceptible to instability.

In-Line

A compressor with all impellers facing in the same direction generates substantial axial forces. The axial pressures generated by each impeller for all the stages are additive. As a result, massive axial loads are transmitted to the fixed bearing. Because of this load, most of these compressors use either a Kingsbury thrust bearing or a balancing piston to resist axial thrusting. Figure 12.3 illustrates a typical balancing piston.

All compressors that use in-line impellers must be monitored closely for axial thrusting. If the compressor is subjected to frequent or constant unloading, the axial clearance will increase due to this thrusting cycle.

Ultimately, this frequent thrust loading will lead to catastrophic failure of the compressor.

Opposed

By reversing the direction of alternating impellers, the axial forces generated by each impeller or stage can be minimized. In effect, the opposed impellers tend to cancel the axial forces generated by the preceding stage. This design is more stable and should not generate measurable axial thrusting. This allows these units to contain a normal float and fixed rolling-element bearing.

Bullgear

The bullgear design uses a direct-driven helical gear to transmit power from the primary driver to a series of pinion-gear-driven impellers that are located around the circumference of the bullgear. Figure 12.4 illustrates a typical bullgear compressor layout.

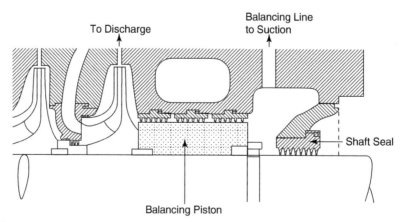

FIGURE 12.3. Balancing piston resists axial thrust from the in-line impeller design of a centerline centrifugal compressor.

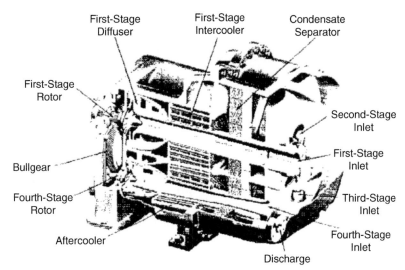

First-Stage Diffuser

First-Stage Intercooler

Condensate Separator

First-Stage Rotor

Bullgear

Fourth-Stage Rotor

Aftercooler

Second-Stage Inlet

First-Stage Inlet

Third-Stage Inlet

Fourth-Stage Inlet

Discharge

FIGURE 12.4. Bullgear centrifugal compressor.

The pinion shafts are typically a cantilever-type design that has an enclosed impeller on one end and a tilting-pad bearing on the other. The pinion gear is between these two components. The number of impeller-pinions (i.e., stages) varies with the application and the original equipment vendor. However, all bullgear compressors contain multiple pinions that operate in series.

Atmospheric air or gas enters the first-stage pinion, where the pressure is increased by the centrifugal force created by the first-stage impeller. The partially compressed air leaves the first stage, passes through an intercooler, and enters the second-stage impeller. This process is repeated until the fully compressed air leaves through the final pinion-impeller, or stage.

Most bullgear compressors are designed to operate with a gear speed of 3,600 rpm. In a typical four-stage compressor, the pinions operate at progressively higher speeds. A typical range is between 12,000 rpm (first stage) and 70,000 rpm (fourth stage).

Because of their cantilever design and pinion rotating speeds, bullgear compressors are extremely sensitive to variations in demand or downstream pressure changes. Because of this sensitivity, their use should be limited to baseload applications.

Bullgear compressors are not designed for, nor will they tolerate, load-following applications. They should not be installed in the same discharge manifold with positive-displacement compressors, especially reciprocating compressors. The standing-wave pulses created by many positive-displacement compressors create enough variation in the discharge manifold to cause potentially serious instability.

In addition, the large helical gear used for the bullgear creates an axial oscillation or thrusting that contributes to instability within the compressor. This axial movement is transmitted throughout the machine-train.

12.2. PERFORMANCE

The physical laws of thermodynamics, which define their efficiency and system dynamics, govern compressed-air systems and compressors. This section discusses both the first and second laws of thermodynamics, which apply to all compressors and compressed-air systems. Also applying to these systems are the Ideal Gas Law and the concepts of pressure and compression.

12.2.1. First Law of Thermodynamics

This law states that energy cannot be created or destroyed during a process, such as compression and delivery of air or gas, although it may change from one form of energy to another. In other words, whenever a quantity of one kind of energy disappears, an exactly equivalent total of other kinds of energy must be produced. This is expressed for a steady-flow open system such as a compressor by the following relationship:

Net energy added to system as heat and work
+ stored energy of mass entering system
− stored energy of mass leaving system = 0

12.2.2. Second Law of Thermodynamics

The second law of thermodynamics states that energy exists at various levels and is available for use only if it can move from a higher to a lower level. For example, it is impossible for any device to operate in a cycle and produce work while exchanging heat only with bodies at a single fixed temperature. In thermodynamics a measure of the unavailability of energy has been devised and is known as entropy. As a measure of unavailability, entropy increases as a system loses heat, but it remains constant when there is no gain or loss of heat as in an adiabatic process. It is defined by the following differential equation:

$$dQ\, dS = T$$

where:

$$T = \text{Temperature (Fahrenheit)}$$

$$Q = \text{Heat added (BTU)}$$

12.2.3. Pressure/Volume/Temperature (PVT) Relationship

Pressure, temperature, and volume are properties of gases that are completely interrelated. Boyle's Law and Charles' Law may be combined into one equation that is referred to as the Ideal Gas Law. This equation is always true for ideal gases and is true for real gases under certain conditions.

$$\frac{P_1 V_1}{T_1} = \frac{P_2 V_2}{T_2}$$

For air at room temperature, the error in this equation is less than 1% for pressures as high as 400 psia. For air at one atmosphere of pressure, the error is less than 1% for temperatures as low as –200°F. These error factors will vary for different gases.

12.2.4. Pressure/Compression

In a compressor, pressure is generated by pumping quantities of gas into a tank or other pressure vessel. Progressively increasing the amount of gas in the confined or fixed-volume space increases the pressure. The effects of pressure exerted by a confined gas result from the force acting on the container walls. This force is caused by the rapid and repeated bombardment from the enormous number of molecules that are present in a given quantity of gas.

Compression occurs when the space is decreased between the molecules. Less volume means that each particle has a shorter distance to travel, thus proportionately more collisions occur in a given span of time, resulting in a higher pressure. Air compressors are designed to generate particular pressures to meet specific application requirements.

12.2.5. Other Performance Indicators

The same performance indicators as those for centrifugal pumps or fans govern centrifugal compressors.

Installation

Dynamic compressors seldom pose serious foundation problems. Since moments and shaking forces are not generated during compressor operation, there are no variable loads to be supported by the foundation. A foundation or mounting of sufficient area and mass to maintain compressor level and alignment and to assure safe soil loading is all that is required. The units may be supported on structural steel if necessary. The principles defined for centrifugal pumps also apply to centrifugal compressors.

It is necessary to install pressure-relief valves on most dynamic compressors to protect them due to restrictions placed on casing pressure, power input, and to keep out of the compressor's surge range. Always install a valve capable of bypassing the full-load capacity of the compressor between its discharge port and the first isolation valve.

Operating Methods

The acceptable operating envelope for centrifugal compressors is very limited. Therefore, care should be taken to minimize any variation in suction supply, backpressure caused by changes in demand, and frequency of unloading. The operating guidelines provided in the compressor vendor's O&M manual should be followed to prevent abnormal operating behavior or premature wear or failure of the system.

Centrifugal compressors are designed to be baseloaded and may exhibit abnormal behavior or chronic reliability problems when used in a load-following mode of operation. This is especially true of bullgear and cantilever compressors. For example, a one-psig change in discharge pressure may be enough to cause catastrophic failure of a bullgear compressor. Variations in demand or backpressure on a cantilever design can cause the entire rotating element and its shaft to flex. This not only affects the compressor's efficiency, but also accelerates wear and may lead to premature shaft or rotor failure.

All compressor types have moving parts, high noise levels, high pressures, and high-temperature cylinder and discharge-piping surfaces.

12.3. POSITIVE DISPLACEMENT

Positive-displacement compressors can be divided into two major classifications: rotary and reciprocating.

12.3.1. Rotary

The rotary compressor is adaptable to direct drive by the use of induction motors or multicylinder gasoline or diesel engines. These compressors are compact, relatively inexpensive, and require a minimum of operating attention and maintenance. They occupy a fraction of the space and weight of a reciprocating machine having equivalent capacity.

Configuration

Rotary compressors are classified into three general groups: sliding vane, helical lobe, and liquid-seal ring.

Sliding Vane

The basic element of the sliding-vane compressor is the cylindrical housing and the rotor assembly. This compressor, which is illustrated in Figure 12.5, has longitudinal vanes that slide radially in a slotted rotor mounted eccentrically in a cylinder. The centrifugal force carries the sliding vanes against the cylindrical case with the vanes forming a number of individual longitudinal cells in the eccentric annulus between the case and rotor. The suction port is located where the longitudinal cells are largest. The size of each cell is reduced by the eccentricity of the rotor as the vanes approach the discharge port, thus compressing the gas.

Cyclical opening and closing of the inlet and discharge ports occurs by the rotor's vanes passing over them. The inlet port is normally a wide opening that is designed to admit gas in the pocket between two vanes. The port closes momentarily when the second vane of each air-containing pocket passes over the inlet port.

When running at design pressure, the theoretical operation curves are identical (see Figure 12.6) to those of a reciprocating compressor. However, there is one major difference between a sliding-vane and a reciprocating compressor. The reciprocating unit has spring-loaded valves that open automatically with small pressure differentials between the outside and inside cylinder. The sliding-vane compressor has no valves.

The fundamental design considerations of a sliding-vane compressor are the rotor assembly, cylinder housing, and the lubrication system.

Housing and Rotor Assembly

Cast iron is the standard material used to construct the cylindrical housing, but other materials may be used if corrosive conditions exist. The rotor is usually a continuous piece of steel that includes the shaft and is made from bar stock. Special materials can be selected for corrosive applications. Occasionally, the rotor may be a separate iron casting keyed to a shaft. On most standard air compressors, the rotor-shaft seals are semimetallic packing in a stuffing box. Commercial mechanical rotary seals can be supplied when needed. Cylindrical roller bearings are generally used in these assemblies.

Vanes are usually asbestos or cotton cloth impregnated with a phenolic resin. Bronze or aluminum also may be used for vane construction. Each vane fits into a milled slot extending the full length of the rotor and slides radially in and out of this slot once per revolution. Vanes are the most maintenance-prone part in the compressor. There are from 8 to 20 vanes on each rotor, depending upon its diameter. A greater number of vanes increase compartmentalization, which reduces the pressure differential across each vane.

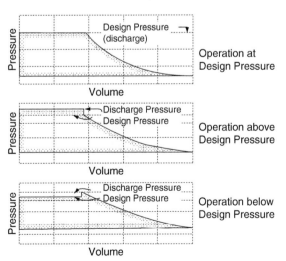

FIGURE 12.6. Theoretical operation curves for rotary compressors with built-in porting.

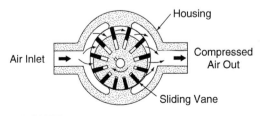

FIGURE 12.5. Rotary sliding-vane compressor.

Lubrication System

AV-belt-driven, force-fed oil lubrication system is used on water-cooled compressors. Oil goes to both bearings and to several points in the cylinder. Ten times as much oil is recommended to lubricate the rotary cylinder as is required for the cylinder of a corresponding reciprocating compressor. The oil carried over with the gas to the line may be reduced 50% with an oil separator on the discharge. Use of an aftercooler ahead of the separator permits removal of 85 to 90% of the entrained oil.

Helical Lobe or Screw

The helical lobe, or screw, compressor is shown in Figure 12.7. It has two or more mating sets of lobe-type rotors mounted in a common housing. The male lobe, or rotor, is usually direct-driven by an electric motor. The female lobe, or mating rotor, is driven by a helical gear set that is mounted on the outboard end of the rotor shafts. The gears provide both motive power for the female rotor and absolute timing between the rotors.

The rotor set has extremely close mating clearance (i.e., about 0.5 mils) but no metal-to-metal contact. Most of these compressors are designed for oil-free operation. In other words, no oil is used to lubricate or seal the rotors. Instead, oil lubrication is limited to the timing gears and bearings that are outside the air chamber. Because of this, maintaining proper clearance between the two rotors is critical.

This type of compressor is classified as a constant volume, variable-pressure machine that is quite similar to the vane-type rotary in general characteristics. Both have a built-in compression ratio.

Helical-lobe compressors are best suited for baseload applications where they can provide a constant volume and pressure of discharge gas. The only recommended method of volume control is the use of variable-speed motors. With variable-speed drives, capacity variations can be obtained with a proportionate reduction in speed. A 50% speed reduction is the maximum permissible control range.

Helical-lobe compressors are not designed for frequent or constant cycles between load and no-load operation. Each time the compressor unloads, the rotors tend to thrust axially. Even though the rotors have a substantial thrust bearing and, in some cases, a balancing piston to counteract axial thrust, the axial clearance increases each time the compressor unloads. Over time, this clearance will increase enough to permit a dramatic rise in the impact energy created by axial thrust during the transient from loaded to unloaded conditions. In extreme cases, the energy can be enough to physically push the rotor assembly through the compressor housing.

Compression ratio and maximum inlet temperature determine the maximum discharge temperature of these compressors. Discharge temperatures must be limited to prevent excessive distortion between the inlet and discharge ends of the casing and rotor expansion. High-pressure units are water-jacketed in order to obtain uniform casing temperature. Rotors also may be cooled to permit a higher operating temperature.

If either casing distortion or rotor expansion occur, the clearance between the rotating parts will decrease, and metal-to-metal contact will occur. Since the rotors typically rotate at speeds between 3,600 and 10,000 rpm, metal-to-metal contact normally results in instantaneous, catastrophic compressor failure.

Changes in differential pressures can be caused by variations in either inlet or discharge conditions (i.e., temperature, volume, or pressure). Such changes can cause the rotors to become unstable and change the load zones in the shaft-support bearings. The result is premature wear and/or failure of the bearings.

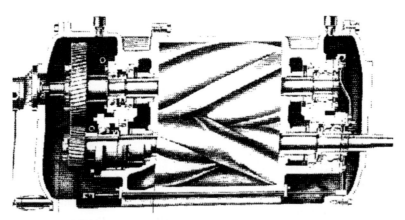

FIGURE 12.7.   Helical lobe, or screw, rotary air compressor.

Always install a relief valve that is capable of bypassing the full-load capacity of the compressor between its discharge port and the first isolation valve. Since helical-lobe compressors are less tolerant to over-pressure operation, safety valves are usually set within 10% of absolute discharge pressure, or 5 psi, whichever is lower.

Liquid-Seal Ring

The liquid-ring, or liquid-piston, compressor is shown in Figure 12.8. It has a rotor with multiple forward-turned blades that rotate about a central cone that contains inlet and discharge ports. Liquid is trapped between adjacent blades, which drive the liquid around the inside of an elliptical casing. As the rotor turns, the liquid face moves in and out of this space due to the casing shape, creating a liquid piston. Porting in the central cone is built-in and fixed, and there are no valves.

Compression occurs within the pockets or chambers between the blades before the discharge port is uncovered. Since the port location must be designed and built for a specific compression ratio, it tends to operate above or below the design pressure (refer back to Figure 12.6).

Liquid-ring compressors are cooled directly rather than by jacketed casing walls. The cooling liquid is fed into the casing where it comes into direct contact with the gas being compressed. The excess liquid is discharged with the gas. The discharged mixture is passed through a conventional baffle or centrifugal-type separator to remove the free liquid. Because of the intimate contact of gas and liquid, the final discharge temperature can be held close to the inlet cooling water temperature. However, the discharge gas is saturated with liquid at the discharge temperature of the liquid.

The amount of liquid passed through the compressor is not critical and can be varied to obtain the desired results. The unit will not be damaged if a large quantity of liquid inadvertently enters its suction port.

Lubrication is required only in the bearings, which are generally located external to the casing. The liquid itself acts as a lubricant, sealing medium, and coolant for the stuffing boxes.

Performance

Performance of a rotary positive-displacement compressor can be evaluated using the same criteria as a positive-displacement pump. As in constant-volume machines, performance is determined by rotation speed, internal slip, and total backpressure on the compressor.

The volumetric output of rotary positive-displacement compressors can be controlled by speed changes. The more slowly the compressor turns, the lower its output volume. This feature permits the use of these compressors in load-following applications. However, care must be taken to prevent sudden, radical changes in speed.

Internal slip is simply the amount of gas that can flow through internal clearances from the discharge back to the inlet. Obviously, internal wear will increase internal slip.

Discharge pressure is relatively constant regardless of operating speed. With the exceptions of slight pressure variations caused by atmospheric changes and backpressure, a rotary positive-displacement compressor will provide a fixed discharge pressure. Backpressure, which is caused by restrictions in the discharge piping or demand from users of the compressed air or gas, can have a serious impact on compressor performance.

If backpressure is too low or demand too high, the compressor will be unable to provide sufficient volume or pressure to the downstream systems. In this instance, the discharge pressure will be noticeably lower than designed.

If the backpressure is too high or demand too low, the compressor will generate a discharge pressure higher than designed. It will continue to compress

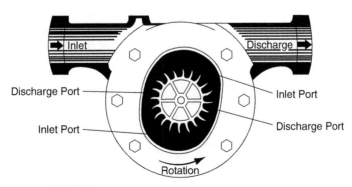

FIGURE 12.8.   Liquid-seal ring rotary air compressor.

the air or gas until it reaches the unload setting on the system's relief valve or until the brake horsepower required exceeds the maximum horsepower rating of the driver.

Installation

Installation requirements for rotary positive-displacement compressors are similar to those for any rotating machine. Review the installation requirements for centrifugal pumps and compressors for foundation, pressure-relief, and other requirements. As with centrifugal compressors, rotary positive-displacement compressors must be fitted with pressure-relief devices to limit the discharge or interstage pressures to a safe maximum for the equipment served.

In applications where demand varies, rotary positive-displacement compressors require a downstream receiver tank or reservoir that minimizes the load-unload cycling frequency of the compressor. The receiver tank should have sufficient volume to permit acceptable unload frequencies for the compressor. Refer to the vendor's O&M manual for specific receiver-tank recommendations.

Operating Methods

All compressor types have moving parts, high noise levels, high pressures, and high-temperature cylinder and discharge-piping surfaces.

Rotary positive-displacement compressors should be operated as baseloaded units. They are especially sensitive to the repeated start-stop operation required by load-following applications. Generally, rotary positive-displacement compressors are designed to unload about every six to eight hours. This unload cycle is needed to dissipate the heat generated by the compression process. If the unload frequency is too great, these compressors have a high probability of failure.

There are several primary operating control inputs for rotary positive-displacement compressors. These control inputs are: discharge pressure, pressure fluctuations, and unloading frequency.

Discharge Pressure

This type of compressor will continue to compress the air volume in the downstream system until (1) some component in the system fails; (2) the brake horsepower exceeds the driver's capacity; or (3) a safety valve opens. Therefore, the operator's primary control input should be the compressor's discharge pressure. If the discharge pressure is below the design point, it is a clear indicator that the total downstream demand is greater than the unit's capacity. If the discharge

pressure is too high, the demand is too low, and excessive unloading will be required to prevent failure.

Pressure Fluctuations

Fluctuations in the inlet and discharge pressures indicate potential system problems that may adversely affect performance and reliability. Pressure fluctuations are generally caused by changes in the ambient environment, turbulent flow, or restrictions caused by partially blocked inlet filters. Any of these problems will result in performance and reliability problems if not corrected.

Unloading Frequency

The unloading function in rotary positive-displacement compressors is automatic and not under operator control. Generally, a set of limit switches, one monitoring internal temperature and one monitoring discharge pressure, are used to trigger the unload process. By design, the limit switch that monitors the compressor's internal temperature is the primary control. The secondary control, or discharge-pressure switch, is a fail-safe design to prevent overloading of the compressor.

Depending on design, rotary positive-displacement compressors have an internal mechanism designed to minimize the axial thrust caused by the instantaneous change from fully loaded to unloaded operating conditions. In some designs, a balancing piston is used to absorb the rotor's thrust during this transient. In others, oversized thrust bearings are used.

Regardless of the mechanism used, none provides complete protection from the damage imparted by the transition from load to no-load conditions. However, as long as the unload frequency is within design limits, this damage will not adversely affect the compressor's useful operating life or reliability. However, an unload frequency greater than that accommodated in the design will reduce the useful life of the compressor and may lead to premature, catastrophic failure.

Operating practices should minimize, as much as possible, the unload frequency of these compressors. Installation of a receiver tank and modification of user-demand practices are the most effective solutions to this type of problem.

12.4. RECIPROCATING

Reciprocating compressors are widely used by industry and are offered in a wide range of sizes and types. They vary from units requiring less than 1 hp to more than 12,000 hp. Pressure capabilities range from low vacuums at intake to special compressors capable of 60,000 psig or higher.

Reciprocating compressors are classified as constant-volume, variable-pressure machines. They are the most efficient type of compressor and can be used for partial-load, or reduced-capacity, applications.

Because of the reciprocating pistons and unbalanced rotating parts, the unit tends to shake. Therefore, it is necessary to provide a mounting that stabilizes the installation. The extent of this requirement depends on the type and size of the compressor.

Because reciprocating compressors should be supplied with clean gas, inlet filters are recommended in all applications. They cannot satisfactorily handle liquids entrained in the gas, although vapors are no problem if condensation within the cylinders does not take place. Liquids will destroy the lubrication and cause excessive wear.

Reciprocating compressors deliver a pulsating flow of gas that can damage downstream equipment or machinery. This is sometimes a disadvantage, but pulsation dampers can be used to alleviate the problem.

12.4.1. Configuration

Certain design fundamentals should be clearly understood before analyzing the operating condition of reciprocating compressors. These fundamentals include frame and running gear, inlet and discharge valves, cylinder cooling, and cylinder orientation.

Frame and Running Gear

Two basic factors guide frame and running gear design. The first factor is the maximum horsepower to be transmitted through the shaft and running gear to the cylinder pistons. The second factor is the load imposed on the frame parts by the pressure differential between the two sides of each piston. This is often called pin load because this full force is directly exerted on the crosshead and crankpin. These two factors determine the size of bearings, connecting rods, frame, and bolts that must be used throughout the compressor and its support structure.

Cylinder Design

Compression efficiency depends entirely upon the design of the cylinder and its valves. Unless the valve area is sufficient to allow gas to enter and leave the cylinder without undue restriction, efficiency cannot be high. Valve placement for free flow of the gas in and out of the cylinder is also important.

Both efficiency and maintenance are influenced by the degree of cooling during compression. The method of cylinder cooling must be consistent with the service intended.

The cylinders and all the parts must be designed to withstand the maximum application pressure. The most economical materials that will give the proper strength and the longest service under the design conditions are generally used.

Inlet and Discharge Valves

Compressor valves are placed in each cylinder to permit one-way flow of gas, either into or out of the cylinder. There must be one or more valve(s) for inlet and discharge in each compression chamber.

Each valve opens and closes once for each revolution of the crankshaft. The valves in a compressor operating at 700 rpm for 8 hours per day and 250 days per year will have cycled (i.e., opened and closed) 42,000 times per hour, 336,000 times per day, or 84 million times in a year. The valves have less than $1/10$ of a second to open, let the gas pass through, and to close.

They must cycle with a minimum of resistance for minimum power consumption. However, the valves must have minimal clearance to prevent excessive expansion and reduced volumetric efficiency. They must be tight under extreme pressure and temperature conditions. Finally, the valves must be durable under many kinds of abuse.

There are four basic valve designs used in these compressors: finger, channel, leaf, and annular ring. Within each class there may be variations in design, depending upon operating speed and size of valve required.

Finger

Figure 12.9 is an exploded view of a typical finger valve. These valves are used for smaller, air-cooled compressors. One end of the finger is fixed and the opposite end lifts when the valve opens.

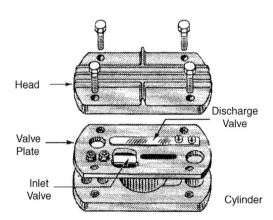

FIGURE 12.9. Finger valve configuration.

Channel

The channel valve shown in Figure 12.10 is widely used in mid- to large-sized compressors. This valve uses a series of separate stainless steel channels. As explained in the figure, this is a cushioned valve, which adds greatly to its life.

Leaf

The leaf valve (see Figure 12.11) has a configuration somewhat like the channel valve. It is made of flat-strip steel that opens against an arched stop plate. This results in valve flexing only at its center with maximum lift. The valve operates as its own spring.

Annular Ring

Figure 12.12 shows exploded views of typical inlet and discharge annular-ring valves. The valves shown have a single ring, but larger sizes may have two or three rings. In some designs, the concentric rings are tied into a single piece by bridges.

The springs and the valve move into a recess in the stop plate as the valve opens. Gas that is trapped in the recess acts as a cushion and prevents slamming. This eliminates a major source of valve and spring breakage. The valve shown was the first cushioned valve built.

Cylinder Cooling

Cylinder heat is produced by the work of compression plus friction, which is caused by the action of the piston and piston rings on the cylinder wall and packing on the rod. The amount of heat generated can be considerable, particularly when moderate to high compression ratios are involved. This can result in undesirably high operating temperatures.

Most compressors use some method to dissipate a portion of this heat to reduce the cylinder wall and discharge gas temperatures. The following are advantages of cylinder cooling:

- Lowering cylinder wall and cylinder head temperatures reduces loss of capacity and horsepower per unit volume due to suction gas preheating during inlet stroke. This results in more gas in the cylinder for compression.
- Reducing cylinder wall and cylinder head temperatures removes more heat from the gas during compression, lowering its final temperature and reducing the power required.
- Reducing the gas temperature and that of the metal surrounding the valves results in longer valve service life and reduces the possibility of deposit formation.
- Reduced cylinder wall temperature promotes better lubrication, resulting in longer life and reduced maintenance.
- Cooling, particularly water-cooling, maintains a more even temperature around the cylinder bore and reduces warpage.

Cylinder Orientation

Orientation of the cylinders in a multistage or multicylinder compressor directly affects the operating dynamics and vibration level. Figure 12.13 illustrates a typical three-piston, air-cooled compressor. Since three pistons are oriented within a 120-degree arc, this type of compressor generates higher vibration levels than the opposed-piston compressor illustrated in Figure 12.14.

Valve closed: A tight seat is formed without slamming or friction, so seat wear is at a minimum. Both channel and spring are precision made to assure a perfect fit. A gas space is formed between the bowed spring and the flat channel.

Valve opening: Channel lifts straight up in the guides without flexing. Opening is even over the full length of the port, giving uniform air velocity without turbulance. Cushioning is effected by the compression and escape of the gas between spring and channel.

Valve wide open: Gas trapped between spring and channel has been compressed and in escaping has allowed channel to float in its stop.

FIGURE 12.10.   Channel valve configuration.

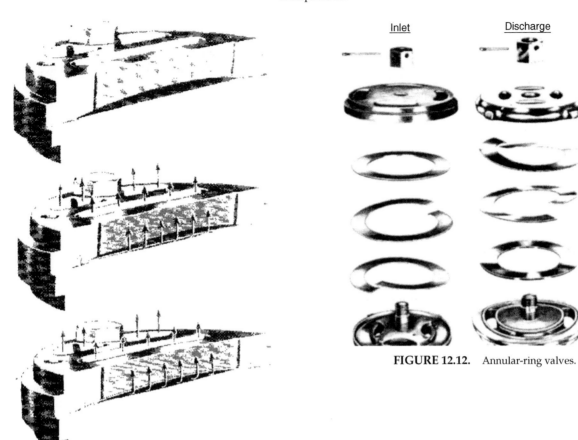

FIGURE 12.11. Leaf spring configuration.

FIGURE 12.12. Annular-ring valves.

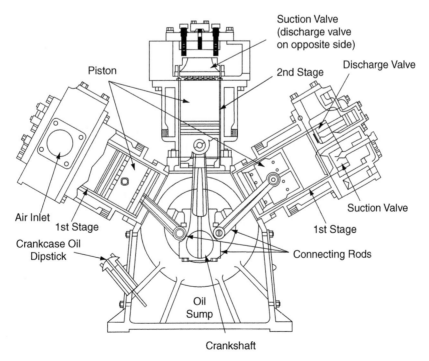

FIGURE 12.13. Three-piston compressor generates higher vibration levels.

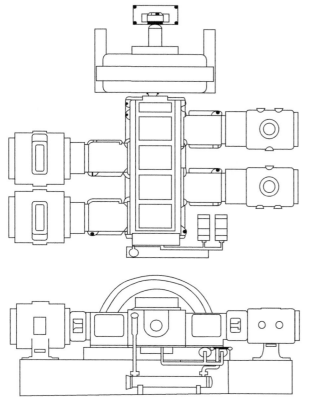

FIGURE 12.14 Opposed-piston compressor balances piston forces.

12.4.2. Performance

Reciprocating-compressor performance is governed almost exclusively by operating speed. Each cylinder of the compressor will discharge the same volume, excluding slight variations caused by atmospheric changes, at the same discharge pressure each time it completes the discharge stroke. As the rotation speed of the compressor changes, so does the discharge volume.

The only other variables that affect performance are the inlet-discharge valves, which control flow into and out of each cylinder. Although reciprocating compressors can use a variety of valve designs, it is crucial that the valves perform reliably. If they are damaged and fail to operate at the proper time or do not seal properly, overall compressor performance will be substantially reduced.

12.4.3. Installation

A carefully planned and executed installation is extremely important and makes compressor operation and maintenance easier and safer. Key components of a compressor installation are location, foundation, and piping.

Location

The preferred location for any compressor is near the center of its load. However, the choice is often influenced by the cost of supervision, which can vary by location. The ongoing cost of supervision may be less expensive at a less-optimum location, which can offset the cost of longer piping.

A compressor will always give better, more reliable service when enclosed in a building that protects it from cold, dusty, damp, and corrosive conditions. In certain locations it may be economical to use a roof only, but this is not recommended unless the weather is extremely mild. Even then, it is crucial to prevent rain and wind-blown debris from entering the moving parts. Subjecting a compressor to adverse inlet conditions will dramatically reduce reliability and significantly increase maintenance requirements.

Ventilation around a compressor is vital. On a motor-driven, air-cooled unit, the heat radiated to the surrounding air is at least 65% of the power input. On a water-jacketed unit with an aftercooler and outside receiver, the heat radiated to the surrounding air may be 15 to 25% of the total energy input, which is still a substantial amount of heat. Positive outside ventilation is recommended for any compressor room where the ambient temperature may exceed 104°F.

Foundation

Because of the alternating movement of pistons and other components, reciprocating compressors often develop a shaking that alternates in direction. This force must be damped and contained by the mounting. The foundation also must support the weight load of the compressor and its driver.

There are many compressor arrangements, and the net magnitude of the moments and forces developed can vary a great deal among them. In some cases, they are partially or completely balanced within the compressors themselves. In others, the foundation must handle much of the force.

When complete balance is possible, reciprocating compressors can be mounted on a foundation just large and rigid enough to carry the weight and maintain alignment. However, most reciprocating compressors require larger, more massive foundations than other machinery.

Depending upon size and type of unit, the mounting may vary from simply bolting to the floor to attaching to a massive foundation designed specifically for the application. A proper foundation must (1) maintain the alignment and level of the compressor and its driver at the proper elevation, and (2) minimize vibration and prevent its transmission to adjacent building structures and machinery. There are five steps to accomplish the first objective:

1. The safe weight-bearing capacity of the soil must not be exceeded at any point on the foundation base.
2. The load to the soil must be distributed over the entire area.
3. The size and proportion of the foundation block must be such that the resultant vertical load due to the compressor, block, and any unbalanced force falls within the base area.
4. The foundation must have sufficient mass and weight-bearing area to prevent its sliding on the soil due to unbalanced forces.
5. Foundation temperature must be uniform to prevent warping.

Bulk is not usually the complete solution to foundation problems. A certain weight is sometimes necessary, but soil area is usually of more value than foundation mass.

Determining if two or more compressors should have separate or single foundations depends on the compressor type. A combined foundation is recommended for reciprocating units since the forces from one unit usually will partially balance out the forces from the others. In addition, the greater mass and surface area in contact with the ground damps foundation movement and provides greater stability.

Soil quality may vary seasonally, and such conditions must be carefully considered in the foundation design. No foundation should rest partially on bedrock and partially on soil; it should rest entirely on one or the other. If placed on the ground, make sure that part of the foundation does not rest on soil that has been disturbed. In addition, pilings may be necessary to ensure stability.

Piping

Piping should easily fit the compressor connections without needing to spring or twist it to fit. It must be supported independently of the compressor and anchored, as necessary, to limit vibration and to prevent expansion strains. Improperly installed piping may distort or pull the compressor's cylinders or casing out of alignment.

Air Inlet

The intake pipe on an air compressor should be as short and direct as possible. If the total run of the inlet piping is unavoidably long, the diameter should be increased. The pipe size should be greater than the compressor's air-inlet connection.

Cool inlet air is desirable. For every 5°F of ambient air temperature reduction, the volume of compressed air generated increases by 1% with the same power consumption. This increase in performance is due to the greater density of the intake air.

It is preferable for the intake air to be taken from outdoors. This reduces heating and air conditioning costs and, if properly designed, has fewer contaminants. However, the intake piping should be a minimum of six feet above the ground and be screened or, preferably, filtered. An air inlet must be free of steam and engine exhausts. The inlet should be hooded or turned down to prevent the entry of rain or snow. It should be above the building eaves and several feet from the building.

Discharge

Discharge piping should be the full size of the compressor's discharge connection. The pipe size should not be reduced until the point along the pipeline is reached where the flow has become steady and nonpulsating. With a reciprocating compressor, this is generally beyond the aftercooler or the receiver. Pipes to handle nonpulsating flow are sized by normal methods, and long-radius bends are recommended. All discharge piping must be designed to allow adequate expansion loops or bends to prevent undue stresses at the compressor.

Drainage

Before piping is installed, the layout should be analyzed to eliminate low points where liquid could collect and to provide drains where low points cannot be eliminated. A regular part of the operating procedure must be the periodic drainage of low points in the piping and separators, as well as inspection of automatic drain traps.

Pressure-Relief Valves

All reciprocating compressors must be fitted with pressure relief devices to limit the discharge or interstage pressures to a safe maximum for the equipment served. Always install a relief valve that is capable of bypassing the full-load capacity of the compressor between its discharge port and the first isolation valve. The safety valves should be set to open at a pressure slightly higher than the normal discharge-pressure rating of the compressor. For standard 100- to 115-psig two-stage air compressors, safety valves are normally set at 125 psig.

The pressure-relief safety valve is normally situated on top of the air reservoir, and there must be no restriction on its operation. The valve is usually of the "huddling chamber" design, in which the static pressure acting on its disk area causes it to open. Figure 12.15

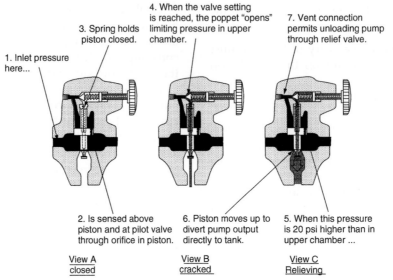

3. Spring holds piston closed.

4. When the valve setting is reached, the poppet "opens" limiting pressure in upper chamber.

7. Vent connection permits unloading pump through relief valve.

1. Inlet pressure here...

2. Is sensed above piston and at pilot valve through orifice in piston.

6. Piston moves up to divert pump output directly to tank.

5. When this pressure is 20 psi higher than in upper chamber ...

View A
closed

View B
cracked

View C
Relieving

FIGURE 12.15.   Illustrates how a safety valve functions.

illustrates how such a valve functions. As the valve pops, the air space within the huddling chamber between the seat and blowdown ring fills with pressurized air and builds up more pressure on the roof of the disk holder. This temporary pressure increases the upward thrust against the spring, causing the disk and its holder to fully pop open.

Once a predetermined pressure drop (i.e., blowdown) occurs, the valve closes with a positive action by trapping pressurized air on top of the disk holder. Raising or lowering the blowdown ring adjusts the pressure-drop setpoint. Raising the ring increases the pressure-drop setting, while lowering it decreases the setting.

12.4.4. Operating Methods

Compressors can be hazardous to work around because they have moving parts. Ensure that clothing is kept away from belt drives, couplings, and exposed shafts. In addition, high-temperature surfaces around cylinders and discharge piping are exposed. Compressors are notoriously noisy, so ear protection should be worn. These machines are used to generate high-pressure gas so, when working around them, it is important to wear safety glasses and to avoid searching for leaks with bare hands. High-pressure leaks can cause severe friction burns.

12.5. TROUBLESHOOTING

Compressors can be divided into three classifications: centrifugal, rotary, and reciprocating. This section identifies the common failure modes for each.

12.5.1. Centrifugal

The operating dynamics of centrifugal compressors are the same as for other centrifugal machine-trains. The dominant forces and vibration profiles are typically identical to pumps or fans. However, the effects of variable load and other process variables (e.g., temperatures, inlet/discharge pressure, etc.) are more pronounced than in other rotating machines. Table 12.1 identifies the common failure modes for centrifugal compressors.

Aerodynamic instability is the most common failure mode for centrifugal compressors. Variable demand and restrictions of the inlet-air flow are common sources of this instability. Even slight variations can cause dramatic changes in the operating stability of the compressor.

Entrained liquids and solids also can affect operating life. When dirty air must be handled, open-type impellers should be used. An open design provides the ability to handle a moderate amount of dirt or other solids in the inlet-air supply. However, inlet filters are recommended for all applications, and controlled liquid injection for cleaning and cooling should be considered during the design process.

12.5.2. Rotary-Type, Positive Displacement

Table 12.2 lists the common failure modes of rotary-type, positive-displacement compressors. This type of compressor can be grouped into two types: sliding vane and rotary screw.

TABLE 12.1.   Common Failure Modes of Centrifugal Compressors

THE CAUSES	Excessive vibration	Compressor surges	Loss of discharge pressure	Low lube oil pressure	Excessive bearing oil drain temp.	Units do not stay in alignment	Persistent unloading	Water in lube oil	Motor trips
Bearing lube oil orifice missing or plugged				●					
Bent rotor (caused by uneven heating and cooling)	●						●		
Build-up of deposits on diffuser		●							
Build-up of deposits on rotor	●	●							
Change in system resistance		●							●
Clogged oil strainer/filter				●					
Compressor not up to speed			●						
Condensate in oil reservoir								●	
Damaged rotor	●								
Dry gear coupling	●								
Excessive bearing clearance	●								
Excessive inlet temperature			●						
Failure of both main and auxiliary oil pumps				●					
Faulty temperature gauge or switch				●	●				●
Improperly assembled parts	●						●		●
Incorrect pressure control valve setting				●					
Insufficient flow		●							
Leak in discharge piping			●						
Leak in lube oil cooler tubes or tube sheet								●	
Leak in oil pump suction piping				●					
Liquid "slugging"	●						●		
Loose or broken bolting	●								

(Continues)

Table 12.1. (*Continued*)

THE CAUSES	Excessive vibration	Compressor surges	Loss of discharge pressure	Low lube oil pressure	Excessive bearing oil drain temp.	Units do not stay in alignment	Persistent unloading	Water in lube oil	Motor trips
Loose rotor parts	●								
Oil leakage				●					
Oil pump suction plugged				●					
Oil reservoir low level				●					
Operating at low speed w/o auxiliary oil pump				●					
Operating in critical speed range	●								
Operating in surge region	●								
Piping strain	●					●	●	●	●
Poor oil condition					●				
Relief valve improperly set or stuck open				●					
Rotor imbalance	●						●		
Rough rotor shaft journal surface					●		●		●
Shaft misalignment	●					●			
Sympathetic vibration	●						●	●	
Vibration					●				
Warped foundation or baseplate							●		●
Wiped or damaged bearings					●				●
Worn or damaged coupling	●								

Sliding Vane Compressors

Sliding-vane compressors have the same failure modes as vane-type pumps. The dominant components in their vibration profile are running speed, vane-pass frequency, and bearing-rotation frequencies. In normal operation, the dominant energy is at the shaft's running speed. The other frequency components are at much lower energy levels. Common failures of this type of compressor occur with shaft seals, vanes, and bearings.

Shaft Seals

Leakage through the shaft's seals should be checked visually once a week or as part of every data-acquisition route. Leakage may not be apparent from the outside

TABLE 12.2. Common Failure Modes of Rotary-Type, Positive-Displacement Compressors

THE CAUSES	No air/gas delivery	Insufficient discharge pressure	Insufficient capacity	Excessive wear	Excessive heat	Excessive vibration and noise	Excessive power demand	Motor trips	Elevated motor temperature	Elevated air/gas temperature
Air leakage into suction piping or shaft seal		•	•			•				
Coupling misaligned				•	•	•	•		•	
Excessive discharge pressure			•	•		•	•	•		•
Excessive inlet temperature/moisture			•							
Insufficient suction air/gas supply		•	•	•		•		•		
Internal component wear	•	•	•							
Motor or driver failure	•									
Pipe strain on compressor casing				•	•	•	•		•	
Relief valve stuck open or set wrong		•	•							
Rotating element binding				•	•	•	•	•	•	
Solids or dirt in inlet air/gas supply				•						
Speed too low		•	•						•	
Suction filter or strainer clogged	•	•	•			•			•	
Wrong direction of rotation	•	•							•	

of the gland. If the fluid is removed through a vent, the discharge should be configured for easy inspection. Generally, more leakage than normal is the signal to replace a seal. Under good conditions, they have a normal life of 10,000 to 15,000 hours and should routinely be replaced when this service life has been reached.

Vanes

Vanes wear continuously on their outer edges and, to some degree, on the faces that slide in and out of the slots. The vane material is affected somewhat by prolonged heat, which causes gradual deterioration. Typical life expectancy of vanes in 100-psig services is about 16,000 hours of operation. For low-pressure applications, life may reach 32,000 hours.

Replacing vanes before they break is extremely important. Breakage during operation can severely damage the compressor, which requires a complete overhaul and realignment of heads and clearances.

Bearings

In normal service, bearings have a relatively long life. Replacement after about six years of operation is generally recommended. Bearing defects are usually displayed in the same manner in a vibration profile as for any rotating machine-train. Inner and outer race defects are the dominant failure modes, but roller spin also may contribute to the failure.

Rotary Screw

The most common reason for compressor failure or component damage is process instability. Rotary-screw compressors are designed to deliver a constant volume and pressure of air or gas. These units are extremely susceptible to any change in either inlet or discharge conditions. A slight variation in pressure, temperature, or volume can result in instantaneous failure. The following are used as indices of instability and potential problems: rotor mesh, axial movement, thrust bearings, and gear mesh.

Rotor Mesh

In normal operation, the vibration energy generated by male and female rotor meshing is very low. As the process becomes unstable, the energy due to the rotor-meshing frequency increases, with both the amplitude of the meshing frequency and the width of the peak increasing. In addition, the noise floor surrounding the meshing frequency becomes more pronounced. This white noise is similar to that observed in a cavitating pump or unstable fan.

Axial Movement

The normal tendency of the rotors and helical timing gears is to generate axial shaft movement, or thrusting. However, the extremely tight clearances between the male and female rotors do not tolerate any excessive axial movement, and therefore, axial movement should be a primary monitoring parameter. Axial measurements are needed from both rotor assemblies. If there is any increase in the vibration amplitude of these measurements, it is highly probable that the compressor will fail.

Thrust Bearings

While process instability can affect both the fixed and float bearings, the thrust bearing is more likely to show early degradation as a result of process instability or abnormal compressor dynamics. Therefore, these bearings should be monitored closely, and any degradation or hint of excessive axial clearance should be corrected immediately.

Gear Mesh

The gear mesh vibration profile also provides an indication of prolonged compressor instability. Deflection of the rotor shafts changes the wear pattern on the helical gear sets. This change in pattern increases the backlash in the gear mesh, results in higher vibration levels, and increases thrusting.

12.5.3. Reciprocating, Positive Displacement

Reciprocating compressors have a history of chronic failures that include valves, lubrication system, pulsation, and imbalance. Table 12.3 identifies common failure modes and causes for this type of compressor.

Like all reciprocating machines, reciprocating compressors normally generate higher levels of vibration than centrifugal machines. In part, the increased level of vibration is due to the impact as each piston reaches top dead center and bottom dead center of its stroke. The energy levels also are influenced by the unbalanced forces generated by nonopposed pistons and looseness in the piston rods, wrist pins, and journals of the compressor. In most cases, the dominant vibration frequency is the second harmonic (2X) of the main crankshaft's rotating speed. Again, this results from the impact that occurs when each piston changes directions (i.e., two impacts occur during one complete crankshaft rotation).

Valves

Valve failure is the dominant failure mode for reciprocating compressors. Because of their high cyclic rate, which exceeds 80 million cycles per year, inlet and discharge valves tend to work harder and crack.

Lubrication System

Poor maintenance of lubrication-system components, such as filters and strainers, typically causes premature failure. Such maintenance is crucial to reciprocating compressors because they rely on the lubrication system to provide a uniform oil film between closely fitting parts (e.g., piston rings and the cylinder wall). Partial or complete failure of the lube system results in catastrophic failure of the compressor.

Pulsation

Reciprocating compressors generate pulses of compressed air or gas that are discharged into the piping

TABLE 12.3.

THE CAUSES	Air discharge temperature above normal	Carbonaceous deposits abnormal	Compressor fails to starts	Compressor fails to unload	Compressor noisy or knocks	Compressor parts overheat	Crankcase oil pressure low	Crankcase water accumulation	Delivery less than rated capacity	Discharge pressure below normal	Excessive compressor vibration	Intercooler pressure above normal	Intercooler pressure below normal	Intercooler safety valve pops	Motor overheating	Oil pumping excessive (single-acting compressor)	Operating cycle abnormally long	Outlet water temperature above normal	Piston ring, piston, cylinder wear excessive	Piston rod or packing wear excessive	Receiver pressure above normal	Receiver safety valve pops	Starts too often	Valve wear and breakage normal
																		THE PROBLEM						
Air discharge temperature too high		●																●						
Air filter defective		●																	●	●				●
Air flow to fan blocked	●	●				●																		
Air leak into pump suction								●																
Ambient temperature too high	●	●				●									●									
Assembly incorrect																								●
Bearings need adjustment or renewal					●	●	●								●									
Belts slipping					●				●	●														
Belts too tight			●			●									●									
Centrifugal pilot valve leaks																●								
Check or discharge valve defective					●																			
Control air filter, strainer clogged				●																				
Control air line clogged																						●		
Control air pipe leaks																						●	●	
Crankcase oil pressure too high																●								
Crankshaft end play too great					●																			
Cylinder, head, cooler dirty	●	●																						
Cylinder, head, intercooler dirty					●													●						
Cylinder (piston) worn or scored		●	●		●	●			●	●	●	●H	●L		●H	●L	●	●	●H		●H			
Detergent oil being used (3)							●																	
Demand too steady (2)																								●
Dirt, rust entering cylinder		●																	●		●			●

(*Continues*)

Table 12.3. (Continued)

THE CAUSES	Air discharge temperature above normal	Carbonaceous deposits abnormal	Compressor fails to start	Compressor fails to unload	Compressor noisy or knocks	Compressor parts overheat	Crankcase oil pressure low	Crankcase water accumulation	Delivery less than rated capacity	Discharge pressure below normal	Excessive compressor vibration	Intercooler pressure above normal	Intercooler pressure below normal	Intercooler safety valve pops	Motor overheating	Oil pumping excessive (single-acting compressor)	Operating cycle abnormally long	Outlet water temperature above normal	Piston ring, piston, cylinder wear excessive	Piston rod or packing wear excessive	Receiver pressure above normal	Receiver safety valve pops	Starts too often	Valve wear and breakage normal
Discharge line restricted	●														●									
Discharge pressure above rating	●	●			●	●			●		●	●		●	●		●	●	●	●	●	●		
Electrical conditions wrong			●												●									
Excessive number of starts															●									
Excitation inadequate			●												●									
Foundation bolts loose					●						●													
Foundation too small											●													
Foundation uneven-unit rocks					●						●													
Fuses blown			●																					
Gaskets leak	●	●			●	●			●	●	● H	● L	● H	● L			●		● H	● H				
Gauge defective							●			●		●	●								●			
Gear pump worn/defective							●																	
Grout, improperly placed											●													
Intake filter clogged	●				●	●			●	●		●			●		●	●						
Intake pipe restricted, too small, too long	●				●	●			●	●		●			●		●	●						
Intercooler, drain more often								●																
Intercooler leaks													●											
Intercooler passages clogged												●		●										
Intercooler pressure too high																		●						
Intercooler vibrating					●																			
Leveling wedges left under compressor											●													
Liquid carry-over					●			●											●	●				●

Table 12.3. (*Continued*)

THE PROBLEM

THE CAUSES	Air discharge temperature above normal	Carbonaceous deposits abnormal	Compressor fails to start	Compressor fails to unload	Compressor noisy or knocks	Compressor parts overheat	Crankcase oil pressure low	Crankcase water accumulation	Delivery less than rated capacity	Discharge pressure below normal	Excessive compressor vibration	Intercooler pressure above normal	Intercooler pressure below normal	Intercooler safety valve pops	Motor overheating	Oil pumping excessive (single-acting compressor)	Operating cycle abnormally long	Outlet water temperature above normal	Piston ring, piston, cylinder wear excessive	Piston rod or packing wear excessive	Receiver pressure above normal	Receiver safety valve pops	Starts too often	Valve wear and breakage normal
Location too humid and damp								●																
Low oil pressure relay open			●																					
Lubrication inadequate	●				●	●									●		●		●	●				●
Motor overload relay tripped			●																					
Motor rotor loose on shaft					●						●													
Motor too small			●												●									
New valve on worn seat																						●		
"Off" time insufficient	●	●			●																			
Oil feed excessive		●		●													●					●		
Oil filter or strainer clogged							●																	
Oil level too high	●	●			●	●									●									
Oil level too low					●	●																		
Oil relief valve defective						●																		
Oil viscosity incorrect		●			●	●	●							●	●		●		●			●		
Oil wrong type															●									
Packing rings worn, stuck, broken																	●							
Piping improperly supported											●													
Piston or piston nut loose					●																			
Piston or ring drain hole clogged															●									
Piston ring gaps not staggered															●									
Piston rings worn, broken, or stuck	●	●			●	●		●	●	●		●H	●L	●H	●L		●	●	●H	●H				
Piston-to-head clearance too small				●																				

(*Continues*)

Table 12.3. (*Continued*)

THE PROBLEM

THE CAUSES	Air discharge temperature above normal	Carbonaceous deposits abnormal	Compressor fails to start	Compressor fails to unload	Compressor noisy or knocks	Compressor parts overheat	Crankcase oil pressure low	Crankcase water accumulation	Delivery less than rated capacity	Discharge pressure below normal	Excessive compressor vibration	Intercooler pressure above normal	Intercooler pressure below normal	Intercooler safety valve pops	Motor overheating	Oil pumping excessive (single-acting compressor)	Operating cycle abnormally long	Outlet water temperature above normal	Piston ring, piston, cylinder wear excessive	Piston rod or packing wear excessive	Receiver pressure above normal	Receiver safety valve pops	Starts too often	Valve wear and breakage normal
Pulley or flywheel loose					●						●													
Receiver, drain more often																							●	
Receiver too small																							●	
Regulation piping clogged				●																				
Resonant pulsation (inlet or discharge)												●	●	●	●									●
Rod packing leaks	●				●	●			●	●														
Rod packing too tight					●																			
Rod scored, pitted, worn																				●				
Rotation wrong	●	●	●																					
Runs too little (2)								●																
Safety valve defective												●	●									●		
Safety valve leaks	●				●				●	●		●		●										
Safety valve set too low													●									●		
Speed demands exceed rating																	●							
Speed lower than rating									●	●														
Speed too high	●	●				●					●				●			●						
Springs broken																								●
System demand exceeds rating	●			●					●	●		●		●			●							
System leakage excessive	●			●					●	●		●		●			●						●	
Tank ringing noise					●																			
Unloader running time too long (1)																●								
Unloader or control defective	●	●	●	●	●	●			●	●	●	●	●	●	●		●		●	●	●	●	●	●

Table 12.3. (*Continued*)

THE CAUSES	Air discharge temperature above normal	Carbonaceous deposits abnormal	Compressor fails to start	Compressor fails to unload	Compressor noisy or knocks	Compressor parts overheat	Crankcase oil pressure low	Crankcase water accumulation	Delivery less than rated capacity	Discharge pressure below normal	Excessive compressor vibration	Intercooler pressure above normal	Intercooler pressure below normal	Intercooler safety valve pops	Motor overheating	Oil pumping excessive (single-acting compressor)	Operating cycle abnormally long	Outlet water temperature above normal	Piston ring, piston, cylinder wear excessive	Piston rod or packing wear excessive	Receiver pressure above normal	Receiver safety valve pops	Starts too often	Valve wear and breakage normal
Unloader parts worn or dirty				•																				
Unloader setting incorrect	•	•	•		•	•			•	•		•	•	•	•		•		•	•	•	•	•	
V-belt or other misalignment					•	•					•													
Valves dirty	•	•				•						•	•											•
Valves incorrectly located	•	•			•	•			•	•		•H	•L	•H	•L		•		•H	•H				
Valves not seated in cylinder	•	•			•	•			•	•		•H	•L	•H	•L		•		•H	•H				
Valves worn or broken	•	•			•	•			•	•		•H	•L	•H	•L		•	•H	•	•H				
Ventilation poor	•	•				•									•									
Voltage abnormally low			•												•									
Water inlet temperature too high	•	•				•			•			•						•						
Water jacket or cooler dirty	•	•																						
Water jackets or intercooler dirty						•						•						•						
Water quantity insufficient	•					•			•			•						•						
Wiring incorrect			•																					
Worn valve on good seat																								•
Wrong oil type		•																	•	•				
(1) Use automatic start/stop control																								
(2) Use constant speed control																								
(3) Change to nondetergent oil																								
H (In high pressure cylinder)																								
L (In low pressure cylinder)																								

Crank arrangements		Forces		Couples	
		Primary	Secondary	Primary	Secondary
Single crank		F' without counterwts. 0.5F' with counterwts.	F''	None	None
Two cranks at 180° In line cylinders		Zero	2F''	F'D without counterwts. F'D/2 with counterwts.	None
Opposed cylinders		Zero	Zero	Nil	Nil
Two cranks at 90°		141F' without counterwts. 0.707F' with counterwts.	Zero	707F'D without counterwts. 0.354F'D with counterwts.	F''D
Two cylinders on one crank Cylinders at 90°		F' without counterwts. Zero with counterwts.	1.41F''	Nil	Nil
Two cylinders on one crank Opposed cylinders		2F' without counterwts. F' with counterwts.	Zero	None	Nil
Three cranks at 120°		Zero	Zero	3.46F'D without counterwts. 1.73F'D with counterwts.	3.46F''D
Four cylinders Cranks at 180°		Zero	4F''	Zero	Zero
Cranks at 90°		Zero	Zero	1.41F'D without counterwts. 0.707F'D with counterwts.	4.0F''D
Six cylinders		Zero	Zero	Zero	Zero
Key					

F' = Primary inertia force in lbs.
F' = .0000284 RN^2W
F'' = Secondary inertia force in lbs.
F'' = R/L F'
R = Crank radius, inches
N = R.P.M.
W = Reciprocating weight of one cylinder, lbs.
L = Length of connecting rod, inches
D = Cylinder center distance

FIGURE 12.16. Unbalanced inertial forces and couples for various reciprocating compressors.

that transports the air or gas to its point(s) of use. This pulsation often generates resonance in the piping system, and pulse impact (i.e., standing waves) can severely damage other machinery connected to the compressed-air system. While this behavior does not cause the compressor to fail, it must be prevented to protect other plant equipment. Note, however, that most compressed-air systems do not use pulsation dampers.

Each time the compressor discharges compressed air, the air tends to act like a compression spring. Because it rapidly expands to fill the discharge piping's available volume, the pulse of high-pressure air can cause serious damage. The pulsation wavelength, λ, from a compressor having a double-acting piston design can be determined by:

$$\lambda = \frac{60a}{2n} = \frac{34{,}050}{n}$$

where:

λ = Wavelength, feet
a = Speed of sound = 1,135 feet/second
n = Compressor speed, revolutions/minute

For a double-acting piston design, a compressor running at 1,200 rpm will generate a standing wave of 28.4 feet. In other words, a shock load equivalent to the discharge pressure will be transmitted to any piping or machine connected to the discharge piping and located within 28 feet of the compressor. Note that for a single-acting cylinder, the wavelength will be twice as long.

Imbalance

Compressor inertial forces may have two effects on the operating dynamics of a reciprocating compressor, affecting its balance characteristics. The first is a force in the direction of the piston movement, which is displayed as impacts in a vibration profile as the piston reaches top and bottom dead center of its stroke. The second effect is a couple, or moment, caused by an offset between the axes of two or more pistons on a common crankshaft. The interrelationship and magnitude of these two effects depend upon such factors as: (1) number of cranks; (2) longitudinal and angular arrangement; (3) cylinder arrangement; and (4) amount of counterbalancing possible. Two significant vibration periods result, the primary at the compressor's rotation speed (X) and the secondary at 2X.

Although the forces developed are sinusoidal, only the maximum (i.e., the amplitude) is considered in the analysis. Figure 12.16 shows relative values of the inertial forces for various compressor arrangements.

13

Gears and Gearboxes*

A gear is a form of disc, or wheel, that has teeth around its periphery for the purpose of providing a positive drive by meshing the teeth with similar teeth on another gear or rack.

FIGURE 13.2. Rack or straight-line gear.

13.1. SPUR GEARS

The *spur gear* might be called the basic gear, since all other types have been developed from it. Its teeth are straight and parallel to the center bore line, as shown in Figure 13.1. Spur gears may run together with other spur gears or parallel shafts, with internal gears on parallel shafts, and with a *rack*. A rack such as the one illustrated in Figure 13.2 is in effect a straight-line gear. The smallest of a pair of gears (Figure 13.3) is often called a pinion.

FIGURE 13.3 Typical spur gears.

FIGURE 13.1. Example of a spur gear.

*Source: Ricky Smith and Keith Mobley, *Industrial Machinery Repair: Best Maintenance Practices Pocket Guide* (Boston: Butterworth–Heinemann, 2003), pp. 283–313.

The involute profile or form is the one most commonly used for gear teeth. It is a curve that is traced by a point on the end of a taut line unwinding from a circle. The larger the circle, the straighter the curvature; and for a rack, which is essentially a section of an infinitely large gear, the form is straight or flat. The generation of an involute curve is illustrated in Figure 13.4.

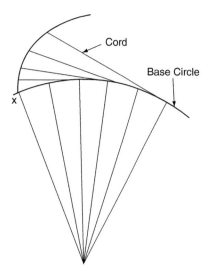

FIGURE 13.4.   Involute curve.

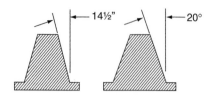

FIGURE 13.6.   Different pressure angles on gear teeth.

The involute system of spur gearing is based on a rack having straight or flat sides. All gears made to run correctly with this rack will run with each other.

The sides of each tooth incline toward the center top at an angle called the *pressure angle,* shown in Figure 13.5.

The 14.5-degree pressure angle was standard for many years. In recent years, however, the use of the 20-degree pressure angle has been growing, and today 14.5-degree gearing is generally limited to replacement work. The principal reasons are that a 20-degree pressure angle results in a gear tooth with greater strength and wear resistance and permits the use of pinions with fewer teeth. The effect of the pressure angle on the tooth of a rack is shown in Figure 13.6.

It is extremely important that the pressure angle be known when gears are mated, as all gears that run together must have the same pressure angle. The pressure angle of a gear is the angle between the line of action and the line tangent to the pitch circles of mating gears. Figure 13.7 illustrates the relationship of the pressure angle to the line of action and the line tangent to the pitch circles.

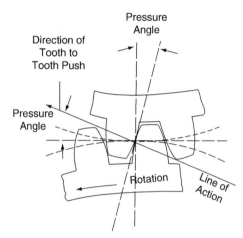

FIGURE 13.7.   Relationship of the pressure angle to the line of action.

13.2. PITCH DIAMETER AND CENTER DISTANCE

Pitch circles have been defined as the imaginary circles that are in contact when two standard gears are in correct mesh. The diameters of these circles are the pitch diameters of the gears. The center distance of the two gears, therefore, when correctly meshed, is equal to one half of the sum of the two pitch diameters, as shown in Figure 13.8.

This relationship may also be stated in an equation, and may be simplified by using letters to indicate the various values, as follows:

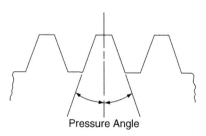

FIGURE 13.5.   Pressure angle.

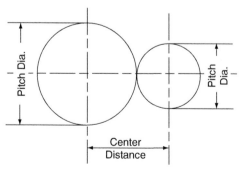

FIGURE 13.8.   Pitch diameter and center distance.

C = center distance
D₁ = first pitch diameter
D₂ = second pitch diameter

$$C = \frac{D_1 + D_2}{2} \quad D_1 = 2C - D_2 \quad D_2 = 2C - D_1$$

Example: The center distance can be found if the pitch diameters are known (illustration in Figure 13.9).

13.3. CIRCULAR PITCH

A specific type of pitch designates the size and proportion of gear teeth. In gearing terms, there are two specific types of pitch: *circular pitch* and *diametrical pitch*. Circular pitch is simply the distance from a point on one tooth to a corresponding point on the next tooth, measured along the pitch line or circle, as illustrated in Figure 13.10. Large-diameter gears are frequently made to circular pitch dimensions.

13.4. DIAMETRICAL PITCH AND MEASUREMENT

The diametrical pitch system is the most widely used, as practically all common-size gears are made to diametrical pitch dimensions. It designates the size and proportions of gear teeth by specifying the number of teeth in the gear for each inch of the gear's pitch diameter. For each inch of pitch diameter, there are

pi (π) inches, or 3.1416 inches, of pitch-circle circumference. The diametric pitch number also designates the number of teeth for each 3.1416 inches of pitch-circle circumference. Stated in another way, the *diametrical pitch* number specifies the number of teeth in 3.1416 inches along the pitch line of a gear.

For simplicity of illustration, a whole-number pitch-diameter gear (4 inches), is shown in Figure 13.11.

Figure 13.11 illustrates that the *diametrical pitch* number specifying the number of teeth per inch of pitch diameter must also specify the number of teeth per 3.1416 inches of pitch-line distance. This may be more easily visualized and specifically dimensioned when applied to the rack in Figure 13.12.

Because the pitch line of a rack is a straight line, a measurement can be easily made along it. In Figure 13.12, it is clearly shown that there are 10 teeth in 3.1416 inches; therefore the rack illustrated is a 10 diametrical pitch rack.

A similar measurement is illustrated in Figure 13.13, along the pitch line of a gear. The diametrical pitch being the number of teeth in 3.1416 inches of pitch line, the gear in this illustration is also a 10 diametrical pitch gear.

In many cases, particularly on machine repair work, it may be desirable for the mechanic to determine the diametrical pitch of a gear. This may be done very easily without the use of precision measuring tools, templates, or gauges. Measurements need not be exact because diametrical pitch numbers are usually whole numbers. Therefore, if an approximate calculation results in a value close to a whole number, that whole number is the diametrical pitch number of the gear.

The following two methods may be used to determine the approximate diametrical pitch of a gear. A common steel rule, preferably flexible, is adequate to make the required measurements.

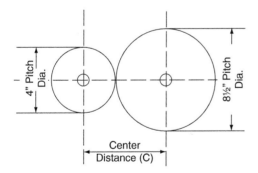

FIGURE 13.9.   Determining center distance.

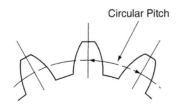

FIGURE 13.10.   Circular pitch.

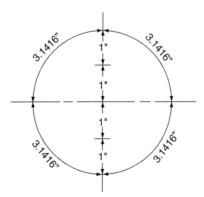

FIGURE 13.11.   Pitch diameter and diametrical pitch.

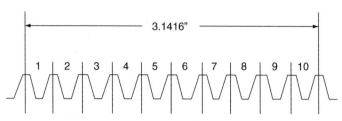

FIGURE 13.12. Number of teeth in 3.1416 inches.

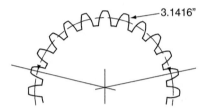

FIGURE 13.13. Number of teeth in 3.1416 inches on the pitch circle.

13.4.1. Method 1

Count the number of teeth in the gear, add 2 to this number, and divide by the outside diameter of the gear. Scale measurement of the gear to the closest fractional size is adequate accuracy.

Figure 13.14 illustrates a gear with 56 teeth and an outside measurement of $5^{13}/_{16}$ inches. Adding 2 to 56 gives 58; dividing 58 by $5^{13}/_{16}$ gives an answer of $9^{31}/_{32}$. Since this is approximately 10, it can be safely stated that the gear is a 10 diametrical pitch gear.

13.4.2. Method 2

Count the number of teeth in the gear and divide this number by the measured pitch diameter. The pitch diameter of the gear is measured from the root or bottom of a tooth space to the top of a tooth on the opposite side of the gear.

Figure 13.15 illustrates a gear with 56 teeth. The pitch diameter measured from the bottom of the tooth space to the top of the opposite tooth is $5^5/_8$ inches. Dividing 56 by $5^5/_8$ gives an answer of $9^{15}/_{16}$ inches, or approximately 10. This method also indicates that the gear is a 10 diametrical pitch gear.

13.5. PITCH CALCULATIONS

Diametrical pitch, usually a whole number, denotes the ratio of the number of teeth to a gear's pitch diameter. Stated another way, it specifies the number of teeth in a gear for each inch of pitch diameter. The relationship of pitch diameter, diametrical pitch, and number of teeth can be stated mathematically as follows.

$$P = \frac{N}{D} \qquad D = \frac{N}{P} \qquad N = D \times P$$

where:

D = pitch diameter
P = diametrical pitch
N = number of teeth

If any two values are known, the third may be found by substituting the known values in the appropriate equation.

Example 1: What is the *diametrical* pitch of a 40-tooth gear with a 5-inch pitch diameter?

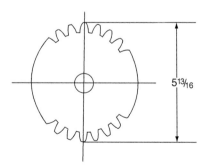

FIGURE 13.14. Using method 1 to approximate the diametrical pitch. In this method the outside diameter of the gear is measured.

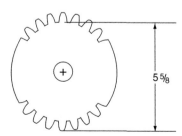

FIGURE 13.15. Using method 2 to approximate the diametrical pitch. This method uses the pitch diameter of the gear.

$$P = \frac{N}{P} \quad \text{or} \quad P = \frac{40}{5} \quad \text{or} \quad P = 8 \text{ diametrical pitch}$$

Example 2: What is the *pitch diameter* of a 12 diametrical pitch gear with 36 teeth?

$$D = \frac{N}{P} \quad \text{or} \quad D = \frac{36}{12} \quad \text{or} \quad D = 3'' \text{ pitch diameter}$$

Example 3: How many teeth are there in a 16 *diametrical pitch* gear with a pitch diameter of $3\frac{3}{4}$ inches?

$$N = D \times P \quad \text{or} \quad N = 3\frac{3}{4} \times 16 \quad \text{or} \quad N = 60 \text{ teeth}$$

Circular pitch is the distance from a point on a gear tooth to the corresponding point on the next gear tooth measured along the pitch line. Its value is equal to the circumference of the pitch circle divided by the number of teeth in the gear. The relationship of the circular pitch to the *pitch-circle circumference, number of teeth,* and the *pitch diameter* may also be stated mathematically as follows:

Circumference of pitch circle = πD

$$p = \frac{\pi D}{N} \qquad D = \frac{pN}{\pi} \qquad N = \frac{\pi D}{p}$$

where:

D = pitch diameter
N = number of teeth
p = circular pitch
π = pi, or 3.1416

If any two values are known, the third may be found by substituting the known values in the appropriate equation.

Example 1: What is the *circular pitch* of a gear with 48 teeth and a *pitch diameter* of 6"?

$$p = \frac{\pi D}{N} \quad \text{or} \quad \frac{3.1416 \times 6}{48} \quad \text{or} \quad \frac{3.1416}{8}$$
$$\text{or} \quad p = 0.3927 \text{ inches}$$

Example 2: What is the *pitch diameter* of a 0.500" *circular-pitch* gear with 128 teeth?

$$D = \frac{pN}{\pi} \quad \text{or} \quad \frac{0.5 \times 128}{3.1416} \qquad D = 20.371 \text{ inches}$$

The list that follows offers just a few names of the various parts given to gears. These parts are shown in Figures 13.16 and 13.17.

- Addendum: Distance the tooth projects above, or outside, the pitch line or circle.
- Dedendum: Depth of a tooth space below, or inside, the pitch line or circle.

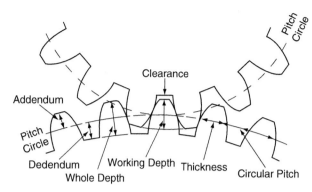

FIGURE 13.16. Names of gear parts.

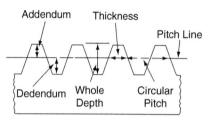

FIGURE 13.17. Names of rack parts.

- Clearance: Amount by which the dedendum of a gear tooth exceeds the addendum of a matching gear tooth.
- Whole Depth: The total height of a tooth or the total depth of a tooth space.
- Working Depth: The depth of tooth engagement of two matching gears. It is the sum of their addendums.
- Tooth Thickness: The distance along the pitch line or circle from one side of a gear tooth to the other.

13.6. TOOTH PROPORTIONS

The full *depth involute system* is the gear system in most common use. The formulas (with symbols) shown below are used for calculating tooth proportions of full-depth involute gears. Diametrical pitch is given the symbol P as before.

$$\text{Addendum, } a = \frac{1}{P}$$

$$\text{Whole depth, } Wd = \frac{20 + 0.002}{P} \text{ (20P or smaller)}$$

$$\text{Dedendum, } Wd = \frac{2.157}{P} \text{ (larger than 20P)}$$

$$\text{Whole depth, } b = Wd - a$$

Clearance, c = b − a

Tooth thickness, $t = \dfrac{1.5708}{P}$

13.7. BACKLASH

Backlash in gears is the play between teeth that prevents binding. In terms of tooth dimensions, it is the amount by which the width of tooth spaces exceeds the thickness of the mating gear teeth. Backlash may also be described as the distance, measured along the pitch line, that a gear will move when engaged with another gear that is fixed or immovable, as illustrated in Figure 13.18.

Normally there must be some backlash present in gear drives to provide running clearance. This is necessary, as binding of mating gears can result in heat generation, noise, abnormal wear, possible overload, and/or failure of the drive. A small amount of backlash is also desirable because of the dimensional variations involved in practical manufacturing tolerances. Backlash is built into standard gears during manufacture by cutting the gear teeth thinner than normal by an amount equal to one-half the required figure. When two gears made in this manner are run together, at standard center distance, their allowances combine, provided the full amount of backlash is required.

On nonreversing drives or drives with continuous load in one direction, the increase in backlash that results from tooth wear does not adversely affect operation. However, on reversing drive and drives where timing is critical, excessive backlash usually cannot be tolerated.

13.8. OTHER GEAR TYPES

Many styles and designs of gears have been developed from the spur gear. While they are all commonly used in industry, many are complex in design and manufacture. Only a general description and explanation of principles will be given, as the field of specialized gearing is beyond the scope of this book.

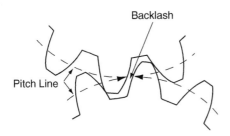

FIGURE 13.18.   Backlash.

Commonly used styles will be discussed sufficiently to provide the millwright or mechanic with the basic information necessary to perform installation and maintenance work.

13.8.1. Bevel and Miter

Two major differences between bevel gears and spur gears are their shape and the relation of the shafts on which they are mounted. The shape of a spur gear is essentially a cylinder, while the shape of a bevel gear is a cone. Spur gears are used to transmit motion between parallel shafts, while bevel gears transmit motion between angular or intersecting shafts. The diagram in Figure 13.19 illustrates the bevel gear's basic cone shape. Figure 13.20 shows a typical pair of bevel gears.

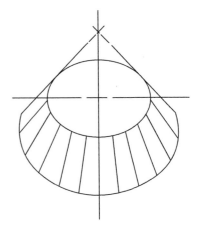

FIGURE 13.19.   Basic shape of bevel gears.

FIGURE 13.20   Typical set of bevel gears.

Special bevel gears can be manufactured to operate at any desired shaft angle, as shown in Figure 13.21. Miter gears are bevel gears with the same number of teeth in both gears operating on shafts at right angles or at 90 degrees, as shown in Figure 13.22.

A typical pair of straight miter gears is shown in Figure 13.23. Another style of miter gears having spiral rather than straight teeth is shown in Figure 13.24. The spiral-tooth style will be discussed later.

The diametrical pitch number, as with spur gears, establishes the tooth size of bevel gears. Because the tooth size varies along its length, it must be measured at a given point. This point is the outside part of the gear where the tooth is the largest. Because each gear in a set of bevel gears must have the same angles and tooth lengths, as well as the same diametrical pitch, they are manufactured and distributed only in mating pairs. Bevel gears, like spur gears, are manufactured in both the 14.5-degree and 20-degree pressure-angle designs.

13.8.2. Helical

Helical gears are designed for parallel-shaft operation like the pair in Figure 13.25. They are similar to spur gears except that the teeth are cut at an angle to the centerline. The principal advantage of this design

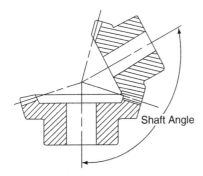

FIGURE 13.21.   Shaft angle, which can be at any degree.

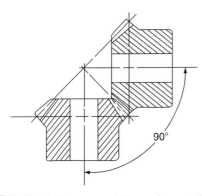

FIGURE 13.22.   Miter gears, which are shown at 90 degrees.

FIGURE 13.23.   Typical set of miter gears.

FIGURE 13.24   Miter gears with spiral teeth.

is the quiet, smooth action that results from the sliding contact of the meshing teeth. A disadvantage, however, is the higher friction and wear that accompanies this sliding action. The angle at which the gear teeth are cut is called the *helix angle* and is illustrated in Figure 13.25.

It is very important to note that the helix angle may be on either side of the gear's centerline. Or if compared to the helix angle of a thread, it may be either a "right-hand" or a "left-hand" helix. The hand of the helix is the same regardless of how it is viewed. Figure 13.26 illustrates a helical gear as viewed from

FIGURE 13.25.   Typical set of helical gears.

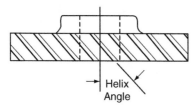

FIGURE 13.26.   Illustrating the angle at which the teeth are cut.

opposite sides; changing the position of the gear cannot change the hand of the tooth's helix angle. A pair of helical gears, as illustrated in Figure 13.27, must have the same pitch and helix angle but must be of opposite hands (one right hand and one left hand).

Helical gears may also be used to connect nonparallel shafts. When used for this purpose, they are often called "spiral" gears or crossed-axis helical gears. This style of helical gearing is shown in Figure 13.28.

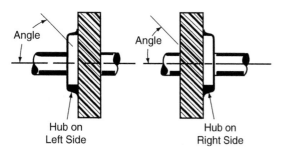

FIGURE 13.27   Helix angle of the teeth the same regardless of side from which the gear is viewed.

FIGURE 13.28.   Typical set of spiral gears.

13.8.3. Worm

The worm and worm gear, illustrated in Figure 13.29, are used to transmit motion and power when a high-ratio speed reduction is required. They provide a steady quiet transmission of power between shafts at right angles. The worm is always the driver and the worm gear the driven member. Like helical gears, worms and worm gears have "hand." The hand is determined by the direction of the angle of the teeth. Thus, in order for a worm and worm gear to mesh correctly, they must be the same hand.

FIGURE 13.29.   Typical set of worm gears.

The most commonly used worms have either one, two, three, or four separate threads and are called single, double, triple, and quadruple thread worms. The number of threads in a worm is determined by counting the number of starts or entrances at the end of the worm. The thread of the worm is an important feature in worm design, as it is a major factor in worm ratios. The ratio of a mating worm and worm gear is found by dividing the number of teeth in the worm gear by the number of threads in the worm.

13.8.4. Herringbone

To overcome the disadvantage of the high end thrust present in helical gears, the herringbone gear, illustrated in Figure 13.30, was developed. It consists simply of two sets of gear teeth, one right-hand and one left-hand, on the same gear. The gear teeth of both hands cause the thrust of one set to cancel out the thrust of the other. Thus, the advantage of helical gears is obtained, and quiet, smooth operation at higher speeds is possible. Obviously they can only be used for transmitting power between parallel shafts.

13.8.5. Gear Dynamics and Failure Modes

Many machine trains utilize gear drive assemblies to connect the driver to the primary machine. Gears and gearboxes typically have several vibration spectra associated with normal operation. Characterization of a gearbox's vibration signature box is difficult to acquire but is an invaluable tool for diagnosing machine-train problems. The difficulty is that (1) it is often difficult to mount the transducer close to the individual gears and (2) the number of vibration sources in a multigear drive

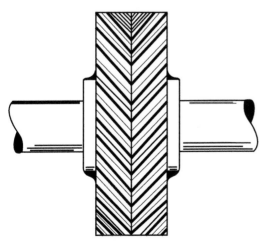

FIGURE 13.30.   Herringbone gear.

results in a complex assortment of gear mesh, modulation, and running frequencies. Severe drive-train vibrations (gearbox) are usually due to resonance between a system's natural frequency and the speed of some shaft. The resonant excitation arises from, and is proportional to, gear inaccuracies that cause small periodic fluctuations in pitch-line velocity. Complex machines usually have many resonance zones within their operating speed range because each shaft can excite a system resonance. At resonance these cyclic excitations may cause large vibration amplitudes and stresses.

Basically, forcing torque arising from gear inaccuracies is small. However, under resonant conditions torsional amplitude growth is restrained only by damping in that mode of vibration. In typical gearboxes this damping is often small and permits the gear-excited torque to generate large vibration amplitudes under resonant conditions.

One other important fact about gear sets is that all gear sets have a designed preload and create an induced load (thrust) in normal operation. The direction, radial or axial, of the thrust load of typical gear sets will provide some insight into the normal preload and induced loads associated with each type of gear.

To implement a predictive maintenance program, a great deal of time should be spent understanding the dynamics of gear/gearbox operation and the frequencies typically associated with the gearbox. As a minimum, the following should be identified.

Gears generate a unique dynamic profile that can be used to evaluate gear condition. In addition, this profile can be used as a tool to evaluate the operating dynamics of the gearbox and its related process system.

Gear Damage

All gear sets create a frequency component, called *gear mesh*. The fundamental gear mesh frequency is equal to the number of gear teeth times the running speed of the shaft. In addition, all gear sets will create a series of sidebands or modulations that will be visible on both sides of the primary gear mesh frequency. In a normal gear set, each of the sidebands will be spaced at exactly the 1X or running speed of the shaft and the profile of the entire gear mesh will be symmetrical.

Normal Profile

In addition, the sidebands will always occur in pairs, one below and one above the gear mesh frequency. The amplitude of each of these pairs will be identical. For example, the sideband pair indicated, as −1 and +1 in Figure 13.31, will be spaced at exactly input speed and have the same amplitude.

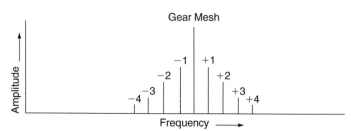

FIGURE 13.31.   Normal profile is symmetrical.

If the gear mesh profile were split by drawing a vertical line through the actual mesh, i.e., the number of teeth times the input shaft speed, the two halves would be exactly identical. Any deviation from a symmetrical gear mesh profile is indicative of a gear problem. However, care must be exercised to ensure that the problem is internal to the gears and induced by outside influences. External misalignment, abnormal induced loads, and a variety of other outside influences will destroy the symmetry of the gear mesh profile. For example, the single reduction gearbox used to transmit power to the mold oscillator system on a continuous caster drives two eccentrics. The eccentric rotation of these two cams is transmitted directly into the gearbox and will create the appearance of eccentric meshing of the gears. The spacing and amplitude of the gear mesh profile will be destroyed by this abnormal induced load.

Excessive Wear

Figure 13.32 illustrates a typical gear profile with worn gears. Note that the spacing between the sidebands becomes erratic and is no longer spaced at the input shaft speed. The sidebands will tend to vary between the input and output speeds but will not be evenly spaced.

In addition to gear tooth wear, center-to-center distance between shafts will create an erratic spacing

and amplitude. If the shafts are too close together, the spacing will tend to be at input shaft speed, but the amplitude will drop drastically. Because the gears are deeply meshed, i.e., below the normal pitch line, the teeth will maintain contact through the entire mesh. This loss of clearance will result in lower amplitudes but will exaggerate any tooth profile defect that may be present.

If the shafts are too far apart, the teeth will mesh above the pitch line. This type of meshing will increase the clearance between teeth and amplify the energy of the actual gear mesh frequency and all of its sidebands. In addition, the load bearing characteristics of the gear teeth will be greatly reduced. Since the pressure is focused on the tip of each tooth, there is less cross-section and strength in the teeth. The potential for tooth failure is increased in direct proportion to the amount of excess clearance between shafts.

Cracked or Broken Tooth

Figure 13.33 illustrates the profile of a gear set with a broken tooth. As the gear rotates, the space left by the chipped or broken tooth will increase the mechanical clearance between the pinion and bullgear. The result will be a low amplitude sideband that will occur to the left of the actual gear mesh frequency. When the next, undamaged teeth mesh, the added clearance will result in a higher energy impact.

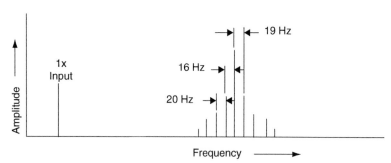

FIGURE 13.32.   Wear or excessive clearance changes sideband spacing.

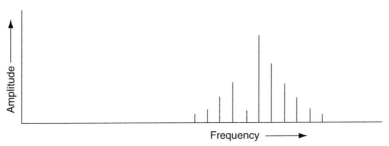

FIGURE 13.33.   A broken tooth will produce an asymmetrical sideband profile.

The resultant sideband, to the right of the mesh frequency, will have much higher amplitude. The paired sidebands will have nonsymmetrical amplitude that represents this disproportional clearance and impact energy.

If the gear set develops problems, the amplitude of the gear mesh frequency will increase, and the symmetry of the sidebands will change. The pattern illustrated in Figure 13.34 is typical of a defective gear set. Note the asymmetrical relationship of the sidebands.

13.8.6. Common Characteristics

You should have a clear understanding of the types of gears generally utilized in today's machinery, how they interact, and the forces they generate on a rotating shaft. There are two basic classifications of gear drives: (1) shaft centers parallel, and (2) shaft centers not parallel. Within these two classifications are several typical gear types.

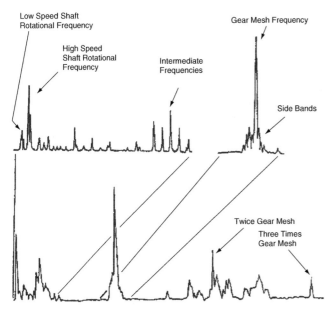

FIGURE 13.34.   Typical defective gear mesh signature.

Shaft Centers Parallel

There are four basic gear types that are typically used in this classification. All are mounted on parallel shafts and, unless an idler gear in also used, will have opposite rotation between the drive and driven gear (if the drive gear has a clockwise rotation, then the driven gear will have a counterclockwise rotation). The gear sets commonly used in machinery include the following.

Spur Gears

The shafts are in the same plane and parallel. The teeth are cut straight and parallel to the axis of the shaft rotation. No more than two sets of teeth are in mesh at one time, so the load is transferred from one tooth to the next tooth rapidly. Usually spur gears are used for moderate- to low-speed applications. Rotation of spur gear sets is opposite unless one or more idler gears are included in the gearbox. Typically, spur gear sets will generate a radial load (preload) opposite the mesh on their shaft support bearings and little or no axial load.

Backlash is an important factor in proper spur gear installation. A certain amount of backlash must be built into the gear drive allowing for tolerances in concentricity and tooth form. Insufficient backlash will cause early failure due to overloading.

As indicated in Figure 13.11, spur gears by design have a preload opposite the mesh and generate an induced load, or *tangential force,* TF, in the direction of rotation. This force can be calculated as:

$$TF = \frac{126,000 * hp}{D_p * rpm}$$

In addition, a spur gear will generate a separating force, S_{TF}, that can be calculated as:

$$S_{TF} = TF * \tan \phi$$

where:

TF = tangential force
hp = input horsepower to pinion or gear
D_p = pitch diameter of pinion or gear
rpm = speed of pinion or gear
φ = pinion or gear tooth pressure angle

Helical Gears

The shafts are in the same plane and parallel but the teeth are cut at an angle to the centerline of the shafts. Helical teeth have an increased length of contact, run more quietly and have a greater strength and capacity than spur gears. Normally the angle created by a line through the center of the tooth and a line parallel to the shaft axis is 45 degrees. However, other angles may be found in machinery. Helical gears also have a preload by design; the critical force to be considered, however, is the thrust load (axial) generated in normal operation; see Figure 13.12.

$$TF = \frac{126,000 * hp}{D_p * rpm}$$

$$S_{TF} = \frac{TF * \tan\phi}{\cos\lambda}$$

$$T_{TF} = TF * \tan\lambda$$

where:

TF = tangential force
S_{TF} = separating force
T_{TF} = thrust force
hp = input horsepower to pinion or gear
D_p = pitch diameter of pinion or gear
rpm = speed of pinion or gear
φ = pinion or gear tooth pressure angle
λ = pinion or gear helix angle

Herringbone Gears

Commonly called "double helical" because they have teeth cut with right and left helix angles, they are used for heavy loads at medium to high speeds. They do not have the inherent thrust forces that are present in helical gear sets. Herringbone gears, by design, cancel the axial loads associated with a single helical gear. The typical loads associated with herringbone gear sets are the radial side-load created by gear mesh pressure and a tangential force in the direction of rotation.

Internal Gears

Internal gears can only be run with an external gear of the same type, pitch, and pressure angle. The preload and induced load will depend on the type of gears used. Refer to spur or helical for axial and radial forces.

13.9. TROUBLESHOOTING

One of the primary causes of gear failure is the fact that, with few exceptions, gear sets are designed for operation in one direction only. Failure is often caused by inappropriate bidirectional operation of the gearbox or backward installation of the gear set. Unless specifically manufactured for bidirectional operation, the "nonpower" side of the gear's teeth is not finished. Therefore, this side is rougher and does not provide the same tolerance as the finished "power" side.

Note that it has become standard practice in some plants to reverse the pinion or bullgear in an effort to extend the gear set's useful life. While this practice permits longer operation times, the torsional power generated by a reversed gear set is not as uniform and consistent as when the gears are properly installed.

Gear overload is another leading cause of failure. In some instances, the overload is constant, which is an indication that the gearbox is not suitable for the application. In other cases, the overload is intermittent and only occurs when the speed changes or when specific production demands cause a momentary spike in the torsional load requirement of the gearbox.

Misalignment, both real and induced, is also a primary root cause of gear failure. The only way to assure that gears are properly aligned is to "hard blue" the gears immediately following installation. After the gears have run for a short time, their wear pattern should be visually inspected. If the pattern does not conform to vendor's specifications, alignment should be adjusted.

Poor maintenance practices are the primary source of real misalignment problems. Proper alignment of gear sets, especially large ones, is not an easy task. Gearbox manufacturers do not provide an easy, positive means to assure that shafts are parallel and that the proper center-to-center distance is maintained.

Induced misalignment is also a common problem with gear drives. Most gearboxes are used to drive other system components, such as bridle or process rolls. If misalignment is present in the driven members (either real or process induced), it also will directly affect the gears. The change in load zone caused by the misaligned driven component will induce misalignment in the gear set. The effect is identical to real misalignment within the gearbox or between the gearbox and mated (i.e., driver and driven) components.

Visual inspection of gears provides a positive means to isolate the potential root cause of gear damage or failures. The wear pattern or deformation of gear teeth provides clues as to the most likely forcing function or cause. The following sections discuss the clues that can be obtained from visual inspection.

TABLE 13.1. Common Failure Modes of Gearboxes and Gear Sets

THE CAUSES	Gear failures	Variations in torsional power	Insufficient power output	Overheated bearings	Short bearing life	Overload on driver	High vibration	High noise levels	Motor trips
Bent shaft				•	•	•	•		
Broken or loose bolts or setscrews				•			•		
Damaged motor						•	•		•
Eliptical gears		•	•			•	•		
Exceeds motor's brake horsepower rating				•			•		
Excessive or too little backlash	•	•							
Excessive torsional loading	•	•	•	•	•	•			•
Foreign object in gearbox	•						•	•	•
Gear set not suitable for application	•		•				•	•	
Gears mounted backward on shafts			•				•	•	
Incorrect center-to-center distance between shafts							•	•	
Incorrect direction of rotation			•			•	•		
Lack of or improper lubrication	•	•		•	•		•	•	•
Misalignment of gears or gearbox	•	•		•	•		•	•	
Overload	•		•	•	•	•			
Process induced misalignment	•	•		•	•				
Unstable foundation		•		•			•	•	
Water or chemicals in gearbox	•								
Worn bearing							•	•	
Worn couplings							•		

Source: Integrated Systems Inc.

13.9.1. Normal Wear

Figure 13.35 illustrates a gear that has a normal wear pattern. Note that the entire surface of each tooth is uniformly smooth above and below the pitch line.

13.9.2. Abnormal Wear

Figures 13.36 through 13.38 illustrate common abnormal wear patterns found in gear sets. Each of these wear patterns suggests one or more potential failure modes for the gearbox.

FIGURE 13.35. Normal wear pattern.

FIGURE 13.36. Wear pattern caused by abrasives in lubricating oil.

FIGURE 13.37.   Pattern caused by corrosive attack on gear teeth.

FIGURE 13.38.   Pitting caused by gear overloading.

Abrasion

Abrasion creates unique wear patterns on the teeth. The pattern varies, depending on the type of abrasion and its specific forcing function. Figure 13.36 illustrates severe abrasive wear caused by particulates in the lubricating oil. Note the score marks that run from the root to the tip of the gear teeth.

Chemical Attack or Corrosion

Water and other foreign substances in the lubricating oil supply also cause gear degradation and premature failure. Figure 13.37 illustrates a typical wear pattern on gears caused by this failure mode.

Overloading

The wear patterns generated by excessive gear loading vary, but all share similar components. Figure 13.38 illustrates pitting caused by excessive torsional loading. The pits are created by the implosion of lubricating oil. Other wear patterns, such as spalling and burning, can also help to identify specific forcing functions or root causes of gear failure.

CHAPTER

14

Packing and Seals*

All machines such as pumps and compressors that handle liquids or gases must include a reliable means of sealing around their shafts so that the fluid being pumped or compressed does not leak. To accomplish this, the machine design must include a seal located at various points to prevent leakage between the shaft and housing. In order to provide a full understanding of seal and packing use and performance, this manual discusses fundamentals, seal design, and installation practices.

14.1. FUNDAMENTALS

Shaft seal requirements and two common types of seals, packed stuffing boxes and simple mechanical seals, are described and discussed in this section. A packed box typically is used on slow- to moderate-speed machinery where a slight amount of leakage is permissible. A mechanical seal is used on centrifugal pumps or other type of fluid handling equipment where shaft sealing is critical.

14.1.1. Shaft Seal Requirements

Figure 14.1 shows the cross-section of a typical end-suction centrifugal pump where the fluid to be pumped enters the suction inlet at the eye of the impeller. Due to the relatively high speed of rotation, the fluid collected within the impeller vanes is held captive because of the close tolerance between the front face of the impeller and the pump housing.

*Source: Ricky Smith and Keith Mobley, *Industrial Machinery Repair: Best Maintenance Practices Pocket Guide* (Boston: Butterworth–Heinemann, 2003), pp. 361–385.

With no other available escape route, the fluid is passed to the outside of the impeller by centrifugal force and into the volute, where its kinetic energy is converted into pressure. At the point of discharge (i.e., discharge nozzle), the fluid is highly pressurized compared to its pressure at the inlet nozzle of the pump. This pressure drives the fluid from the pump and allows a centrifugal pump to move fluids to considerable heights above the centerline of the pump.

This highly pressurized fluid also flows around the impeller to a lower pressure zone where, without an adequate seal, the fluid will leak along the drive shaft to the outside of the pump housing. The lower pressure results from a pump design intended to minimize the pressure behind the impeller. Note that this design element is specifically aimed at making drive shaft sealing easier.

Reducing the pressure acting on the fluid behind the impeller can be accomplished by two different methods, or a combination of both, on an open-impeller unit. One method is where small pumping vanes are cast on the backside of the impeller. The other method is for balance holes to be drilled through the impeller to the suction eye.

In addition to reducing the driving force behind shaft leakage, decreasing the pressure differential between the front and rear of the impeller using one or both of the methods described above greatly decreases the axial thrust on the drive shaft. This decreased pressure prolongs the thrust bearing life significantly.

14.1.2. Sealing Devices

Two sealing devices are described and discussed in this section: packed stuffing boxes and simple mechanical seals.

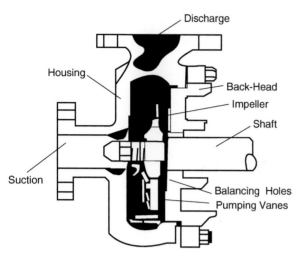

FIGURE 14.1.   Cross-section of a typical end-suction centrifugal pump.

Packed Stuffing Boxes

Before the development of mechanical seals, a soft pliable material or packing placed in a box and compressed into rings encircling the drive shaft was used to prevent leakage. Compressed packing rings between the pump housing and the drive shaft, accomplished by tightening the gland-stuffing follower, formed an effective seal.

Figure 14.2 shows a typical packed box that seals with rings of compressed packing. Note that if this packing is allowed to operate against the shaft without adequate lubrication and cooling, frictional heat eventually builds up to the point of total destruction of the packing and damage to the drive shaft. Therefore, all packed boxes must have a means of lubrication and cooling.

Lubrication and cooling can be accomplished by allowing a small amount of leakage of fluid from the machine or by providing an external source of fluid. When leakage from the machine is used, leaking fluid is captured in collection basins that are built into the machine housing or baseplate. Note that periodic maintenance to recompress the packing must be carried out when leakage becomes excessive.

Packed boxes must be protected against ingress of dirt and air, which can result in loss of resilience and lubricity. When this occurs, packing will act like a grinding stone, effectively destroying the shaft's sacrificial sleeve, and cause the gland to leak excessively. When the sacrificial sleeve on the drive shaft becomes ridged and worn, it should be replaced as soon as possible. In effect, this is a continuing maintenance program that can readily be measured in terms of dollars and time.

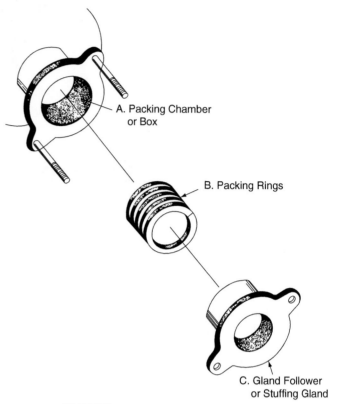

FIGURE 14.2.   Typical packed stuffing box.

Uneven pressures can be exerted on the drive shaft due to irregularities in the packing rings, resulting in irregular contact with the shaft. This causes uneven distribution of lubrication flow at certain locations, producing acute wear and packed-box leakages. The only effective solution to this problem is to replace the shaft sleeve or drive shaft at the earliest opportunity.

Simple Mechanical Seals

Mechanical seals, which are typically installed in applications where no leakage can be tolerated, are described and discussed in this section. Toxic chemicals and other hazardous materials are primary examples of applications where mechanical seals are used.

Components and Assembly

Figure 14.3 shows the components of a simple mechanical seal, which is made up of the following:

- Coil spring
- O-ring shaft packing
- Seal ring

The seal ring fits over the shaft and rotates with it. The spring must be made from a material that is compatible with the fluid being pumped so that it will withstand

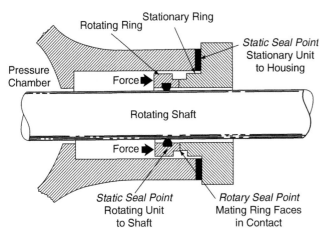

FIGURE 14.3. Simple mechanical seal.

corrosion. Likewise, the same care must be taken in material selection of the O-ring and seal materials. The insert and insert O-ring mounting are installed in the bore cavity provided in the gland ring. This assembly is installed in a pump-stuffing box, which remains stationary when the pump shaft rotates.

A carbon graphite insertion ring provides a good bearing surface for the seal ring to rotate against. It is also resistive to attack by corrosive chemicals over a wide range of temperatures.

Figure 14.4 depicts a simple seal that has been installed in the pump's stuffing box. Note how the coil spring sits against the back of the pump's impeller, pushing the packing O-ring against the seal ring. By doing so, it remains in constant contact with the stationary insert ring.

As the pump shaft rotates, the shaft packing rotates with it due to friction. (In more complex mechanical seals, the shaft-packing element is secured to the rotat-

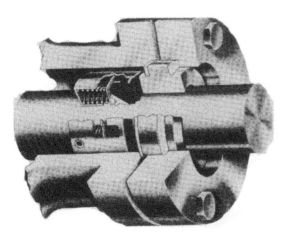

FIGURE 14.4. Pump stuffing box seal.

ing shaft by Allen screws.) There is also friction between the spring, the impeller, and the compressed O-ring. Thus, the whole assembly rotates together when the pump shaft rotates. The stationary insert ring is located within the gland bore. The gland itself is bolted to the face of the stuffing box. This part is held stationary due to the friction between the O-ring insert mounting and the inside diameter (ID) of the gland bore as the shaft rotates within the bore of the insert.

How It Prevents Leakage

Having discussed how a simple mechanical seal is assembled in the stuffing box, we must now consider how the pumped fluid is stopped from leaking out to the atmosphere.

In Figure 14.4, the O-ring shaft packing blocks the path of the fluid along the drive shaft. Any fluid attempting to pass through the seal ring is stopped by the O-ring shaft packing. Any further attempt by the fluid to pass through the seal ring to the atmospheric side of the pump is prevented by the gland gasket and the O-ring insert. The only other place where fluid can potentially escape is the joint surface, which is between the rotating carbon ring and the stationary insert. (Note: The surface areas of both rings must be machined-lapped perfectly flat, measured in light bands with tolerances of one-millionth of an inch.)

Sealing Area and Lubrication

The efficiency of all mechanical seals is dependent upon the condition of the sealing area surfaces. The surfaces remain in contact with each other for the effective working life of the seal and are friction-bearing surfaces.

As in the compressed packing gland, lubrication also must be provided in mechanical seals. The sealing area surfaces should be lubricated and cooled with pumped fluid (if it is clean enough) or an outside source of clean fluid. However, much less lubrication is required with this type of seal because the frictional surface area is smaller than that of a compressed packing gland, and the contact pressure is equally distributed throughout the interface. As a result, a smaller amount of lubrication passes between the seal faces to exit as leakage.

In most packing glands there is a measurable flow of lubrication fluid between the packing rings and the shaft. With mechanical seals, the faces ride on a microscopic film of fluid that migrates between them, resulting in leakage. However, leakage is so slight that if the temperature of the fluid is above its saturation point at atmospheric pressure, it flashes off to vapor before it can be visually detected.

Advantages and Disadvantages

Mechanical seals offer a more reliable seal than compressed packing seals. Because the spring in a mechanical seal exerts a constant pressure on the seal ring, it automatically adjusts for wear at the faces. Thus, the need for manual adjustment is eliminated. Additionally, because the bearing surface is between the rotating and stationary components of the seal, the shaft or shaft sleeve does not become worn. Although the seal will eventually wear out and need replacing, the shaft will not experience wear.

However, much more precision and attention to detail must be given to the installation of mechanical seals compared to conventional packing. Nevertheless, it is not unusual for mechanical seals to remain in service for many thousands of operational hours if they have been properly installed and maintained.

14.2. MECHANICAL SEAL DESIGNS

Mechanical seal designs are referred to as friction drives, or single-coil spring seals, and positive drives.

14.2.1. Single-Coil Spring Seal

The seal shown back in Figure 14.4 depicts a typical friction drive or single-coil spring seal unit. This design is limited in its use because the seal relies on friction to turn the rotary unit. Because of this, its use is limited to liquids such as water or other nonlubricating fluids. If this type of seal is to be used with liquids that have natural lubricating properties, it must be mechanically locked to the drive shaft.

Although this simple seal performs its function satisfactorily, there are two drawbacks that must be considered. Both drawbacks are related to the use of a coil spring that is fitted over the drive shaft:

- One drawback of the spring is the need for relatively low shaft speeds because of a natural tendency of the components to distort at high surface speeds. This makes the spring push harder on one side of the seal than the other, resulting in an uneven liquid film between the faces. These cause excessive leakage and wear at the seal.
- The other drawback is simply one of economics. Because pumps come in a variety of shaft sizes and speeds, the use of this type of seal requires inventorying several sizes of spare springs, which ties up capital.

Nevertheless, the simple and reliable coil spring seal has proven itself in the pumping industry and is often selected for use despite its drawbacks. In regulated industries, this type of seal design far exceeds the capabilities of a compressed packing ring seal.

14.2.2. Positive Drive

There are two methods of converting a simple seal to positive drive. Both methods, which use collars secured to the drive shaft by setscrews, are shown in Figure 14.5. In this figure, the end tabs of the spring are bent at 90 degrees to the natural curve of the spring. These end tabs fit into notches in both the collar and the seal ring. This design transmits rotational drive from the collar to the seal ring by the spring. Figure 14.5 also shows two horizontally mounted pins that extend over the spring from the collar to the seal ring.

14.3. INSTALLATION PROCEDURES

This section describes the installation procedures for packed stuffing boxes and mechanical seals.

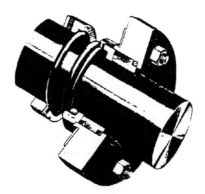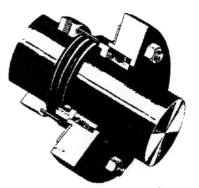

FIGURE 14.5. Conversion of a simple seal to positive drive.

14.3.1. Packed Stuffing Box

This procedure provides detailed instructions on how to repack centrifugal pump packed stuffing boxes or glands. The methodology described here is applicable to other gland sealed units such as valves and reciprocating machinery.

Tool List

The following is a list of the tools needed to repack a centrifugal pump gland:

- Approved packing for specific equipment
- Mandrel sized to shaft diameter
- Packing ring extractor tool
- Packing board
- Sharp knife
- Approved cleaning solvent
- Lint-free cleaning rags

Precautions

The following precautions should be taken in repacking a packed stuffing box:

- Ensure coordination with operations control.
- Observe site and area safety precautions at all times.
- Ensure equipment has been electrically isolated and suitably locked out and tagged.
- Ensure machine is isolated and depressurized, with suction and discharge valves chained and locked shut.

Installation

The following are the steps to follow in installing a gland:

1. Loosen and remove nuts from the gland bolts.
2. Examine threads on bolts and nuts for stretching or damage. Replace if defective.
3. Remove the gland follower from the stuffing box and slide it along the shaft to provide access to the packing area.
4. Use packing extraction tool to carefully remove packing from the gland.
5. Keep the packing rings in the order they are extracted from the gland box. This is important in evaluating wear characteristics. Look for rub marks and any other unusual markings that would identify operational problems.
6. Carefully remove the lantern ring. This is a grooved, bobbin-like spool piece that is situated exactly on the centerline of the seal water inlet connection to the gland (Figure 14.6). Note: It is most important to place the lantern ring under the seal water inlet connection to ensure the water is properly distributed within the gland to perform its cooling and lubricating functions.

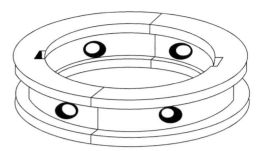

FIGURE 14.6.   Lantern ring or seal cage.

7. Examine the lantern ring for scoring and possible signs of crushing. Make sure the lantern ring's outside diameter (OD) provides a sliding fit with the gland box's internal dimension. Check that the lantern ring's ID is a free fit along the pump's shaft sleeve. If the lantern ring does not meet this simple criterion, replace it with a new one.
8. Continue to remove the rest of the packing rings as previously described. Retain each ring in the sequence that it was removed for examination.
9. Do not discard packing rings until they have been thoroughly examined for potential problems.
10. Turn on the gland seal cooling water slightly to ensure there is no blockage in the line. Shut the valve when good flow conditions are established.
11. Repeat Steps 1 through 10 with the other gland box.
12. Carefully clean out the gland stuffing boxes with a solvent-soaked rag to ensure that no debris is left behind.
13. Examine the shaft sleeve in both gland areas for excessive wear caused by poorly lubricated or over-tightened packing. Note: If the shaft sleeve is ridged or badly scratched in any way, the split housing of the pump may have to be split and the impeller removed for the sleeve to be replaced. Badly installed and maintained packing causes this.
14. Check total indicated runout (TIR) of the pump shaft by placing a magnetic base-mounted dial indicator on the pump housing and a dial stem on the shaft. Zero the dial and rotate the pump shaft one full turn. Record reading (Figure 14.7). Note: If the TIR is greater than ±0.002 inches, the pump shaft should be straightened.
15. Determine the correct packing size before installing using the following method (Figure 14.8):

    Measure the ID of the stuffing box, which is the OD at the packing (*B*), and the diameter of the shaft (*A*). With this data, the packing cross-section size is calculated by:

$$\text{Packing Cross Section} = \frac{B - A}{2}$$

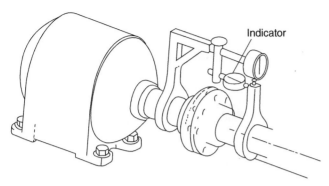

FIGURE 14.7.   Dial indicator check for runout.

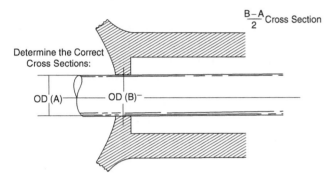

FIGURE 14.8.   Selecting correct packing size.

The packing length is determined by calculating the circumference of the packing within the stuffing box. The centerline diameter is calculated by adding the diameter of the shaft to the packing cross-section that was calculated in the preceding formula. For example, a stuffing box with a 4" ID and a shaft with a 2" diameter will require a packing cross section of 1". The centerline of the packing would then be 3".

Therefore, the approximate length of each piece of packing would be:

$$\text{Packing Length} = \text{Centerline Diameter} \times 3.1416$$
$$= 3.0 \times 3.1416 = 9.43 \text{ inches}$$

The packing should be cut approximately ¼" longer than the calculated length so that the end can be bevel cut.

16. Controlled leakage rates easily can be achieved with the correct size packing.

17. Cut the packing rings to size on a wooden mandrel that is the same diameter as the pump shaft. Rings can be cut either square (butt cut) or diagonally (approximately 30 degrees). Note: Leave at least a 1/6" gap between the butts regardless of the type of cut used. This permits the packing rings to move under compression or temperature without binding on the shaft surface.

18. Ensure that the gland area is perfectly clean and is not scratched in any way before installing the packing rings.

19. Lubricate each ring lightly before installing in the stuffing box. Note: When putting packing rings around the shaft, use an "S" twist. *Do not bend open.* See Figure 14.9.

20. Use a split bushing to install each ring, ensuring that the ring bottoms out inside the stuffing box. An offset tamping stick may be used if a split bushing is not available. *Do not use a screwdriver.*

21. Stagger the butt joints, placing the first ring butt at 12 o'clock; the second at 6 o'clock; the third at 3 o'clock; the fourth at 9 o'clock; etc., until the packing box is filled (Figure 14.10). Note: When the last ring has been installed, there should be enough room to insert the gland follower 1/8 to 3/16 inches into the stuffing box (Figure 14.11).

22. Install the lantern ring in its correct location within the gland. Do not force the lantern ring into position (Figure 14.12).

23. Tighten up the gland bolts with a wrench to seat and form the packing to the stuffing box and shaft.

24. Loosen the gland nuts one complete turn and rotate the shaft by hand to get running clearance.

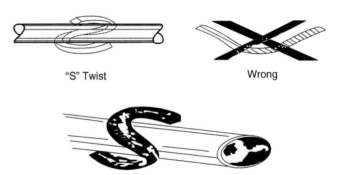

FIGURE 14.9.   Proper and improper installation of packing.

FIGURE 14.10.   Stagger butt joints.

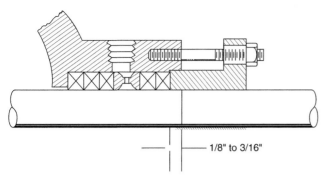

1/8" to 3/16"

FIGURE 14.11   Proper gland follower clearance.

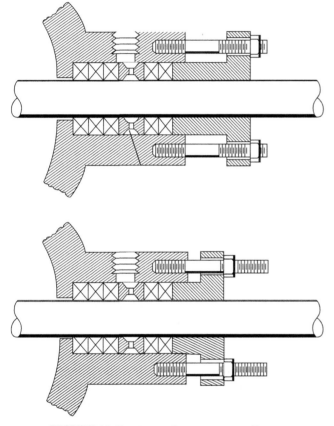

FIGURE 14.12.   Proper lantern ring installation.

25. Retighten the nuts finger tight only. Again, rotate the shaft by hand to make sure the packing is not too tight.
26. Contact operations to start the pump and allow the stuffing box to leak freely. Tighten the gland bolts one flat at a time until the desired leakage is obtained and the pump runs cool.
27. Clean up the work area and account for all tools before returning them to the tool crib.
28. Inform operations of project status and complete all paperwork.

29. After the pump is in operation, periodically inspect the gland to determine its performance. If it tends to leak more than the allowable amount, tighten by turning the nuts one flat at a time. Give the packing enough time to adjust before tightening it more. If the gland is tightened too much at one time, the packing can be excessively compressed, causing unnecessary friction and subsequent burnout of the packing.

14.3.2. Mechanical Seals

A mechanical seal's performance depends on the operating condition of the equipment where it is installed. Therefore, inspection of the equipment before seal installation can potentially prevent seal failure and reduce overall maintenance expenses.

Equipment Checkpoints

The pre-installation equipment inspection should include the following: stuffing box space, lateral or axial shaft movement (end play), radial shaft movement (whip or deflection), shaft runout (bent shaft), stuffing box face squareness, stuffing box bore concentricity, driver alignment, and pipe strain.

Stuffing Box Space

To properly receive the seal, the radial space and depth of the stuffing box must be the same as the dimensions shown on the seal assembly drawing.

Lateral or Axial Shaft Movement (Endplay)

Install a dial indicator with the stem against the shoulder of the shaft. Use a soft hammer or mallet to lightly tap the shaft on one end and then on the other. Total indicated endplay should be between 0.001 and 0.004 inches. A mechanical seal cannot work properly with a large amount of endplay or lateral movement. If the hydraulic condition changes (as frequently happens), the shaft could "float," resulting in sealing problems. Minimum endplay is a desirable condition for the following reasons:

• Excessive endplay can cause pitting, fretting, or wear at the point of contact between the shaft packing in the mechanical seal and the shaft or sleeve OD. As the mechanical seal-driving element is locked to the shaft or sleeve, any excessive endplay will result in either overloading or underloading of the springs, causing excessive wear or leaks.
• Excessive endplay as a result of defective thrust bearings can reduce seal performance by disturbing both the established wear pattern and the lubricating film.

- A floating shaft can cause chattering, which results in chipping of the seal faces, especially the carbon element. Ideal mechanical seal performance requires a uniform wear pattern and a liquid film between the mating contact faces.

Radial Shaft Movement (Whip or Deflection)

Install the dial indicator as close to the radial bearing as possible. Lift the shaft or exert light pressure at the impeller end. If more than 0.002 to 0.003 inches of radial movement occurs, investigate bearings and bearing fits (especially the bore) for the radial bearing fit. An oversized radial bearing bore caused by wear, improper machining, or corrosion will cause excessive radial shaft movement resulting in shaft whip and deflection. Minimum radial shaft movement is important for the following reasons:

- Excessive radial movement can cause wear, fretting, or pitting of the shaft packing or secondary sealing element at the point of contact between the shaft packing and the shaft or sleeve OD.
- Extreme wear at the mating contact faces will occur when excessive shaft whip or deflection is present due to defective radial bearings or bearing fits. The contact area of the mating faces will be increased, resulting in increased wear and the elimination or reduction of the lubricating film between the faces, further shortening seal life.

Shaft Runout (Bent Shaft)

A bent shaft can lead to poor sealing and cause vibration. Bearing life is greatly reduced, and the operating conditions of both radial and thrust bearings can be affected.

Clamp the dial indicator to the pump housing and measure the shaft runout at two or more points on the OD of the impeller end of the shaft. Also measure the shaft runout at the coupling end of the shaft. If the runout exceeds 0.002 inches, remove the shaft and straighten or replace it.

Square Stuffing Box Face

With the pump stuffing box cover bolted down, clamp the dial indicator to the shaft with the stem against the face of the stuffing box. The total indicator runout should not exceed 0.003 inches.

When the face of the stuffing box is "out-of-square," or not perpendicular to the shaft axis, the result can be serious malfunction of a mechanical seal for the following reasons:

- The stationary gland plate that holds the stationary insert or seat in position is bolted to the face of the stuffing box. Misalignment will cause the gland to cock, resulting in cocking of the stationary element.

This results in seal wobble or operation in an elliptical pattern. This condition is a major factor in fretting, pitting, and wearing of the mechanical seal shaft packing at the point of contact with the shaft or sleeve.

- A seal that is wobbling on the shaft can also cause wear on the drive pins. Erratic wear on the face contact causes poor seal performance.

Stuffing Box Bore Concentricity

With the dial indicator set up as described above, place the indicator stem well into the bore of the stuffing box. The stuffing box should be concentric to the shaft axis to within a 0.005-inch total indicator reading.

Eccentricity alters the hydraulic loading of the seal faces, reducing seal life and performance. If the shaft is eccentric to the box bore, check the slop, or looseness, in the pump bracket fits. Rust, atmospheric corrosion, or corrosion from leaking gaskets can cause damage to these fits, making it impossible to ensure a stuffing box that is concentric with the shaft. A possible remedy for this condition is welding the corroded area and remachining to proper dimensions.

Driver Alignment and Pipe Strain

Driver alignment is extremely important, and periodic checks should be performed. Pipe strain can also damage pumps, bearings, and seals.

In most plants, it is customary to blind the suction and discharge flanges of inactive pumps. These blinds should be removed before the pump driver alignment is made, or the alignment job is incomplete.

After the blinds have been removed and as the flanges on the suction and discharge are being connected to the piping, check the dial indicator reading on the outside diameter of the coupling half and observe movement of the indicator dial as the flanges are being secured. Deviation indicates pipe strain. If severe strain exists, corrective measures should be taken, or damage to the pump and unsatisfactory seal service can result.

Seal Checkpoints

The following are important seal checkpoints:

- Ensure that all parts are kept clean, especially the running faces of the seal ring and insert.
- Check the seal rotary unit, and make sure the drive pins and spring pins are free in the pinholes or slots.
- Check the setscrews in the rotary unit collar to see that they are free in the threads. Setscrews should be replaced after each use.

- Check the thickness of all gaskets against the dimensions shown on the assembly drawing. Improper gasket thickness will affect the seal setting and the spring load imposed on the seal.
- Check the fit of the gland ring to the equipment. Make sure there is no interference or binding on the studs or bolts or other obstructions. Be sure the gland ring pilot, if any, enters the bore with a reasonable guiding fit for proper seal alignment.
- Make sure all rotary unit parts of the seal fit over the shaft freely.
- Check both running faces of the seal (seal ring and insert) and be sure there are no nicks or scratches. Imperfections of any kind on either of these faces will cause leaks.

Installing the Seal

The following steps should be taken when installing a seal:

- Instruction booklets and a copy of the assembly drawing are shipped with each seal. Be sure each is available, and read the instructions before starting installation.
- Remove all burrs and sharp edges from the shaft or shaft sleeve, including sharp edges of keyways and threads. Worn shafts or sleeves should be replaced.
- Check the stuffing box bore and face to ensure they are clean and free of burrs.
- The shaft or sleeve should be lightly oiled before the seal is assembled to allow the seal parts to move freely over it. This is especially desirable when assembling the seal collar because the bore of the collar usually has only a few thousandths of an inch clearance. Care should be taken to avoid getting the collar cocked.
- Install the rotary unit parts on the shaft or sleeve in the proper order.
- Be careful when passing the seal gland ring and insert over the shaft. Do not bring the insert against the shaft because it might chip away small pieces from the edge of the running face.
- Wipe the seal faces clean and apply a clean oil film before completing the equipment assembly. A clean finger, which is not apt to leave lint, will do the best job when giving the seal faces the final wiping.
- Complete the equipment assembly, taking care when compressing the seal into the stuffing box.
- Seat the gland ring and gland ring gasket to the face of the stuffing box by tightening the nuts or bolts evenly and firmly. Be sure the gland ring is not cocked. Tighten the nuts or bolts only enough to affect a seal at the gland ring gasket, usually finger tight and ½ to ¾ of a turn with a wrench. Excessively tightening the gland ring nut or bolt will cause distortion that will be transmitted to the running face, resulting in leaks.

If the seal assembly drawing is not available, the proper seal setting dimension for inside seals can be determined as follows:

- Establish a reference mark on the shaft or sleeve flush with the face of the stuffing box.
- Determine how far the face of the insert will extend into the stuffing box bore. This dimension is taken from the face of the gasket.
- Determine the compressed length of the rotary unit by compressing the rotary unit to the proper spring gap.
- This dimension added to the distance the insert extends into the stuffing box will give the seal setting dimension from the reference mark on the shaft or sleeve to the back of the seal collar.
- Outside seals are set with the spring gap equal to the dimension stamped on the seal collar.
- Cartridge seals are set at the factory and installed as complete assemblies. These assemblies contain spacers that must be removed after the seal assembly is bolted into position and the sleeve collar is in place.

Installation of Environmental Controls

Mechanical seals are often chosen and designed to operate with environmental controls. If this is the case, check the seal assembly drawing or equipment drawing to ensure that all environmental control piping is properly installed. Before equipment startup, all cooling and heating lines should be operating and remain so for at least a short period after equipment shutdown.

Before startup, all systems should be properly vented. This is especially important on vertical installations where the stuffing box is the uppermost portion of the pressure-containing part of the equipment. The stuffing box area must be properly vented to avoid a vapor lock in the seal area that would cause the seal to run dry.

On double seal installations, be sure the sealing liquid lines are connected, the pressure control valves are properly adjusted, and the sealing liquid system is operating before starting the equipment.

Seal Startup Procedures

When starting equipment with mechanical seals, make sure the seal faces are immersed in liquid from the beginning so they will not be damaged from dry operation. The following recommendations for seal startup apply to most types of seal installations and will improve seal life if followed:

- Caution the electrician not to run the equipment dry while checking motor rotation. A slight turnover will not hurt the seal, but operating full speed for several minutes under dry conditions will destroy or severely damage the rubbing faces.

- The stuffing box of the equipment, especially centrifugal pumps, should always be vented before startup. Even though the pump has a flooded suction, it is still possible that air may be trapped in the top of the stuffing box after the initial liquid purge of the pump.
- Check installation for need of priming. Priming might be necessary in applications with a low or negative suction head.
- Where cooling or bypass recirculation taps are incorporated in the seal gland, piping must be connected to or from these taps before startup.

These specific environmental control features must be used to protect the organic materials in the seal and to ensure its proper performance. Cooling lines should be left open at all times or whenever possible. This is especially true when a hot product might be passing through standby equipment while it is not online. Many systems provide for product to pass through the standby equipment, so the need for additional product volume or an equipment change is only a matter of pushing a button.

- With hot operational equipment that is shut down at the end of each day, it is best to leave the cooling water on at least long enough for the seal area to cool below the temperature limits of the organic materials in the seal.
- Face lubricated–type seals must be connected from the source of lubrication to the tap openings in the seal gland before startup. This is another predetermined environmental control feature that is mandatory for proper seal function. Where double seals are to be operated, it is necessary that the lubrication feed lines be connected to the proper ports for both circulatory or dead-end systems before equipment startup. This is very important because all types of double seals depend on the controlled pressure and flow of the sealing fluid to function properly. Even before the shaft is rotated, the sealing liquid pressure must exceed the product pressure opposing the seal. Be sure a vapor trap does not prevent the lubricant from reaching the seal face promptly.
- Thorough warm-up procedures include a check of all steam piping arrangements to be sure that all are connected and functioning, as products that will solidify must be fully melted before startup. It is advisable to leave all heat sources on during shutdown to ensure a liquid condition of the product at all times. Leaving the heat on at all times further facilitates quick start-ups and equipment switchovers that may be necessary during a production cycle.

- Thorough chilling procedures are necessary on some installations, especially liquefied petroleum gases (LPG) applications. LPG must always be kept in a liquid state in the seal area, and startup is usually the most critical time. Even during operation, the recirculation line piped to the stuffing box might have to be run through a cooler in order to overcome frictional heat generated at the seal faces. LPG requires a stuffing box pressure that is greater than the vapor pressure of the product at pumping temperature (25 to 50 psi differential is desired).

14.4. TROUBLESHOOTING

Failure modes that affect shaft seals are normally limited to excessive leakage and premature failure of the mechanical seal or packing. Table 14.1 lists the common failure modes for both mechanical seals and packed boxes. As the table indicates, most of these failure modes can be directly attributed to misapplication, improper installation, or poor maintenance practices.

14.4.1. Mechanical Seals

By design, mechanical seals are the weakest link in a machine train. If there is any misalignment or eccentric shaft rotation, the probability of a mechanical seal failure is extremely high. Most seal tolerances are limited to no more than 0.002 inches of total shaft deflection or misalignment. Any deviation outside of this limited range will cause catastrophic seal failure.

Physical misalignment of a shaft will either cause seal damage, permitting some leakage through the seal, or it will result in total seal failure. Therefore, it is imperative that good alignment practices be followed for all shafts that have an installed mechanical seal.

Process and machine-induced shaft instability also create seal problems. Primary causes for this failure mode include aerodynamic or hydraulic instability, critical speeds, mechanical imbalance, process load changes, or radical speed changes. These can cause the shaft to deviate from its true centerline enough to result in seal damage.

Chemical attack (i.e., corrosion or chemical reaction with the liquid being sealed) is another primary source of mechanical seal problems. Generally, two primary factors cause chemical attack: misapplication or improper flushing of the seal.

Misapplication is another major cause of premature seal failure. Little attention is generally given to the selection of mechanical seals. Most plants rely on the vendor to provide a seal that is compatible with the application. Too often there is a serious breakdown

TABLE 14.1.   Common Failure Modes of Packing and Mechanical Seals

	THE CAUSES	Excessive leakage	Continuous stream of liquid	No leakage	Shaft hard to turn	Shaft damage under packing	Frequent replacement required	Bellows spring failure	Seal face failure
Packed box — Nonrotating	Cut ends of packing not staggered	•	•				•		
	Line pressure too high	•							
	Not packed properly				•	•	•		
	Packed box too loose	•	•						
	Packing gland too loose	•	•						
	Packing gland too tight	•	•		•	•	•		
Packed box — Rotating	Cut end of packing not staggered		•						
	Line pressure too high	•							
	Mechanical damage (seals, seat)	•	•	•			•		
	Noncompatible packing	•	•			•			
	Packing gland too loose	•							
	Packing gland too tight				•	•	•		
Mechanical seal — Internal flush	Flush flow/pressure too low							•	•
	Flush pressure too high	•	•					•	•
	Improperly installed	•						•	•
	Induced misalignment	•							
	Internal flush line plugged							•	•
	Line pressure too high							•	•
	Physical shaft misalignment	•							
	Seal not compatible with application	•							
Mechanical seal — External flush	Contamination in flush liquid	•							•
	External flush line plugged							•	•
	Flush flow/pressure too low							•	•
	Flush pressure too high	•	•					•	•
	Improperly installed	•							•
	Induced misalignment	•						•	•
	Line pressure too high							•	•
	Physical shaft misalignment	•						•	•
	Seal not compatible with application	•							•

Source: Integrated Systems Inc.

in communications between the end user and the vendor on this subject. Either the procurement specification does not provide the vendor with appropriate information, or the vendor does not offer the option of custom ordering the seals. Regardless of the reason, mechanical seals are often improperly selected and used in inappropriate applications.

Seal Flushing

When installed in corrosive chemical applications, mechanical seals must have a clear water flush system to prevent chemical attack. The flushing system must provide a positive flow of clean liquid to the seal and also provide an enclosed drain line that removes the flushing liquid. The flow rate and pressure of the flushing liquid will vary depending on the specific type of seal but must be enough to assure complete, continuous flushing.

14.4.2. Packed Boxes

Packing is used to seal shafts in a variety of applications. In equipment where the shaft is not continuously rotating (e.g., valves), packed boxes can be used successfully without any leakage around the shaft. In rotating applications, such as pump shafts, the application must be able to tolerate some leakage around the shaft.

Nonrotating Applications

In nonrotating applications, packing can be installed tightly enough to prevent leakage around the shaft. As long as the packing is properly installed and the stuffing-box gland is properly tightened, there is very little probability that seal failure will occur. This type of application does require periodic maintenance to ensure that the stuffing-box gland is properly tightened or that the packing is replaced when required.

Rotating Applications

In applications where a shaft continuously rotates, packing cannot be tight enough to prevent leakage. In fact, some leakage is required to provide both flushing and cooling of the packing. Properly installed and maintained packed boxes should not fail or contribute to equipment reliability problems. Proper installation is relatively easy, and routine maintenance is limited to periodic tightening of the stuffing-box gland.

15

Electric Motors

Electric motors are the most common source of motive power for machine trains. As a result, more of them are evaluated using microprocessor-based vibration monitoring systems than any other driver. The vibration frequencies of the following parameters are monitored to evaluate operating condition. This information is used to establish a database.

- Bearing frequencies
- Imbalance
- Line frequency
- Loose rotor bars
- Running speed
- Slip frequency
- V-belt intermediate drives

15.1. BEARING FREQUENCIES

Electric motors may incorporate either sleeve or rolling-element bearings. A narrowband window should be established to monitor both the normal rotational and defect frequencies associated with the type of bearing used for each application.

15.2. IMBALANCE

Electric motors are susceptible to a variety of forcing functions that cause instability or imbalance. The narrowbands established to monitor the fundamental and other harmonics of actual running speed are useful in identifying mechanical imbalance, but other indices also should be used.

One such index is line frequency, which provides indications of instability. Modulations, or harmonics, of line frequency may indicate the motor's inability to find and hold a magnetic center. Variations in line frequency also increase the amplitude of the fundamental and other harmonics of running speed.

Axial movement and the resulting presence of a third harmonic of running speed is another indication of instability or imbalance within the motor. The third harmonic is present whenever there is axial thrusting of a rotating element.

15.3. LINE FREQUENCY

Many electrical problems, or problems associated with the quality of the incoming power and internal to the motor, can be isolated by monitoring the line frequency. Line frequency refers to the frequency of the alternating current being supplied to the motor. In the case of 60-cycle power, monitoring of the fundamental or first harmonic (60 Hertz), second harmonic (120 Hz), and third harmonic (180 Hz) should be performed.

15.4. LOOSE ROTOR BARS

Loose rotor bars are a common failure mode of electric motors. Two methods can be used to identify them. The first method uses high-frequency vibration components that result from oscillating rotor bars. Typically, these frequencies are well above the normal maximum frequency used to establish the broadband signature. If this is the case, a high-pass filter, such as high-frequency domain, can be used to monitor the condition of the rotor bars.

The second method uses the slip frequency to monitor for loose rotor bars. The passing frequency created by this failure mode energizes modulations associated with slip. This method is preferred since these frequency components are within the normal bandwidth used for vibration analysis.

15.5. RUNNING SPEED

The running speed of electric motors, both alternating current (ac) and direct current (dc), varies. Therefore, for monitoring purposes, these motors should be classified as variable-speed machines. A narrowband window should be established to track the true running speed.

15.6. SLIP FREQUENCY

Slip frequency is the difference between the synchronous speed and the actual running speed of the motor. A narrowband filter should be established to monitor electrical line frequency. The window should have enough resolution to clearly identify the frequency and the modulations, or sidebands, that represent slip frequency. Normally, these modulations are spaced at the difference between synchronous and actual speed, and the number of sidebands is equal to the number of poles in the motor.

15.7. V-BELT INTERMEDIATE DRIVES

Electric motors with V-belt intermediate drives display the same failure modes as those described previously. However, the unique V-belt frequencies should be monitored to determine if improper belt tension or misalignment is evident.

In addition, electric motors used with V-belt intermediate drive assemblies are susceptible to premature wear on the bearings. Typically, electric motors are not designed to compensate for the sideloads associated with V-belt drives. In this type of application, special attention should be paid to monitoring motor bearings.

15.8. ELECTRIC MOTOR ANALYSIS

The primary data-measurement point on the inboard bearing housing should be located in the plane opposing the induced load (sideload), with the secondary point at 90 degrees. The outboard primary data-measurement point should be in a plane opposite the inboard bearing with the secondary at 90 degrees.

Both radial (x- and y-axis) measurements should be taken at the inboard and outboard bearing housings. Orientation of the measurements is determined by the anticipated induced load created by the driven units. The primary (x-axis) radial measurement should be positioned in the same plane as the worst anticipated shaft displacement. The secondary (y-axis) radial should be positioned at 90 degrees in the direction of rotation to the primary point and oriented to permit vector analysis of actual shaft displacement.

Horizontal motors rely on a magnetic center generated by its electrical field to position the rotor in the axial (z-axis) plane between the inboard and outboard bearings. Therefore, most electric motors are designed with two float bearings instead of the normal configuration incorporating one float and one fixed bearing. Vertical motors should have an axial (z-axis) measurement point at the inboard bearing nearest the coupling and oriented in an upward direction. This data point monitors the downward axial force created by gravity or an abnormal load.

Electric motors are not designed to absorb sideloads, such as those induced by V-belt drives. In applications where V-belts or other radial loads are placed on the motor, the primary radial transducer (x-axis) should be oriented opposite the direction of induced load and the secondary radial (y-axis) point should be positioned at 90 degrees in the direction of rotation. If, for safety reasons, the primary transducer cannot be positioned opposite the induced load, the two radial transducers should be placed at 45 degrees on either side of the load plane created by the side load.

Totally enclosed, fan-cooled, and explosion-proof motors create some difficulty in acquiring data on the outboard bearing. By design, the outboard bearing housing is not accessible. The optimum method of acquiring data is to permanently mount a sensor on the outboard-bearing housing and run the wires to a convenient data-acquisition location. If this is not possible, the x-y data points should be as close as possible to the bearing housing. Ensure that there is a direct mechanical path to the outboard bearing. The use of this approach results in some loss of signal strength from motor-mass damping. Do not obtain data from the fan housing.

Reliability Articles

16.1. TOP FIVE REASONS WHY COMPANIES DON'T MEASURE RELIABILITY: IT SEEMS LIKE EVERYONE HAS AN EXCUSE AS TO WHY THEY DON'T MEASURE RELIABILITY*

Most companies don't measure mean time between failures (MTBF), even though it's the most basic measurement that quantifies reliability. MTBF is the average time an asset functions before it fails. So, why don't they measure MTBF?

16.1.1. Reason 1

Work orders don't capture all emergency work. Many companies have rules such as, "A work order will be written only if the equipment is down for more than one hour." This rule doesn't make sense. Let's say, for example, a circuit overload on a piece of equipment trips 100 times in a month. Many times, small problems lead to major asset failure. Don't wait until a small problem becomes a big one. Start tracking MTBF and you'll be on the road to reliability. Eventually, you'll learn to manage your assets proactively according to their health. Then, you'll see your MTBF improve dramatically.

16.1.2. Reason 2

Not every asset is loaded into the CMMS/EAM. This is a problem that makes writing an emergency work order impossible. If you're not tracking every asset down to the component level, you can't possibly

*Source: Ricky Smith, CMRP, "Top Five Reasons Why Companies Don't Measure Reliability," *Plant Services Management* (November 2005). Reprinted by permission of the publisher.

identify any true reliability issue. Think about it this way; if 20% of your assets eat up 80% of your resources, wouldn't you want to identify that 20%, the bad actors? Put all of your assets in your CMMS/EAM, track the MTBF and the bad actors will become obvious.

16.1.3. Reason 3

It isn't important to measure MTBF because other metrics provide equivalent value. Yes, you can get asset reliability from other metrics, but keep it simple by using MTBF. Count the number of breakdowns (the number of emergency work orders) for an asset during a given time interval. That's all it takes to learn how long the equipment runs (on average) before it fails.

16.1.4. Reason 4

The maintenance organization is in such a reactive mode that there's no time to generate any metrics. They're constantly scrambling merely to react to the latest crisis. But, taking a small step in the right direction—tracking just one measure of reliability—will reveal the 20% of the assets that are burning 80% of the resources. If you start with the worst actor, you'll be surprised at how quickly you can rise out of the reactivity quagmire.

For example, a plant manager who recently measured the MTBF for what he called his "Top 10 Critical Assets" was shocked at the results. He expected the combined MTBF for these assets would be around eight hours to nine hours. In the first month of this initiative, he found that the actual MTBF was 0.7 hours. You may find yourself in the same situation. You'll never know the true reliability status on your plant floor until you begin measuring it.

16.1.5. Reason 5

There are too many other problems to worry about right now without being pressured to measure reliability, too. I've heard this many times and what it tells me is that the organization is in total reactive mode. This organization deals only with the problem of the hour. If 20% of your assets are taking 80% of your resources, dig yourself out of the problem by attacking the assets that cause the most pain—the "high payoff assets" that will respond to a reliability improvement initiative. We've got to stop fighting fires. The characteristics of adept firefighters include:

- High turnover of personnel (mostly in production).
- Maintenance costs that continue to rise.
- Maintenance costs that are capped before the month ends ("Don't spend any more money this month. We're over budget.")
- Every day is a new day of problems and chaos.
- Maintenance is blamed for missing the production goals.

It isn't easy to fight fires and initiate reliability improvement at the same time, but it can be done. Start measuring MTBF and attack the high-payoff assets. You can't change a company's culture from reactive to proactive overnight, but you can eliminate reliability problems one major system at a time. That's where you'll find a rapid return on investment. Change people's activities and behaviors slowly and you'll transition to a proactive culture.

Asset reliability is the key to keeping a company profitable, increasing its capacity and reducing its maintenance cost. In a future column, we'll present some reliability improvement ideas.

16.2. CREATING A CULTURE CHANGE IN YOUR MAINTENANCE DEPARTMENT: IS YOUR MAINTENANCE CREW IN A REACTIVE MINDSET? CHECK OUT A LIST OF QUALIFIERS TO FIND OUT AND THEN LEARN HOW TO CHANGE IT*

It's difficult to manage a maintenance crew effectively in a reactive environment. We're either unaware we're in a reactive mode or we don't know how to get out of it. The following list of qualifiers determines if your crew is reactive:

*Source: Ricky Smith, CMRP, "Creating a Culture Change in Your Maintenance Department," *Plant Services Management* (March 2006). Reprinted by permission of the publisher.

- PM labor hours stay the same (or increase) and emergency labor hours trend upward.
- PM work orders lack specifications, procedures and other data.
- Yesterday's maintenance problems and reliability issues consume 90% of daily maintenance meetings.
- The maintenance supervisor is a hero one day, a no good the next.
- The maintenance supervisor must work late at least twice a week.
- Maintenance crews don't know what equipment they'll be working on tomorrow.
- The maintenance supervisor routinely expedites parts for emergency work.
- Equipment reliability issues prevent the plant from operating at targeted capacity.

If these points seem too close to home, you're probably operating in a reactive maintenance environment. The challenges and obstacles you face are many. I know, I was there and faced these issues on a daily basis. The toughest challenge was a cultural one—my maintenance crew was resistant to change.

Now, as a consultant, I find that in many plants neither Production nor Maintenance feels responsible or accountable for equipment reliability. Instead, the maintenance department focuses its effort on time-based PMs that don't work anyway, doing too much too soon and doing too little too late. Remember these words of wisdom: You know you're in reactive mode when you continue to perform preventive maintenance on equipment that continues to fail. Also, I've found some plants never meet capacity projections. In fact, I've seen management formally reduce production projections and even change the name from "projections" to "stretch goals." Shouldn't we admit that if we have a stretch goal, it really means we don't believe we'll ever meet it? Nevertheless, as maintenance and manufacturing costs continue to rise for no apparent reason, maintenance comes under pressure to do something quickly.

So, quick fixes are tried and tried again, but they never really work reliably.

So how do you get out of a downward spiral and move your crew from reactive to proactive? I changed my crew's behavior by convincing and proving to them that there was a better way—being proactive in maintenance. And, I had to convince them there was an easy way to get there, and they would benefit personally from the change.

Before starting this culture change initiative, I had to gain support and sponsorship from plant management. The only way to get that prerequisite is to develop a compelling business case. Believe me, it will be compelling—a reactive environment leaves big dollars sitting on the table.

Moving from reactive to proactive will reduce maintenance costs by at least 20%, depending on the severity of the problems. In addition, capacity will increase because you're improving reliability. I've seen asset reliability raise capacity by as much as 10% to 15%.

Then, armed with management support, I needed true believers. I had to prove to my crew that life was better in a proactive environment. To make the biggest impact quickly, we took one of our worst performing assets and focused on changing our process to improve its reliability. We changed the day-to-day activities and behaviors of the people in Maintenance and Operations and ensured that people understood what to do.

The people who operated and maintained the asset owned and executed the asset reliability program, conducting proactive inspections at designated frequencies. We got some expert help in developing the asset reliability program (there are many work identification methodologies available). We didn't invest in heavy statistical analyses, nor did we use an abundance of additional predictive technologies, but we validated those we had in place when we developed the asset reliability program to ensure we were focused on the right work.

Within six months, we had tuned up the reliability and performance of that asset. We put key performance indicators in place to manage the process and kept it going. The team knew they were successful and they felt great. The change had occurred. Sure, it was only one asset, but now others wanted a ride on our success train.

You don't have to tolerate managing maintenance in a reactive mode. Developing a proactive asset reliability program and focusing on a process to implement it is the key to success in changing from a reactive organization to a proactive one.

16.3. EXTERMINATE LUBE PROBLEMS: GREASE AND OIL EXPERTISE CAN BE A SERIOUS COMPETITIVE EDGE*

An enormous amount of productivity is lost because the correct oil or grease is not properly installed at the right time. Lubrication is a critical responsibility, but in many organizations effective techniques and the technicians who know and do them get little respect. Building the role of lubrication experts—your "men

*Source: Ricky Smith, CMRP, contributing editor, "Exterminate Lube Problems," *Plant Services Management* (November 2005). Reprinted by permission of the publisher.

in black"—is a relatively low-cost way to materially improve reliability.

Studies have shown that 70% to 85% of equipment failures are self-induced, meaning that maintenance practices and processes are directly responsible for the failures. A recent survey I conducted online shows that poor lubrication practices represent about 40% of maintenance-related self-induced failures. In the same study, more than 80% of respondents indicated they consider lubrication to be a significant problem in their operation.

Lubrication plays a role in the operation of most equipment—gear reducers, electric motors, chain drives, air compressors, bearings and more—so it's obvious that doing it properly is key to the success of capital-intensive companies.

One of the main reasons companies struggle with lubrication effectiveness is they overrely on standard original equipment manufacturer (OEM) recommendations. Instead, lubrication activities should be driven by asset health and the true lubrication needs of the asset. The combination of the right lubrication activities and proper practices creates a significant opportunity to improve plant reliability.

16.3.1. Big, Bad, and Ugly

More than 200 maintenance professionals participated in the August 2005 survey on lubrication and its impact on reliability. The results may shock you, or they may simply validate what you are seeing in your operation. It's clear that although some companies are doing things right when it comes to lubrication, most are not.

Do you consider lubrication to be a problem? Responses to this first question clearly show the significance of lubrication with more than 80% saying it's a problem in their operation (Figure 16.1).

What percentage of your equipment downtime is related to lubrication? More than 18% of companies

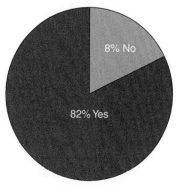

FIGURE 16.1. Asked "Do you consider lubrication to be a problem?", more than 80% of survey respondents said yes.

report that more than 20% of their equipment failures are directly related to lubrication problems (Figure 16.2). The first step in solving a problem is knowing you have a problem, and the next step is knowing how large the problem is.

Companies tell me they don't have money to improve anything. If their total annual sales are $60 million, total downtime is 10% and 25% of downtime is due to improper lubrication, the lost opportunity cost due to lubrication is $1.5 million. With numbers like these, the money is there—it's just that no one in the operation knows it or can measure the losses.

In what area do you have the most lubrication problems? In a reactive environment, we do not focus on the real problems but on the problems that face us on a specific day. I used to work in that type of environment, but later transitioned to a very proactive environment. Based on my experience as a maintenance practitioner, I thought motors and gear reducers would be the biggest problems.

But as one can see (Figure 16.3), respondents say bearings are the largest problem in most organizations, with gear reducers a distant second. Only about 3% report motors as the biggest problem. It could be the motor rewind shops are not telling us the whole story— I visited a large motor rewind facility and was shown numerous motors that had failed as a result of either lubrication bypassing a sealed bearing or being pushed through the bearing and into the motor windings.

Do you have a person dedicated to lubrication? I have seen lubricators do a great job and I have seen them do a very bad job. One thing I found consistent among the ones that did a bad job was they were all trained on the job by their predecessor. Formal training is the key to solving this problem.

Half the respondents say they have a dedicated person and half say they do not, but more than 80% have lubrication problems so apparently having a person

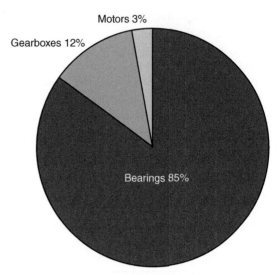

FIGURE 16.3. Bearings lead the list of lubrication problem areas.

dedicated to lubrication does not ensure it is performed correctly. To perform their job to standard, lubricators must be trained to the prerequisites of the job and then held to that standard.

Do you have a well-defined lubrication program? Having a well-defined lubrication program is key and the first step to ensuring success. Some 48% of respondents said they did; 52% said they did not.

We saw that more than 30% of all respondents say at least 10% of their equipment downtime is related to lubrication issues. It seems likely they are the ones that do not have a well-defined lubrication program. Maybe it is time to invest in one.

At what skill level is your maintenance staff in lubrication? More than 40% stated that the lubrication skill level of their maintenance staff is below 3 on a scale of 1 to 10 where 10 is highest (Figure 16.4). Part of the solution to the downtime issue is to train your people and make lubrication training an ongoing event.

16.3.2. Make Lube Expertise a Specialty

The results of the survey point to three conclusions. First, training is an issue. Most companies either do not look at lubrication training as important or they use what I call "check the box" lubrication training: If asked, "Do you train your people in lubrication?" they answer yes, but the training is not focused enough to change behavior.

Second, lubrication procedures have either not been developed or are not followed. You can have the best lubrication procedures in the world but if no one follows them, they're useless. Management must ensure proper procedures are written, then ensure they are followed.

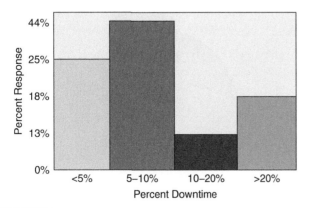

FIGURE 16.2. About 18% of companies report more than 20% of their equipment failures are lubrication-related.

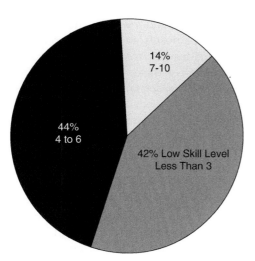

FIGURE 16.4. Asked to rate the lubrication skill level of their maintenance staff on scale of 1 to 10, fully 41% say less than 4. No one gave a 10.

Finally, it seems that organizations do not understand the relationships between lubrication and reliability. Many failures attributed to normal wear or faulty components are actually caused by poor lubrication practices (Table 16.1).

In some cases, lubrication is blamed when storage or installation may actually be the root cause. You must look for the true root cause of any failure before coming to a conclusion, or you'll end up treating a symptom and not the problem.

For example, lubrication is often blamed when bearings have failed due to improper storage. If bearings are allowed to lay open in a storeroom, are stored in an environment with vibration, or large bearings are not rotated on a scheduled basis, premature failure will occur.

Lubrication is also wrongly blamed for problems due to improper bearing installation. Best practices for bearing installation include never handling them with bare hands, never rotate an unlubricated bearing, and always heat the bearing to the manufacturer's specifications before installing to an interference fit.

One must also follow the manufacturer's recommendations when installing a new or rebuilt gear reducer. For example, most manufacturers will tell you that the gear reducer must have the oil changed within 24 to 48 hours of operation. This removes foreign matter that may be in the gear reducer. I followed this process as a maintenance supervisor back in the 1980s and never had a gear reducer fail after installation—those gear reducers operated without problems for many years.

Understanding lubrication is more than understanding how to follow procedures. It also requires understanding the fundamentals of storage and installation. Only then can you connect the fundamentals of lubrication to actual lubrication failures.

TABLE 16.1.   Poor Lubrication Practices

Problem	Root Cause
Bearing failure due to contamination of grease with dirt, dust or silica	Failure to wipe grease fitting or the end of the grease gun nozzle clean
Bearing failure due to contamination of grease by dirt, dust or silica	Seal not holding due to overlubrication
Bearing failure due to lubricant not providing barrier to prevent metal-to-metal contact	Wrong grease/oil or heat reduced viscosity due to temperature rise beyond range of lubricant
Gear Reducers	
Failed bearings and damaged gear teeth due to contamination causing interference between gears thus overloading bearings	Gear oil added to gearbox through a dirty funnel or dirty container or bucket
Failed bearings due to contamination of lubricant with dirt, dust or silica	Seal leaking due to overpressurization of gearbox caused by blocked air intake on housing
Electric Motors	
Bearing failure due to contamination of the grease with dirt, dust or silica	Failure to wipe grease fitting or the end of the grease gun nozzle clean
Bearing failure due to contamination of the grease with dirt, dust or silica	Seal not holding due to overlubrication
Windings failed because of grease buildup inside the motor	Relief plug not removed before introducing grease into zerk fitting
Windings failed because of grease buildup inside the motor	Sealed bearings—grease cannot enter the bearing

16.3.3. Get the Job Done

So implementing effective lubrication practices is important. The necessary steps depend on where you stand. First, review the current lubrication practices. If time-based preventive maintenance procedures are followed, consider whether or not the reliability of the equipment can be monitored based on condition instead of time.

Figure 16.5 illustrates the typical room for improvement. It validates our survey in showing that reactive maintenance is the norm: too little, too late. A "best in class" organization will monitor bearing condition based on oil sampling, heat gain, vibration analysis, current draw, ultrasonic and other predictive technologies.

An effective monitoring program will manage condition data with alarms set in a CMMS/EAM to tell maintenance when action is required. There are software programs sold on the market today that do just that.

There will always be lubrication practices that require time-based PMs, but consider implementing condition-based PMs. I am not recommending you run out and implement a condition-monitoring program, but think about how you might improve the way you lubricate your equipment.

When you have considered and, if necessary, modified your lubrication practices, implement an effective training and monitoring program.

Track performance of your lubrication program through agreed-upon metrics. Possibilities include MTBF (root causes of failures will have to be determined to identify lubrication failures), production losses, maintenance costs associated with each problem piece of equipment, and replacement parts costs.

Start with a baseline of the metrics before you implement the program, and me asure afterward on a weekly basis. Trend the results and post without comment for four weeks. After four weeks, allow your maintenance staff to comment on the results. See what is working and what is not, and understand what is a process problem and not a people problem. People problems can be solved through training and enforcement of the standard.

Keep everyone involved and interested with a Top Five list of lubrication failures: post the top five problems of equipment that has failed due to lubrication issues. Post charts of the metrics, and have the maintenance staff identify on them what actions were taken to correct problems.

In short, preventing failures via proper lubrication depends on two things: defining and documenting procedures with specifications, and having the discipline to follow those procedures. An educated staff is more able to understand the reasoning behind the procedures, and thus more likely to be proactive in following them. Create and support expert "men in black" to rid your plant of lubrication problems.

16.4. WHAT IT TAKES TO MAKE THE CLIMB FROM REACTIVE TO RCM*

Plant Services decided to measure the degree of implementation of RCM, the prevailing attitude toward it, and the differences between the RCM haves and have-nots.

*Source: Ricky Smith, CMRP, "What It Takes to Make the Climb from Reactive to RCM," *Plant Services Management* (December 2006). Reprinted by permission of the publisher.

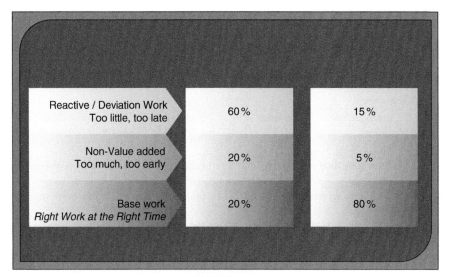

FIGURE 16.5.

It's great to talk and write about best practices, reliability-centered maintenance (RCM) and how important they are to a plant's ability to compete in our global economy, and we've been doing it for years. We know many facilities have embraced the principles and some have even implemented them. We also know many have not.

We decided to take a measure of the degree of implementation, the prevailing attitude toward RCM, and the differences between the RCM haves and have-nots. We invited maintenance professionals of all stripes to tell us via a Web-based survey (see below) not only where their plant's actual practices are on the scale from reactive to reliability, but also how well their departments cooperate, their management's attitude toward maintenance, and some key factors that drive the necessary culture changes.

Some 272 industry professionals participated in the survey, and more than two-thirds are maintenance and reliability managers. A significant number of senior managers as well as plant and production managers participated.

The results may be surprising to many and, I hope, eye-opening to those who have them closed. Bear in mind that responses are voluntary, and people are more willing to volunteer good news than bad.

16.4.1. Waving the Flag

Starting off on a high note, every single respondent stated that asset reliability is a significant concern to them ("Survey Results" sidebar, question 1). I'm not surprised. Most reliability professionals I know are always questioning themselves whether or not they have optimal asset reliability at optimal cost, and this concern resonates with senior management.

1. What is your position or title?

	Response Percent	Response Total
Maintenance Management	46.7%	127
Engineering Management	1.8%	5
Maintenance/Engineering Management	19.5%	53
Production Management	1.1%	3
Plant Management	5.5%	15
Senior Management	2.6%	7
Other (please specify)	22.8%	62
	Total Respondents	272
	(skipped this question)	2

2. Do you consider reliability of your assets to be a significant concern to you?

	Response Percent	Response Total
Yes	100%	271
No	0%	0
	Total Respondents	271
	(skipped this question)	3

3. Does senior management understand the significance of reliability in your facility/plant?

	Response Percent	Response Total
Yes	77.9%	211
No	22.1%	60
	Total Respondents	271
	(skipped this question)	3

4. Has your facility/plant implemented Lean or Six Sigma (or both)?

	Response Percent	Response Total
Yes	29.6%	80
No	**48.1%**	**130**
In progress at this time	22.2%	60
	Total Respondents	**270**
	(skipped thist question)	**4**

5. In you have a Lean initiative, does it include a focus on asset reliability?

	Response Percent	Response Total
Yes	44.1%	98
No	**55.9%**	**124**
	Total Respondents	**222**
	(skipped this question)	**52**

6. Is your plant at risk to being shutdown or downsized due to cost or other factors?

	Response Percent	Response Total
Yes	28.3%	76
No	**71.7%**	**193**
	Total Respondents	**269**
	(skipped this question)	**5**

7. With what metrics do you measure the reliability of your assets? (SELECT THE TOP TWO)

	Response Percent	Response Total
MTBF (Mean Time Between Failure)	23.3%	60
Asset availability	27.9%	72
Asset utilization	20.9%	54
Equipment downtime	**53.5%**	**138**
Equipment uptime	27.1%	70
OEE	14.3%	37
Cost	31.8%	82
# of potential functional failures identified within a specified time period	9.3%	24
We do not measure reliability	13.6%	35
Other (please specify)	1.6%	4
	Total Respondents	**258**
	(skipped this question)	**16**

8. Do you manage your maintenance and reliability process through the use of leading KPIs (Key Performance Indicators)?

	Response Percent	Response Total
Yes	41.2%	106
No	**45.5%**	**117**

	Response Percent	Response Total
Uncertain as to what you are talking about	13.2%	34
	Total Respondents	257
	(skipped this question)	17

9. Who owns reliability in your facility/plant?

	Response Percent	Response Total
Maintenance Department	**46.3%**	**119**
Production/Operations Department	6.6%	17
Both	25.3%	65
Senior Management	1.6%	4
Everyone	20.2%	52
	Total Respondents	257
	(skipped this question)	17

10. Do your operators assist in basic maintenance, lubrication, or inspection of their equipment as it relates to maintaining or inspecting reliability?

	Response Percent	Response Total
Yes	**51%**	**131**
No	49%	126
	Total Respondents	257
	(skipped this question)	17

11. Do you have all the elements of your maintenance and reliability process mapped and defined in a work flow process?

	Response Percent	Response Total
Yes	23.8%	61
No	**71.5%**	**183**
Uncertain as to what you are talking about	4.7%	12
	Total Respondents	256
	(skipped this question)	18

12. Has your facility/plant ranked all of your assets based on their failure consequence and risk to the business?

	Response Percent	Response Total
Yes	25.7%	65
Neither	**52.2%**	**132**
Just failure consequence	12.6%	32
Just risk to the business	9.5%	24
	Total Respondents	253
	(skipped this question)	21

13. What percentage of your maintenance work is "too little, too late"?

	Response Percent	Response Total
Less than 10%	22.9%	58
More than 10% but less than 20%	27.3%	69
More than 20% but less than 50%	**30%**	**76**
Between 50 and 70%	17%	43
Above 80%	2.8%	7
	Total Respondents	**253**
	(skipped this question)	**21**

14. What percentage of your maintenance work is "too much too early"?

	Response Percent	Response Total
Less than 10%	**51%**	**127**
More than 10% but less than 20%	36.9%	92
More than 20% but less than 50%	11.2%	28
Between 50 and 70%	0.8%	2
Above 80%	0%	0
	Total Respondents	**249**
	(skipped this question)	**25**

15. What percentage of your maintenance work is the "right maintenance work at the right time"?

	Response Percent	Response Total
Less than 10%	13.5%	34
More than 10% but less than 20%	20.2%	51
More than 20% but less than 50%	**29.8%**	**75**
Between 50 and 70%	27.4%	69
Above 80%	9.1%	23
	Total Respondents	**252**
	(skipped this question)	**22**

16. Do you have an effective PM (Preventive Maintenance) Program?

	Response Percent	Response Total
Yes	**69.8%**	**176**
No	30.2%	76
	Total Respondents	**252**
	(skipped this question)	**22**

17. Are breakdowns the norm (reactive maintenance is in full affect)?

	Response Percent	Response Total
Yes	43.7%	111
No	**56.3%**	**143**
	Total Respondents	**254**
	(skipped this question)	**20**

18. Do you use some type of PdM (Predictive Maintenance) technology?

	Response Percent	Response Total
Yes	**73.1%**	**185**
No	26.9%	68
	Total Respondents	**253**
	(skipped this question)	**21**

19. Would you consider your PdM Program effective in predicting failures far enough in advance so the corrective work can be scheduled and planned properly and does this actually occur?

	Response Percent	Response Total
Yes	**50.8%**	**128**
No	25.8%	65
Do not have a PdM Program	23.4%	59
Do not understand the question	0%	0
	Total Respondents	**252**
	(skipped this question)	**22**

20. Does your company apply principles of the PF Interval?

	Response Percent	Response Total
Yes	11.2%	27
No	**46.3%**	**112**
Not sure	42.6%	103
	Total Respondents	**242**
	(skipped this question)	**32**

21. How well would you say you understand the definition of failure modes, equipment functions, total functional failures, and partial functional failure? (1 = unknown, 10 = expert)

	Response Percent	Response Total
No at all (1)	9.1%	22
A little (3)	21.4%	52
Somewhat (6)	**45.3%**	**110**
Yes, I very much understand the above definitions very well (10)	24.3%	59
	Total Respondents	**243**
	(skipped this question)	**31**

22. What CMMS/EAM do you use?

	Response Percent	Response Total
SAP	14.5%	33
Oracle	5.3%	12
Datastream	16.3%	37

Response Percent	Response Total	
MRO	6.6%	15
Other (please specify)	57.3%	130
	Total Respondents	227
	(skipped this question)	47

23. Will your CMMS/EAM provide information such as MTBF (mean time between failure) or a "bad actors" report?

	Response Percent	Response Total
Yes	**43.1%**	**100**
No	30.6%	71
Uncertain	26.3%	61
	Total Respondents	232
	(skipped this question)	42

24. What percentage of the time does a functional failure or breakdown have a work order written for it?

	Response Percent	Response Total
None	8.8%	21
Less than 50%	20.6%	49
More than 50% less than 70%	13%	31
More than 70% less than 90%	23.1%	55
Close to 100%	**34.5%**	**82**
	Total Respondents	238
	(skipped this question)	36

25. How often is a RCFA (Root Cause Failure Analysis) or RCA (Root Cause Analysis) applied to an equipment failure?

	Response Percent	Response Total
None	20.3%	48
Once in a while	**36%**	**85**
Only on critical assets	28.8%	68
For most failures	14.8%	35
	Total Respondents	236
	(skipped this question)	38

26. If you were to rate the reliability of your assets on a scale of 1 to 10 what would it be? (1–Real Bad, 10–World Class)

	Response Percent	Response Total
1	0.4%	1
2	1.3%	3
3	6.4%	15
4	10.6%	25
5	12.7%	30
6	13.6%	32
7	22%	52

Response Percent	Response Total	
8	25%	59
9	6.8%	16
10	1.3%	3
	Total Respondents	**236**
	(skipped this question)	**38**

27. How much money does your company lose per year (your estimate, take your best guess) because of reliability issues?

	Response Percent	**Response Total**
None	0.4%	1
Less than 100,000 dollars	35.4%	81
More than 100,000 less than 1 million dollars	**38%**	**87**
1 million to 10 million dollars	21.4%	49
Over 10 million dollars but less than 50 million dollars	3.5%	8
More than 50 million dollars	1.3%	3
	Total Respondents	**229**
	(skipped this question)	**45**

28. Do you use some type of reliability software to monitor the health of your assets?

	Response Percent	**Response Total**
Yes	24%	56
No	**76%**	**177**
	Total Respondents	**233**
	(skipped this question)	**41**

29. Do you currently have a successful proactive asset reliability improvement initiative underway in your plant/facility?

	Response Percent	**Response Total**
Yes	33.5%	78
No	**66.5%**	**155**
	Total Respondents	**233**
	(skipped this question)	**41**

30. If so, when will the initiative pay for itself?

	Response Percent	**Response Total**
1–6 months	18.4%	16
6–12 months	17.2%	15
1–3 years	**46%**	**40**
More than 3 years	18.4%	16
	Total Respondents	**87**
	(skipped this question)	**187**

31. How long do you expect this initiative to last?

	Response Percent	Response Total
Forever	37%	51
1–5 years	7.2%	10
More than 5 years however it does have an end date	4.3%	6

Response Percent	Response Total	
Have no idea	**51.4%**	**71**
	Total Respondents	**138**
	(skipped this question)	**136**

32. Has your facility/plant used RCM or some other type of failure analysis methodology (RCM II, MTA, FMEA, etc.) to determine what must be done to prevent or predict failure of your assets? (at least critical assets)

	Response Percent	Response Total
Yes	34.1%	76
No	**59.2%**	**132**
Uncertain as to what you are talking about	6.7%	15
	Total Respondents	**223**
	(skipped this question)	**51**

33. If you used any of the above methodologies was it effective?

	Response Percent	Response Total
Yes	**64.5%**	**69**
No	35.5%	38
	Total Respondents	**107**
	(skipped this question)	**167**

34. What percentage of your assets did you use a form of failure analysis (RCM II, RCM Turbo, etc.)?

	Response Percent	Response Total
Less than 5%	**62%**	**98**
More than 5% but less than 10%	9.5%	15
More than 10% but less than 20%	12%	19
More than 20% but less than 50%	8.9%	14
More than 50% but less than 70%	3.8%	6
More than 70%	3.8%	6
	Total Respondents	**158**
	(skipped this question)	**116**

35. Did you "template" the results to "like" equipment using a form of failure analysis?

	Response Percent	Response Total
Yes	27.2%	46
No	72.8%	123
	Total Respondents	169
	(skipped this question)	105

36. Would you like a copy of the results once the survey is complete?

	Response Percent	Response Total
Yes	67.6%	140
No	32.4%	67
	Total Respondents	207
	(skipped this question)	67

37. If yes please provide the following information:

	Response Percent	Response Total
Name:	99.3%	139
Title:	97.9%	137
Company Name:	97.1%	136
Location:	97.1%	136
Email address:	99.3%	139
	Total Respondents	140
	(skipped this question)	134

But is reliability really under control—and is it sustainable for the future? No matter how good a grip you think you have, never underestimate the need to keep looking for ways to get better by ensuring the reliability of your capital assets, measuring reliability and continuously improving. The impact of asset reliability on asset utilization and performance dictates that we pay constant attention to this critical process.

16.4.2. Does Management Understand?

In most plants surveyed, senior management seems to understand the significance of reliability (question 2). One of the questions I constantly hear is, "If senior management understands the significance of reliability, why don't they support a reliability initiative?"

Most senior management cannot and won't accept a reliability initiative if it is not supported by a business case, points out Jack Nicholas, a world-renowned reliability expert. In the business plan, senior management wants to see:

1. The value a reliability initiative will bring in hard dollars through:
   a. Increased capacity, asset availability.
   b. Reduced maintenance cost.
2. Other outputs (not typically captured in hard dollars).
   a. Decreased risk of environmental incidents.
   b. Decreased asset life-cycle cost.
   c. Decreased capital maintenance (replacing equipment because it is "worn out" or "old").
3. The time to value (from when the initiative starts to when the company will start realizing results).
4. The cost of the initiative (nothing is free).
5. Amount of internal and external resources required.
6. A plan with a timeline.
7. Key performance indicators (KPIs) that will be used to manage the initiative (leading and lagging KPIs).
8. Length of time for total return on investment (must be validated by the company's financial expert).

If these items can be delivered in a professional manner, it's hard for management not to accept and support the initiative. In fact, we want senior management to be the sponsors of any reliability initiative. Top leadership has control over the destiny of a plant. In particular, if a plant is at risk of closure, projects such as reliability improvement initiatives can be game-changers.

Speaking of closure, 28% of respondents report their plant or operation was at risk of being downsized or shut down (question 3). Numerous government reports say that in the next three to five years, 25% to 30% of companies will be downsized or shut down. For example, Ford and General Motors are closing plants and laying off thousands of employees.

"Business conditions that used to change every seven to nine years now change every seven to nine months," Andy Harshaw, vice president, Dofasco Steel, was recently quoted as saying. "Companies must be flexible to change or face the fact that they may shut their doors." Harshaw went on to say that managing asset reliability was important to Dofasco's strategic goal and survivability of his company.

16.4.3. Who Owns Reliability?

More than 46% of respondents say the maintenance department owns the reliability of their plant/facility. From numerous discussions and my own experience as a maintenance manager, I know most companies blame asset reliability issues on maintenance. I say, "In the best companies in the world, everyone owns reliability." Not surprisingly, only 20% of respondents gave what I consider the best answer (question 4). Until production accepts a partnership with maintenance to care for assets and keep them reliable, the plant will probably never reach the level of optimized reliability at optimal cost that is required for the company to reach its business goals.

Production/Operations should be the number one believer and driver of an effective preventive maintenance (PM) program. If they don't own the reliability of the assets, the PM program probably won't be effective. Almost 70% of survey respondents stated they had an effective PM program, but 44% indicate that equipment breakdowns are the norm (question 5). A preventive maintenance program cannot be effective if equipment breakdowns are the norm.

An interesting correlation is that 44% of respondents say breakdowns are the norm, and about the same percentage assess the reliability of their assets as ranking between being 1 and 5 on a scale where 1 is "real bad" and 10 is "world-class." I conclude without surprise that an effective operations-driven preventive maintenance program improves equipment reliability, reduces reactive maintenance, and adds value to a company.

16.4.4. Informal versus Formal PM Programs

Looking at the situation more closely, I must ask, "How are companies developing PM programs?" In my experience, PM programs are typically developed informally, based on manufacturers' suggestions, work requests (largely reactive), or simply on work that has always been done that way. When your PM program isn't technically based and not connected to a reliability-based maintenance strategy, typically more than 80% of the work you are executing is reactive, creating the defects we know as equipment failures. Progressive environments use a formal, technically sound process where work orders can be traced back to the failure analysis that found the problem and created the task (Figure 16.6).

Only 34% of respondents say they use a formal methodology of looking at failures to determine the maintenance strategy to prevent and predict failures (question 6). A similar percentage (35%) rank the reliability of their assets between 8 and 10 on the scale where 1 is real bad and 10 is world-class. I can assume the 35% of companies who have high reliability also use some type of failure analysis methodology to develop their maintenance strategy. The analysis they perform is most likely RCM, failure modes and effects analysis (FMEA), maintenance task analysis (MTA), or some other proven methodology.

16.4.5. To Measure Is to Manage

Most people have heard of Dr. W. Edward Deming and his manufacturing philosophies. Perhaps his most famous quote, which all successful companies believe (and unsuccessful ones tend to forget), is, "You cannot manage something you cannot measure."

The survey results on measurements point to some interesting findings. Fully 41% of respondents say they manage using leading KPIs (question 7). Leading KPIs are the only effective way an organization can manage their reliability process. "Leading KPIs lead to results," says Ron Thomas, a reliability leader at Dofasco Steel.

The results are tracked by lagging KPIs such as cost, asset downtime, number of failures, etc. Some 45% of respondents say they don't manage with leading indicators, so at best, we assume they try to manage with lagging indicators. But decisions need to be made based on problems in the asset reliability process before they impact results. An example may be that scheduled compliance (a leading KPI) is off-target. If this situation isn't corrected, the result could be higher production cost because maintenance work isn't

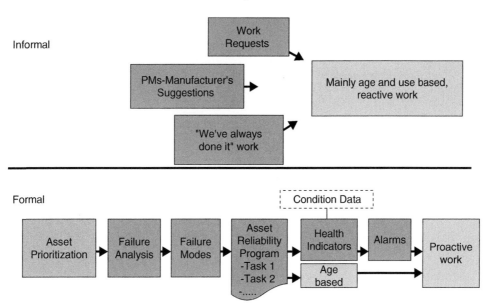

FIGURE 16.6. Progressive plants use a formal process where work orders can be traced back to the failure analysis that found the problem and created the task.

being accomplished on time with the right amount of resources, which causes excessive equipment downtime.

Many people ask me, "What is the first step to develop leading and lagging KPIs for my organization?" Most really don't want to hear my answer, because everyone is looking for the silver bullet or a quick fix. If you want to effectively manage an asset reliability process, you must have the process elements (such as work identification, planning, scheduling, work execution, etc.) mapped and defined with tasks, roles and responsibilities; leading and lagging KPIs; etc. In the survey, just 23% of respondents say they have mapped and defined their reliability process. Figure 16.7 shows an example of a process map, in this case for procured materials and services.

Using the right KPI is critical to knowing where you are in a process. When we asked, "With what metrics do you measure the reliability of your assets?" only 23% of respondents state they used mean time between failures (MTBF). (MTBF is simply dividing the number of asset failures into time—for example, if you have three functional failures in 24 hours, the MTBF is 24 divided by 3, or 8 hours.) MTBF is one of the most fundamental measures of reliability. Other measurements may be affected by reliability, but MTBF's only focus is measuring asset reliability.

MTBF becomes less important as reliability increases, so then a company may begin focusing on, say, the number of potential functional failures identified in a specified period of time. In the survey, 9% indicate they are currently using this metric to measure asset reliability. These are probably the plants you would want to visit to learn how they do it.

Interestingly, even though only 23% of companies measure MTBF to manage reliability, 43% say their CMMS/EAM can provide this information. The real problem is that most companies cannot measure reliability of their assets because they currently don't collect the data in a manner that would make this KPI valid.

Only 34% of respondents say that a work order is written close to 100% of the time for a functional failure or breakdown. Almost 30% say they either don't write a work order, or write one less than 50% of the time. I believe that you cannot improve something you cannot measure, and all successful managers agree with this philosophy. Another is that managing with bad data leads to bad decisions.

16.4.6. Depth of Understanding

The great Winston Churchill said, "I am always willing to learn, however I do not always like to be taught." This is true in the world of reliability. Most managers are willing to learn, however, they aren't willing to be taught something new so they can understand the basics of reliability.

More than 90% of managers are intimidated by the word reliability because they do not understand reliability, says Terrence O'Hanlon, CMRP, of

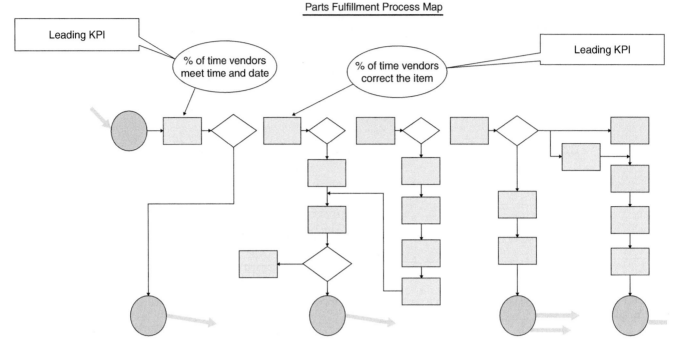

FIGURE 16.7. Understand how to apply leading and lagging indicators by mapping the process. Then see where measurements can spot performance problems before they affect reliability.

ReliabilityWeb.com and *Reliability* and *Uptime* magazines. The survey shows a serious gap between what people think they know about reliability and their actual knowledge of reliability fundamentals. Most managers don't understand nor apply the basic principle of reliability.

For example, only 11% of respondents say their company applies the principles of the PF Interval, and 46% state they don't use this basic concept at all (question 8). The PF Interval is one of the foundational principles of asset reliability, which focuses on detecting failures far enough in advance that a proactive task can be implemented to mitigate the failure. This is the foundation of an effective preventive and predictive maintenance program. I always say, "It isn't what you know that will kill you—it is the things you don't know." This is definitely true in the world of asset reliability.

Question 9 asked how well respondents understand the definitions of failure modes, equipment functions, total functional failure, and partial functional failure. These are some of the most important foundational elements of reliability, and must be understood to develop a proactive maintenance strategy. Only 24% of respondents say they understand these fundamental elements, while 30% either know nothing or very little about them.

Malcolm Forbes says, "The goal of education is to replace an empty mind with an open mind." Once a manager is educated in the basic principles of reliability, their world will change. They will feel like they have suddenly seen the sunlight after having lived under a mushroom all their life.

16.4.7. Indicated Actions

This survey helped identify serious gaps in many companies' relationship to reliability. At the same time, it indicates a path to understanding how we can optimize asset reliability. A reliability initiative will be supported and can be successful if you have the business case—essentially a financial improvement plan for your company.

I have seen many companies try some type of initiative to improve reliability. Usually it either didn't provide the value expected, or took too long to see the gains. Most reliability improvement initiatives deliver some return, but to make a quick impact to the bottom line—to achieve what I call performance breakthrough and a rapid payback—we need sustainable change (Figure 16.8). That change can only occur when managers and floor-level personnel see success and participate.

The survey found that 33% of companies have a "successful" reliability improvement initiative currently in place, and 36% of those companies say the

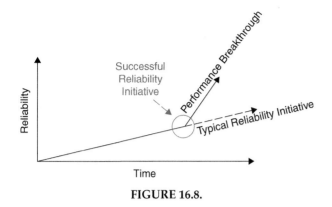

FIGURE 16.8.

initiative will pay for itself in one year or less (question 10). All told, 82% say the initiative will pay for itself in less than three years.

More than 37% say the reliability initiative will last forever. It's so important to understand that a true proactive asset reliability initiative is a continuous improvement process that lasts forever. As assets age, as the company experiences equipment failures, and as its business changes, reliability must be continually optimized. Continuous improvement must be embedded into the maintenance and reliability process.

The maintenance and reliability model in Figure 16.9 is a perfect example of how continuous improvement becomes part of the maintenance and reliability process. This model is known as the "Proactive Asset Reliability Process" and is used by some of the most successful companies in the world.

16.4.8. Lessons Are Simple

A few simple lessons must be learned if you want a successful reliability improvement initiative. These aren't options, but principles which must be followed or reliability will be at risk.

- Executive sponsorship is required. A company needs a committed champion at the executive level to take ownership and responsibility of the initiative.
- Floor-level operators and maintainers must be part of the design and share in the success of the new maintenance and reliability strategy.
- Everyone from the floor level to the boardroom must have some level of education in reliability. For change to occur, people need to understand why they need to change. If you need to educate everyone in reliability, contact me and I will provide resources to help you develop and execute effective reliability training.
- Develop a balanced scorecard for all levels of the operation, from the floor to the board room. Establish targets and goals for most KPIs on this scorecard. People want to know their score in the game.
- Be successful by developing a plan and following it. With respect to meeting financial targets and deadlines associated with the plan, remember the saying, "undersell, overdeliver."

Finally, here are the steps, based on best practices, to implement a successful asset reliability process:

Step 1: Develop a business case to identify the financial opportunity. The business case must identify the projected financial outcome in hard dollars. The

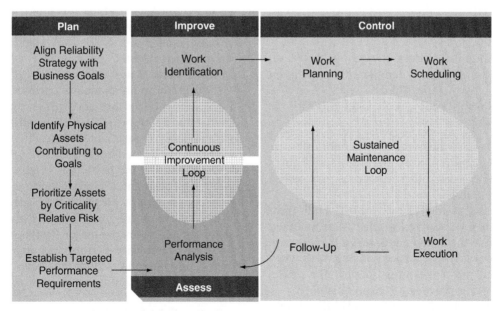

FIGURE 16.9. The Proactive Asset Reliability Process shows the role of continous improvement, and is used by some of the most successful companies in the world.

financial outcome may be found by increased capacity, reduced maintenance labor and material cost, increase asset utilization, and more. The plant management team developing the business case must include a finance person (comptroller, chief financial officer, etc.).

Step 2: Assets should be ranked based on risk to the business and their condition. Knowing your critical assets is so important to ensuring success of this initiative. More than 48% of survey respondents have ranked their critical assets. You will need to execute this initiative one asset at a time and focus first on the asset that provides the quickest payback. People will only change if they see change occur and believe in it. Taking the right step at the right time is so important to a successful reliability initiative.

Step 3: RCM methodology (RCM, FMEA, or MTA) must be applied to the asset with a joint team of operators and maintainers working together to design a proactive maintenance strategy for the asset. In the survey, the numbers of respondents who say they have a successful reliability initiative and who say they use RCM methodology is the same—a big hint.

Step 4: Use reliability software to assist in managing the health data of your assets. Reliability will now be managed based on the health of the assets, not breakdowns. Less than 28% of respondents say they have a successful reliability initiative in place, and only 20% use reliability software to collect and disseminate health data from their assets. In a typical plant, you could be managing as many as 60,000 to 80,000 data points coming from visual inspections, PLCs, predictive maintenance tools such as vibration monitors, and other sources. It's also very important that reliability software be linked to a CMMS/EAM to reduce human error and integrate continuous improvement into your reliability initiative.

Step 5: Continue the process throughout the plant, at least on critical assets, and template the results on like equipment wherever possible.

Step 6: Establish Leading and Lagging KPIs to manage the process.

To be successful when improving and optimizing reliability at optimal cost, you need four things in harmony with each other: practices, processes, technology, and people.

A proactive asset reliability process must be followed. Best practices must be adopted, applied, and followed for each element of the maintenance and reliability process. An example of a best practice noted in this survey is that successful companies must identify the proactive work that will improve and sustain reliability.

Methodologies such as RCM, MTA, and FMEA should be used. Technology including a CMMS/EAM system, reliability software, and PdM tools are the enablers.

Of course, people are the heart of all initiatives, no matter what the domain. We need proactive senior management that will sponsor and drive reliability projects. We need middle management that will champion projects and support employees in the midst of cultural change from reactive to proactive. Finally, we need Maintenance, Operations, and Engineering employees empowered to care for their assets to optimize reliability and embrace the change, because it really does mean a better way of life.

16.5. PUT A PLANT-WIDE FOCUS ON FUNCTIONAL FAILURES*

Short-term and long-term solutions can increase reliability, asset availability, utilization, quality, and capacity, but they're not easy or simple. Look at the barriers that hold a plant back.

The United States lost 40,000 manufacturing jobs in August. Plants are shutting down every day. Companies struggle to find the nonexistent "silver bullet" for increasing reliability and equipment performance while reducing costs. Short-term and long-term solutions can increase reliability, asset availability, utilization, quality, and capacity, but they're not easy or simple. Look at the barriers that hold a plant back.

First is the belief that a failure means the equipment is broken. That's wrong. A true failure of an asset is when it no longer meets the function required of it at some known rate or standard. For example, if a convey that's supposed to operate at 200 meters per minute can't, it has failed functionally, thus affecting revenue.

Second, in many cases, when an asset fails, no one in maintenance seems to understand that it failed. Maintenance doesn't get involved when quality or production rate issues arise.

Last, many maintenance departments don't know the equipment's performance targets and don't understand why it's important to know them. This isn't a maintenance department failure. It's a breakdown revealing misaligned goals.

Overcoming these barriers is essential to rapid reliability improvement. An understanding of and focus on functional failure by all plant personnel leads to rapid results. Focus on aligning the total plant on meeting each asset's performance target. Post the targets and current rates to publicize performance gaps to both maintenance and operations. Both organizations must accept responsibility for eliminating the performance gap.

*Source: Ricky Smith, CMRP, "Put a Plant-wide Focus on Functional Failures," *Plant Services Management* (October 2006). Reprinted by permission of the publisher.

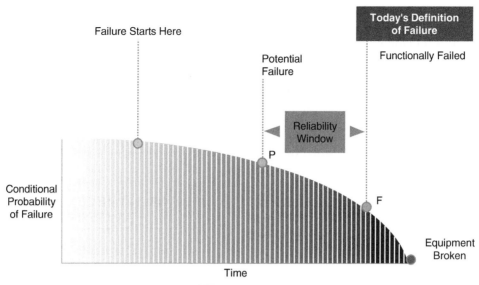

FIGURE 16.10.

Achieving rapid breakthroughs requires you to define failures properly. A reactive company might say failure means the conveyor is broken or stopped because of mechanical problem. A proactive company says the conveyor no longer performs the function required. A partial functional failure occurs when it's supposed to operate at 200 meters per minute, but can only run 160 meters per minute. A total functional failure indicates it stopped because of a mechanical problem (see Figure 16.10).

Uncover your "hidden plant" that provides the gains you need. If your equipment operates below defined performance targets, establish a policy of immediately engaging the maintenance department. Focusing on equipment performance targets is the key to rapid asset reliability gains.

The old saying, "It's what you don't know that kills you," is applicable to reliability issues. Getting to the rapid breakthroughs in plant performance is just the beginning of a long journey. Don't be satisfied with the reliability results you gain by following my advice. Continue onward by applying Reliability Centered Maintenance (RCM) methodology to zero in on "optimal reliability at optimal cost." I didn't say use RCM; I said RCM methodology, which could be RCM II, Streamlined RCM, FMEA or MTA. They're different methodologies, so learn about them and how they can foster rapid performance gains.

Just remember, some people have made short-term, rapid performance gains. Most, however, have only sustained their performance. If you would like further information about rapid breakthroughs in performance, send me an email. My final plea is that America should wake up before it's too late.

16.6. RELIABILITY IS WORTH A SECOND LOOK: STATISTICAL ANALYSIS AND TIME-BASED PREVENTIVE MAINTENANCE DON'T REALLY ADDRESS THE ABILITY TO PERFORM—IT'S TIME TO GET FAMILIAR WITH THE DEFINITION OF RELIABILITY*

The word "reliability" unnecessarily intimidates many maintenance professionals. In my opinion, most people associate reliability with reliability centered maintenance (RCM) and they're unclear about what it actually means. The definition is simple: Reliability is the ability of an item to perform a required function under a stated set of conditions for a specific time period.

This definition isn't at all intimidating, but having worked with more than 400 maintenance organizations, I found that most still focus on fixing failed equipment, not on ensuring reliability and avoiding failure.

A common reason is there's no time available to investigate the true requirements that ensure equipment reliability. Yet, there's a growing awareness among reactive maintenance organizations of the consequences of poor equipment performance:

• Higher maintenance costs
• More failures
• Downtime, safety and environmental issues

It's time for reactive maintenance organizations to admit there's no silver bullet to the equipment performance problem. Lean manufacturing and world class

*Source: Ricky Smith, CMRP, "It's Time to Get Familiar with the Definition of Reliability," *Plant Services Management* (June 2006). Reprinted by permission of the publisher.

manufacturing don't directly address optimal asset reliability. Forget the silver bullet and focus on asset reliability. The results will follow.

The steel company Dofasco needed a corporate fact-finding mission to change its focus to asset reliability. A team of more than 60 key employees spent more than two years researching the world's best maintenance organizations. They found that a focus on reliability gave the biggest return with the longest lasting results. Today, Dofasco Steel is a top-notch North American steel producer, and the company won many awards for its reliability-focused maintenance.

Companies like Dofasco that understand reliability typically have the best performing plants. A reliability-focused organization takes a holistic approach to asset management, focusing on people and culture. Common characteristics include:

• The goal is optimal asset health at an optimal cost
• A focus on processes—what people do to achieve results
• They measure each process step for effectiveness, not just results
• PM programs focus mainly on monitoring and managing asset health
• PM programs are technically sound, with each task linked to a specific failure mode—formal practices and tools identify the work required to ensure reliability

Don't focus your entire maintenance effort on a PM program that has little to do with meeting the actual equipment reliability needs. Statistical analysis techniques such as Weibull only help to identify assets for which reliability is a problem. You don't need engineering resources to figure out that your MTBF is too small. Besides, it's easy to identify bad actors. Rather than measuring failure frequency, figure out how to improve reliability.

Use statistical analysis to set frequencies for time-based PMs, which should account for a very small portion of your PMs. Here are some sobering facts that will make you think twice about the effectiveness of a time-based PM program.

Less than 20% of asset failures are age related, so how does one identify the PM frequency? My findings indicate that 98% of companies don't have good failure history data.

Most reliability studies say that 80% of asset failures are random. You can detect early signs of random failure by monitoring the right health indicators to determine whether the asset is degrading. The PF interval is the time between the detection of a potential failure (P) and functional failure (F). A maintenance organization needs to know the PF curve for critical equipment. This approach allows time for corrective action, in a scheduled and proactive manner, before functional failure occurs.

Take a step back and review the way you manage equipment performance. If equipment continues to fail after preventive maintenance or overhauls, then something must change. Focus on ensuring plant asset reliability. Everyone in a plant should understand the definition of reliability and what it means to the success of the company. Make reliability your plant's collective buzzword.

16.7. WHEN PREVENTIVE MAINTENANCE DOESN'T WORK*

Sometimes, preventive maintenance just doesn't work. That's when it becomes time to stop the collective insanity and start learning from others.

For many years, I've performed and managed preventive maintenance (PM) on every type of equipment, never asking myself why the equipment still fails, even after I've performed PM. My compliance rates were always high, but so were the number of recurring equipment breakdowns.

Finally, I asked myself how it is possible that a maintenance professional could perform the same PMs on equipment that continues to fail.

I know now that my PM program was flawed because it was essentially a reactive maintenance program that relied mainly on time-based PM tasks following manufacturers' suggestions and stuff we learned along the way. I had no technical justification for any task other than "we always do it this way," or "it's the latest predictive technology," so "we can't stop doing it now or we'll risk more failures." If that isn't the definition of insanity, I don't know what is.

The research that changed the way I think about failures and PM actually started more than 30 years ago, yet many plants are still falling apart today. It's time to stop the collective insanity. If you face the same problems on a daily basis (sometimes with little hope in sight), then read on, because I found a solution—and you can, too.

Research on equipment failures during the past 30-plus years has proven that more than 80% of failures aren't related to equipment age or use. The implication of the finding is that less than 20% of our proactive maintenance tasks should be driven by time, equipment age or usage. The majority (more than 80%) should be predictive and detective forms of proactive maintenance. Predictive maintenance is the use

*Source: Ricky Smith, CMRP, contributing editor, "When Preventive Maintenance Doesn't Work," *Plant Services Management* (February 2006). Reprinted by permission of the publisher.

of technology or some form of condition monitoring to predict equipment failure. Detective maintenance refers to work that determines whether a failure has already occurred, and applies well to hidden failures that aren't (at least initially) evident when they occur.

With this new understanding of failures, I migrated my department from operating in reactive mode to operating in proactive mode. The key difference is that our programs now focused on monitoring asset health and letting that determine the maintenance work to be performed proactively.

The research further showed that, once we truly understand an asset's failure modes (or causes), our program will look more like best-in-class. Here's an example of a maintenance program that transformed itself from reactive, time-based PMs to a proactive maintenance program.

Before:

- Clean the pump strainer once a month.
- Take oil samples from the reservoir, which tells whether to replace or merely filter the oil.
- Inspect pressure gauges to ensure the pump is developing sufficient head.

After:

- Watch for early signs of specific failure modes (reservoir temperature or excessive pressure fluctuations).
- Use electronic predictive checks to watch for early signs of specific failure modes (pressure and flow).
- Use predictive technologies to catch early signs of specific failure modes (e.g., oil sampling for a specific particle types).

There's a significant difference between these two maintenance programs. The new program produces far better asset reliability.

I've seen equipment failures reduced by 30%, 50%, and more. The business impact of a well-defined proactive maintenance program is huge. You'll increase equipment reliability, reduce capital replacement cost, achieve higher equipment availability and reduce maintenance costs. The soft benefits are a motivated workforce, a less-stressed management team, more time at home, and so on.

While the numbers will get management to support a project to prove the benefits on just one asset, once they see the size of the opportunity and the soft benefits, the next question will be, "What is your plan to roll this out on the rest of our critical assets?" Allow management to work with you to develop the plan. They'll feel some ownership of the process.

After running a compliant PM program for years, I found I couldn't rely on time-based maintenance alone. Research and experience in applying that research has proven that there's a better way to run the business of maintenance. The properly balanced use of predictive, detective and time-based maintenance forms a successful proactive maintenance program.

With such a huge potential to improve business competitiveness, maintenance managers have a great vehicle for generating interest and support among senior management, all of whom are looking for rapid return.

16.8. THE TOP FOUR REASONS WHY PREDICTIVE MAINTENANCE FAILS AND "WHAT TO DO ABOUT IT"*

Many companies adopt some form of Predictive Maintenance (PdM) technology as the first step in the path to improved plant reliability. However, the returns from these initial PdM investments often fail to meet the expectations of management. Many of you have seen the ineffective use of predictive maintenance where failures occur even though you are using some type of PdM monitoring. I lived in this world as a maintenance supervisor and it frustrated me that I could not define the use of PdM more effectively. I wrote this article in order to share my experiences with you based on my successes and failures. So let's look at the top four reasons why PdM has failed to meet management's expectations as I have seen.

In order to define why Predictive Maintenance fails let's first understand the definitions of "Predictive Maintenance" and "Predictive Maintenance Technologies" or PdM Technologies.

Predictive Maintenance is the monitoring of an asset's health in order to anticipate the opportunities to proactively perform maintenance to preserve an asset from failure or to protect it in some way. PdM Technologies are the instruments or technologies used to collect asset health data.

The purpose of Predictive Maintenance is to maximize, at optimal cost, the likelihood that a given asset will deliver the performance necessary to support the plant's business goals. By "optimal cost" we imply that if it is feasible, and economically sensible to perform a task that detects a failure far enough in advance to make intervention practical, then we will

---

*Source: Ricky Smith, CMRP, "The Top Four Reasons Why Predictive Maintenance Fails and 'What to Do about It,' *Plant Services Management* (November 2006). Reprinted by permission of the publisher.

have avoided the far greater costs of equipment downtime, secondary damage, as human injury, environmental impact, quality and others.

In order to use PdM technology one must understand how equipment fails. Through studies we know 20% of failures are time based and 80% of failures are random in nature and cannot be effectively correlated to time or operating hours. PdM provides one of the major tools to predict failure of an asset. PdM use for random failures must focus on the health of the asset (through monitoring indicators such as temperature, ultrasonic sound waves, vibration, etc.) in order to determine where an asset is on the degradation or PF Curve. Point "P" is the first point at which we can detect degradation. Point "F," the true definition of failure, is the point at which the asset fails to perform at the required functional level. In the past, we defined "Failure" as the point at which the equipment broke down. You can see points P and F and the two different definitions of failure by referring back to Figure 16.10.

16.8.1. PF Curve

The amount of time that elapses between the detection of a potential failure (P) and its deterioration to functional failure (F) is known as the PF interval. A maintenance organization needs to know the PF Curve on critical equipment in order to maintain reliability at the level required to meet the needs of the plant. Without this knowledge how can one truly understand how to manage the reliability of the asset?

PdM should be used to define where on the PF interval is the health of the asset. Define the point of failure in the PF interval far enough in advance that the asset can have planned and scheduled maintenance performed to restore the asset. As you can now see, understanding failures is very important to understand how to use PdM technologies to its full potential.

Let's now look at the four main reasons that PdM has failed to deliver expected value.

16.8.2. Reason 1: The Collection of PdM Data Is Not Viewed as Part of the Total Maintenance Process (Refer Back to Figure 16.9)

Many organizations, at least initially, view PdM as a separate activity from the core role of the maintenance function, and so it is not covered in the maintenance process. Some organizations start down the PdM path by "trying it out" on a contract basis. The contractor's role is to email or snail-mail the resulting predictive data to the plant. In other companies, a PdM resource (often seen as the Reliability Technician) is assigned the predictive role, or a PdM Team is formed. When these individuals or teams are not seen as an integral part of the maintenance department, their value is unlikely to be realized. Also, quite often the predictive data will be supplied to the maintenance organization, but the technician who collected the data is not consulted on the results, so the potential for well-informed data-driven decisions is limited. If PdM is disconnected from the maintenance process, the PdM program will likely fail because the value cannot be identified.

For example, have you ever seen a case where a maintenance employee becomes the new PdM technician? He may be the lucky one picked to operate the brand new $50,000 thermography equipment. In an immature reliability environment, the new role usually comes with a title that includes the word "Reliability." This new Reliability Technician goes out and starts snapping pictures of assets that show interesting heat profiles (when your only tool is a hammer, everything looks like a nail). But for most of these assets, a reasonably sound failure analysis, if performed, would not identify excessive overheating as the best predictor of failure. Or, potentially even worse, after the failure of a particular asset is determined to be "overheating," the Reliability Technician is assigned to produce thermographic profiles of every similar asset in the plant— regardless of probability of failure, frequency of failure, failure consequence, etc. Is it any wonder that production and maintenance personnel see limited value in the Reliability Technician's data?

To get the most out of PdM, I recommend that you make it an essential part of the Work Identification and Work Execution elements in your maintenance process. The steps in the Work Identification should clearly identify failure modes, and the best techniques for predicting those failures. PdM tasks are identified as part of a complete asset maintenance program, so we understand why we are doing the work and we are not doing unnecessary work. Work Execution conducts the work specified in the asset maintenance program in the most efficient manner possible. Tasks should be grouped in routes and handheld devices used where the PdM technology requires human intervention.

Involve production, maintenance, and PdM personnel in failure analyses and the resulting work execution. In this way, we ensure that the prescription for failure management applies our PdM capabilities where they are most valuable. The involvement of these groups also ensures that the predictive data will be welcomed and seen as valuable as it arrives.

16.8.3. Reason 2: The Collected PdM Data Arrives Too Late to Prevent Equipment Failures

In this scenario, maintenance and operations management ask "Why did we not see this equipment failure coming?" Yet the PdM Technician can often point to a chart or spreadsheet logged days ago and say "I told you so." Management's perception is that the information was received too late. Yet, in reality, the data was there, but was not visible when it would be most valuable. Predictive maintenance activities generate massive amounts of data related to the health of the equipment. To be of real value to maintenance and operations, the data must be visible to maintenance, effectively analyzed while it's still current, compared against defined "normal" states and the analysis communicated in a real-time manner.

We know from a reliability standpoint you cannot see or predict all equipment failures. However, most degradation in equipment performance can be observed well in advance with the integration of PdM technologies and techniques. Using handheld data collectors, operators and trades people can record real-time and time-stamped health indicators and feed that data into a computerized reliability system. The amount of data collected in any 8-hour shift is likely to be overwhelming if it was to be managed manually. And yet with appropriate computerization, the normal and non-normal state information creates the opportunity to selectively focus on only the handful of data points that are relevant to each shift—where the asset health degradation is evident in the data. This form of data management can lead to the ultimate use of PdM capability, where management can easily make critical maintenance intervention decisions—driven by real-time data, before it is too late.

16.8.4. Reason 3: Many Companies Fail to Take Advantage of Data from PLCs (Programmable Logic Controllers) and DCSs (Distributive Control Systems)

PLCs and DCSs can provide important production data such as pressure, flow, and temperature that can also be useful for assessing asset health. Most of us think of PdM in the traditional sense; vibration analysis or oil analysis. Yet the production data available in most companies is quite extensive. We need to selectively tap into this valuable resource.

A cautionary word about production data; like other forms of PdM information, it's only valuable if used in the context of a failure analysis. Most thorough failure analyses will point to production data as appropriate for understanding indicators of certain failure modes, while the majority of failure modes will rely on the collection of data through human senses. So hooking up a data-rich production database to a CMMS/EAM will only result in increasing the amount of useless data in making the right decision at the right time and help capture equipment historical data which is typically not accurate in most plants.

These more advanced PdM programs recognize where production data can add value, and they take advantage of the fact that it's readily accessible electronically (for production use). Using this data, maintenance can better predict the degradation of equipment performance to determine the opportune time to intervene with proactive maintenance activities.

An example I was very familiar with as a maintenance supervisor was we had a DCS (Distributive Control System) which main function was to monitor the parameters of our production process and production equipment. We managed our production process using Statistical Process Control. Our DCS managed a lot of data and did a great job of it for production. What we missed was using specific data in this system to help maintenance make asset reliability decisions. I will use our rotary press as an example; the rotary press (calendar system) pressed two 300 CM rolls to form a matted product from woven fibers through this drum type press at speeds of over 500 meters per minute. The pressure of this rotary press had to stay constant in order to make the desired product. A complex hydraulic servo system was used to maintain the pressure required on this press in order to deliver the product required. Our DCS monitored the hydraulic servo valve milliamp output as part of their process control measures. We checked (visual inspection by an electrician) daily the milliamp signals from all servos. We did not plot the data and relate the date to the PF Curve and thus the decisions we made on this system were either made too early or most of the time too late. Reliability software now allows for continuous monitoring of the milliamp signals coming from these servos. This data could have been collected real time, plotted and assisted in determining where on the PF Curve we needed to make a decision to change out a servo valve (based on the values from one servo valve controller) or the change out of the hydraulic pump (based on the values from numerous servo valve controllers) using reliability technology and methodology. With this new technology available a milliamp signal would be connected from the DCS to the Reliability Software where a decision can be made based on data with an alarm to the maintenance planner who plan and schedule a change out of a servo valve or pump far enough in advance that failures could have been kept at a minimum. This reliability software can be connected directly to the CMMS/EAM so that planning and scheduling of the work would be seamless and allow accurate history to be documented on the equipment.

16.8.5. Reason 4: Most PdM Data Is Dispersed in Too Many Non-Integrated Databases

Separate software systems are usually employed to manage the many specialized sources of PdM data: contractors have their data, the PdM team has several separate databases for each PdM technology, the production PLCs and DCSs also store required data. In addition, reliability engineers collect condition and state data from a variety of sources (typically as a result of a formal work identification process like RCM) and apply rules and calculations manually (day-after-day). Maintenance and operations personnel, themselves, are collecting and managing an increasing number of condition-based proactive tasks in their own databases or often still on paper checksheets (Figure 16.11). To act on this disjointed information from a variety of sources, it becomes impossible to realize significant value.

So what should you do about it?

With today's technologies, all of these data sources can be integrated to enable timely maintenance decisions. Quite often, the best indicator of health is built using rules or calculations that combine data from multiple sources (Figure 16.12).

With a well integrated solution, maintenance can use real-time data to focus on defining the right proactive work to be performed at the right time.

Utilize systems that sort through normal and non-normal data, and display the results in ways that are easy to understand, and utilize.

Here is an example of a system that eliminates the sifting through piles of data. The plant, all of its assets and failure-mode-specific health indicators are displayed in a Health Indicator Panel, a two-panel screen showing the entire plant hierarchy and all assets on the left side, and relevant health indicators on the right side. This panel allows you to monitor asset condition and, at a glance, see any indications of impending failures—before the failures occur. As non-normal values are recorded, alarms are triggered and displayed, drawing attention to only the few data points that currently signal the potential for equipment failure. These flashing alarms are displayed when assets are moving closer to functional failure and alarm severities are readily understood based on the type of icon displayed. Here corrective maintenance decisions can be made based on asset health and risk to the business.

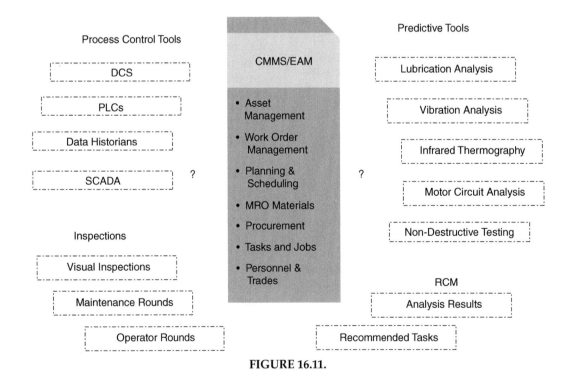

FIGURE 16.11.

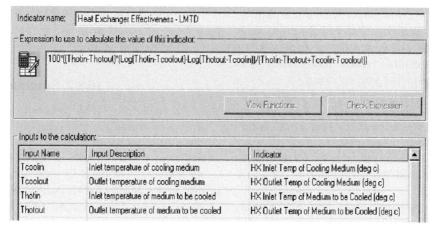

FIGURE 16.12.   Use software that eliminates manual calculations—and captures the knowledge of your experts.

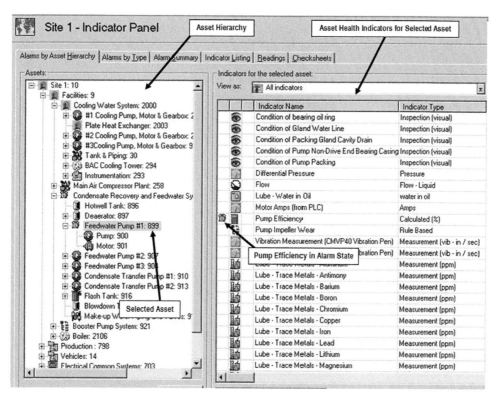

FIGURE 16.13.   Identifying condition alarms.

16.8.6. Some Simple Guidelines Will Help to Get You Moving in the Right Direction

1. Do not stop what you currently are doing in Predictive Maintenance but evolve your PdM strategy into your maintenance program trading the ad hoc wrong work at the wrong time to the "right work at the right time." Do this by aligning your PdM work with the maintenance process required to keep your equipment reliable.

2. Identify the most critical assets (those which are at highest risk to your plant) and focus on putting in place a PdM strategy within the context of a complete maintenance program for these assets. If you want to make your PdM more effective, you need to know which assets are more important to monitor. Your

PdM program will make an impact within the plant as quickly as possible and be a true contributor to asset reliability. When this new strategy is implemented you want "rapid results" which immediately gets people excited about what you are doing.

3. Establish performance targets for these highest risk assets (focus on just one asset at a time) and measure the success of your new strategy. Performance targets must be in production terms: increased capacity, decreased downtime, etc. and in maintenance terms such as increased Mean Time Between Failure (MTBF).

4. Work with operators, maintainers, and PdM technicians to assist in identifying known and likely failure modes on the highest priority assets. Develop a complete asset maintenance program of which PdM is an integral element.

5. Implement this new PdM strategy within the context of a complete maintenance program on one asset at a time and monitor the results. If this process has been followed properly you should see results in a short period of time.

16.8.7. Summary

An effective PdM program must be integrated into a company's asset reliability process so that the right decisions can be made at the right time utilizing accurate data which is fed into a reliability software which is in turn seamlessly linked to an effective CMMS/EAM. Being able to make reliability decisions far enough in advance to plan and schedule maintenance work will drive an organization from being reactive to proactive quickly thus allowing the company to meet it's business goals 100% of the time.

The time has come for change. The best time to begin this new journey is "now."

17

MTBF User Guide: Measuring Mean Time between Failures

Most companies do not measure mean time between failure (MTBF). Yet, it is one of the first and most basic measurements you can use to measure reliability. MTBF is the average time an asset will function before it fails.

17.1. UNDERSTANDING DEFINITIONS

It is important to have definitions because, in reality, the true meaning of a term is not always what the typical industry believes. Everyone in an organization must have a common understanding of a definition. Here are the definitions of some key terms related to MTBF:

- *Bad actors* are pieces of equipment or assets that typically have long-standing reliability issues. Some companies identify "bad actors" by the amount of maintenance dollars spent on assets in labor and material, not production losses. MTBF is a simple measurement to pinpoint these poorly performing assets. Note that the most systematic, technically based method of determining if an asset is critical is to conduct an assessment based on consequence of failure and risk of failure to the business.
- *Total equipment failure* occurs when an asset completely fails or breaks down and is not operating at all.
- *Functional failure* is the inability of an asset to fulfill one or more of its functions (for example, it no longer produces a product that meets quality standards).

- *Partial equipment failure* occurs when equipment continues to run to a standard but some component of the asset is in failure mode. For example, the equipment may now be operating at a reduced speed.
- *MTBF* is the average time an asset will function before it fails.
- *Emergency work order* is a formal document written any time an asset has failed and a maintenance person is called to make a repair.
- *Reliability* is the ability of an item to perform a required function under stated conditions for a stated period of time.

17.2. THE MTBF PROCESS

Step 1. Ensure all emergency work is covered by a work order no matter how minor the equipment failure and that the asset information is captured in the CMMS/EAM by asset number.

Step 2. Begin tracking MTBF, focusing on one production area or asset group. Calculate on a daily basis the mean time between failures:

$$\text{MTBF} = \frac{\text{Time}}{\text{Number of emergency work orders (for all failures, total and functional)}}$$

Step 3. Trend the data you find in this production area or asset group daily on a line graph and post it for everyone to see. (Many people may not like to see this data or even believe it, but it provides knowledge of how the equipment has been performing to date and increases the need to find a solution to improve reliability.)

Step 4. Once you start tracking MTBF, another useful metric to track is the percentage change in MTBF. This allows you to set a target or goal and work toward this goal. This approach often gains support by management for improving reliability.

$$\text{MTBF \% change} = \frac{\text{Current MTBF}}{\text{Previous MTBF}}$$

Step 5. Trend the percentage change.

Step 6. Once you feel comfortable tracking and trending MTBF for this one production area or asset group, begin stepping down to the next level in your asset group. This group is typically called the child in your equipment hierarchy. What you have been measuring thus far is what I call the *father* or *parent* in the equipment hierarchy. You may define the hierarchy differently but in general the message is understood.

Continue the process throughout your organization's production areas and assets.

17.3. EXAMPLE

- Asset: Packaging area for a plant I define as the production area.
- Number of emergency work orders in the past 24 hours = 8: Total equipment failure = 3, functional equipment failure = 5.

   (Note: do not worry about the exact definition of each type of failure. An emergency work order needs to be written any time an asset has a problem and a maintenance person is called to the asset to investigate or make a repair.)
- Time: 24 hours
- Calculation:

$$\text{MTBF} = \frac{24 \text{ hours}}{8 \text{ emergency work orders}} = 3 \text{ hours}$$

Trending this information is valuable when identifying whether an asset's health is improving or getting worse. An example of this trending is seen in Figure 17.1.

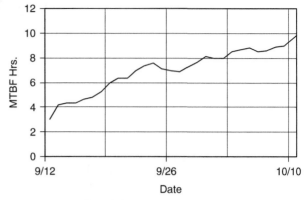

FIGURE 17.1.   Example trend.

17.3.1. MTBF Percentage Change

Over a five day period the MTBF improved from having failures every 3 hours to having a failure every 4.2 hours.

- Time: 5 days
- Previous MTBF: Day 1 = 3 hours
- Current MTBF: Day 5 = 4.2 hours
- Calculation:

$$\text{MTBF \% change} = \frac{4.2 \text{ hours (current MTBF, Day 5)}}{3 \text{ hours (MTBF, Day 1)}}$$

$$= 1.4 \text{ or } 40\%$$

17.3.2. Total Plant MTBF

MTBF calculated for all assets in a plant is an indicator of the total plant reliability.

17.4. SUMMARY

The process of MTBF is the most basic measurement in understanding the current status of reliability of all of your company assets. From here it can be determined which assets are in need of improved reliability.

A

Workflow for Planning

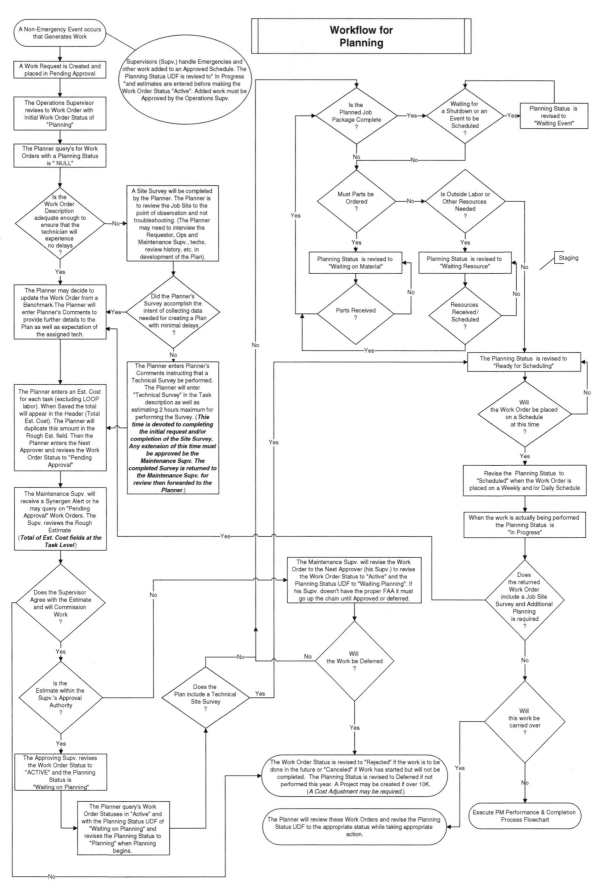

B

Checklists and Forms

Maintenance Operation Checklist	
Is the inventory of skills to support the LEAN program available?	
Is training planned to fill skill and technical shortcomings?	
Does the training support the development of predictive analytical skills?	
Does the training support LEAN management and supervisory skills?	
Are the documentation, procedures, and work practices capable of supporting Lean Maintenance?	
Are the responsibilities for systems and equipment defined and assigned?	
Are the maintenance history data and results distributed to proper users?	
Is there a feedback system in place for continuous maintenance program improvement?	
Is root-cause failure analysis in use and effective?	
Are failed components subject to post-failure examination and results recorded?	
Are predictive forecasts tracked and methods modified based on experience?	
Are PM task and CM monitoring periodicities adjusted based on experience?	
Does the CMMS fully support the maintenance program?	
Are maintenance cost, cost avoidance, and cost savings data collected, analyzed, and disseminated?	
Is baseline condition and performance data updated following major repair or replacement of equipment?	
Are appropriate measures of maintenance performance (metrics) in use?	

FIGURE B.1. Overall maintenance operation: Checklist.

TABLE B.1.   Recommended Predictive Technology Application by Equipment Type

Equipment Item	Recommended Predictive Technologies	Optional Predictive Technologies
Batteries	Battery Impedance Test	Infrared Thermography
Boilers	Hydrostatic Test Airborne Ultrasonic Test Thermodynamic Performance Tests	Infrared Thermography
Breakers	Contact Resistance Test Insulation Resistance Test	Airborne Ultrasonic Test Power Factor Test Insulation Oil Test High Voltage Test Breaker Timing Test Infrared Thermography
Cables	Insulation Resistance Test	Airborne Ultrasonic Test Power Factor Test High Voltage Test
Compressors	Vibration Analysis Balance Test and Measurement Alignment (Laser preferred) Lubricating Oil Test Thermodynamic Performance Tests	Hydraulic Oil Test
Cranes	Vibration Analysis Balance Test and Measurement Alignment (Laser preferred) Lubricating Oil Test Mechanical Performance Tests	Insulation Resistance Test Hydraulic Oil Test
Fans	Vibration Analysis Balance Test and Measurement Alignment (Laser preferred) Lubricating Oil Test Thermodynamic Performance Tests	
Gearboxes	Vibration Analysis Hydraulic Oil Test Lubricating Oil Test	
Heat Exchangers	Hydrostatic Test Airborne Ultrasonic Test Thermodynamic Performance Tests	Infrared Thermography

(Continues)

TABLE B.1. Recommended Predictive Technology Application by
Equipment Type—*(Continued)*

Equipment Item	Recommended Predictive Technologies	Optional Predictive Technologies
HVAC Ducts	Operational Test Ductwork Leakage Test	
Motor Control	Airborne Ultrasonic Test	Insulation Resistance Test
Centers	Infrared Thermography	
Switchgear	Airborne Ultrasonic Test Insulation Resistance Test Infrared Thermography	Contact Resistance Test High Voltage Test Power Factor Test
Motors	Vibration Analysis Balance Test and Measurement Alignment (Laser preferred) Power Factor Test	Infrared Thermography Insulation Resistance Test Motor Circuit Evaluation Test High Voltage Test
Piping Systems	Hydrostatic Test Thermodynamic Performance Tests	Airborne Ultrasonic Test Pulse Ultrasonic Test Infrared Thermography
Pumps	Vibration Analysis Balance Test and Measurement Alignment (Laser preferred) Lubricating Oil Test Thermodynamic Performance Tests	Hydraulic Oil Test
Roofs, Walls, and Insulation	Infrared Thermography	Airborne Ultrasonic Test
Steam Traps	Airborne Ultrasonic Test	
Transformers	Airborne Ultrasonic Test Power Factor Test Insulation Oil Test Infrared Thermography Turns Ratio Test	Contact Resistance Test Insulation Resistance Test High Voltage Test
Valves	Hydrostatic Test	Airborne Ultrasonic Test Thermodynamic Performance Tests Infrared Thermography

Failure Analysis Form

Reliability Centered Maintenance

Equipment Identification: Item No. _____ Equip. Type _____

                              Ser. No. _____ Location _____

Name of Person(s) Responding: _____

Time of Equipment Failure: Date: _____   Time: _____

Time Equipment Returned to Service: Date: _____   Time: _____

Brief Description of Failure: _____

_____

_____

Probable Cause of Failure: _____

_____

_____

Corrective Action Taken: _____

_____

_____

Parts Replaced: _____

_____

Previous Failures (review CMMS): _____

Date Last PM was Performed: _____ **Associated W.O. No.:** _____

Direct Cost Data:	In-House	Contract	Subtotal
Labor:	$_____	$_____	$_____
Material:	$_____	$_____	$_____
		Total Cost:	$_____

Failure Analysis Report Completed By: _____   **Date:** _____

FIGURE B.2.   Failure analysis form.

Sample CMMS Data Collection Form

Initials: _____

Film/Roll/Frame: _____   Date: _____

Mechanical Component: Equipment #: _____ Belongs to: _____
Equipment: _____
Location: _____ Floor: _____ Room _____
MFG: _____

Model, ID, Spec, No.: _____ Size: _____
Model, ID, Spec, No.: _____
Model, ID, Spec, No.: _____ Type: _____
Model, ID, Spec, No.: _____

SN # 1: _____ SN # 2: _____
SN # 3: _____ SN # 4: _____

RPM: _____ GPM: _____ CFM/FPM: _____ MWP@°F: _____

Oil, Air, Fuel Filter Type: _____ MN, PN, Size: _____ Qty.: _____

Oil, Air, Fuel Filter Type: _____ MN, PN, Size: _____ Qty.: _____

Oil, Air, Fuel Filter Type: _____ MN, PN, Size: _____ Qty.: _____

Oil, Air, Fuel Filter Type: _____ MN, PN, Size: _____ Qty.: _____

Oil, Air, Fuel Filter Type: _____ MN, PN, Size: _____ Qty.: _____

Lubricant: Type Oil _____ Type Grease _____

Drive Type: Belt, Chain, Gear, Coupling, Clutch, Direct: _____

Belt Type/Size/Qty. ( ): _____ Coupling Type/Size: _____

Remarks: _____
_____
_____

Electrical Component: Equipment #: _____ Belongs to: _____
Equipment: _____
Location: _____ Floor: _____ Room: _____
MFG: _____

Model, ID, Spec, Cat: _____
Model, ID, Spec, Cat: _____
Model, ID, Spec, Cat: _____
Model, ID, Spec, Cat: _____

SN # 1: _____ SN # 2: _____
SN # 3: _____ SN # 4: _____
HP: _____ Kva/kW: _____ RPM: _____ Amps: _____ V/Ph AC DC: _____ Frame: _____
HP: _____ Kva/kW: _____ RPM: _____ Amps: _____ V/Ph AC DC: _____ Frame: _____
Bearings: Greased: _____ Sealed: _____ Oil Fed: _____ Type Oil: _____ Type Grease: _____
Remarks:

FIGURE B.3.   Sample CMMS Data Collection Form.

			Enter name of company being certified here						
LEAN MAINTENANCE PREPARATION, IMPLEMENTATION AND EXECUTION AUDIT						SCORING				
PHASE	CATEGORY	SUB CATEGORY	ITEM NAME	MAXIMUM SCORE	THIS SCORE	LOW	LOW AVERAGE	AVERAGE	HIGH AVERAGE	HIGH
1	-	-	**PREPARATION AND PLANNING PHASE**	25.00	18.75					
	1.1	-	**Management Commitment**	4.00	3.00				x	
	1.2	-	**Project Manager Assignment**	4.00	3.00				x	
	1.3	-	**Planning**							
		1.3.1	Planning Meeting							
		1.3.1.1	Appropriate Attendees	4.00	3.00				x	
		1.3.1.2	Agenda and Action Items	4.00	3.00				x	
		1.3.2	Master Plan Approval / Published							
		1.3.2.1	Schedule and Milestones	4.00	3.00				x	
		1.3.2.2	Mission Statement	4.00	3.00				x	
		1.3.2.3	Objectives and Goals	4.00	3.00				x	
		1.3.2.4	Assignment of Responsibilities	4.00	3.00				x	
		1.3.2.5	Completeness of Plan	4.00	3.00				x	
	1.4	-	**Pilot Project Selected and Planned**	4.00	3.00				x	
	1.5	-	**Project Selling Campaign / Training Established**	4.00	3.00				x	
2	-	-	**IMPLEMENTATION PHASE**	28.57	21.43					
	2.1	-	**Organization**							
		2.1.1	Structure / Infrastructure Planned	4.00	3.00				x	
		2.1.2	Work Flow Planning	4.00	3.00				x	
	2.2	-	**CMMS**							
		2.2.1	CMMS Selection	4.00	3.00				x	
		2.2.2	CMMS Implementation							
		2.2.2.1	Equipment Inventory Complete and Accurate	4.00	3.00				x	
		2.2.2.2	Technical / Procedural Documentation in CMMS	4.00	3.00				x	
		2.2.2.3	Reports Generation Complete	4.00	3.00				x	
		2.2.2.4	Work Order Sequence Established and Followed	4.00	3.00				x	
	2.3	-	**Total Productive Maintenance**							
		2.3.1	PM / PdM / Condition Monitoring & Testing Program	4.00	3.00				x	
		2.3.2	5-S / Visual Responsibility & Planning	4.00	3.00				x	
		2.3.3	Work Order System Complete	4.00	3.00				x	
		2.3.4	Planner / Scheduler							
		2.3.4.1	Qualified Assignee	4.00	3.00				x	
		2.3.4.2	Training	4.00	3.00				x	
		2.3.5	Value Stream Process Planning	4.00	3.00				x	
	2.4	-	**Maintenance Engineering (Est. & Assign Responsibilities)**							
		2.4.1	Responsibilities Assigned / Understood / Practiced	4.00	3.00				x	
	2.5	-	**MRO Storeroom**							
		2.5.1	Lean Policies and Controls in Place	4.00	3.00				x	
		2.5.2	Reorganization completed	4.00	3.00				x	
		2.5.3	Supplier	.00	3.00				x	
	2.6	-	**Training**							
		2.6.1	General Lean Maintenance Training Completed	4.00	3.00				x	
		2.6.2	JTA and SA in progress or completed	4.00	3.00				x	
3	-	-	**EXECUTION PHASE**	46.43	34.82					
	3.1	-	**Organization**							
		3.1.1	Integration	4.00	3.00				x	
		3.1.2	Communications	4.00	3.00				x	
		3.1.3	Work Flow Discipline	4.00	3.00				x	
	3.2	-	**CMMS**							
		3.2.1	Complete and Effective Use	4.00	3.00				x	
		3.2.2	Reporting Effectiveness	4.00	3.00				x	
		3.2.3	Work Order Discipline	4.00	3.00				x	
	3.3	-	**Total Productive Maintenance (TPM)**							
		3.3.1	Use of Documentation	4.00	3.00				x	
		3.3.2	CMMS Implementation							
		3.3.2.1	Equipment Inventory	4.00	3.00				x	
		3.3.2.2	Technical / Procedural Documentation	4.00	3.00				x	
		3.3.2.3	Report Generation	.00	3.00				x	
		3.3.2.4	Effective Scheduler Usage	4.00	3.00				x	
		3.3.3	5-S / Visual Deployment	4.00	3.00				x	
		3.3.4	Work Measurement	4.00	3.00				x	
		3.3.5	Work Order Usage	.00	3.00				x	
		3.3.6	Schedule Compliance	4.00	3.00				x	
		3.3.7	Value Stream Mapping Process	4.00	3.00				x	
	3.4	-	**Maintenance Engineering**							
		3.4.1	Knowledge of Lean Maintenance Tools	4.00	3.00				x	
		3.4.2	Predictive Maintenance / Condition Monitor & Testing	4.00	3.00				x	
		3.4.3	Continuous PM Evaluation & Improvement	4.00	3.00				x	
		3.4.4	Planning & Scheduling Process	4.00	3.00				x	
	3.5	-	**MRO Storeroom**							
		3.5.1	Inventory Control	4.00	3.00				x	
		3.5.2	Storeroom Organization	4.00	3.00				x	
		3.5.3	CMMS Integration / Work Order Use	4.00	3.00				x	
		3.5.4	Stockouts	4.00	3.00				x	
		3.5.5	Use of JIT Vendors	4.00	3.00				x	
	3.6	-	**Training**							
		3.6.1	Skill needs / Skills availability assessed	4.00	3.00				x	
		3.6.2	Focused Training Program	4.00	3.00				x	
		3.6.3	Qualification / Certification Program	4.00	3.00				x	
		3.6.4	Training Effectiveness Evaluation	4.00	3.00				x	
	3.7	-	**Lean Sustainment Established**	4.00	3.00				x	
			TOTALS	100.00	75.00					

FIGURE B.4.   Lean Maintenance Preparation, Implementation, and Execution Audit.

LEAN MAINTENANCE PRACTICES AUDIT					SCORING				
CATEGORY	SUB CATEGORY	ITEM NAME	MAXIMUM SCORE	THIS SCORE	LOW	LOW AVERAGE	AVERAGE	HIGH AVERAGE	HIGH
1	-	PROGRAM FOUNDATION	21.62	10.14					
	1a	Management Commitment	4.00	3.00				x	
	1b	Master Plan	4.00	2.00			x		
	1c	Mission Statement	4.00	0.00	x				
	1d	Objectives and Goals	4.00	2.00			x		
	1e	Management Reporting	4.00	1.00		x			
2	-	ORGANIZATION	10.81	5.41					
	2a	Structure	4.00	1.00		x			
	2b	Work Flow	4.00	2.00			x		
	2c	Communication	4.00	3.00				x	
3	-	TOTAL PRODUCTIVE MAINTENANCE	29.73	17.57					
	3a	Standardized Work Practices	4.00	2.00			x		
	3b	CMMS Implementation							
	3b(1)	Equipment Inventory	4.00	3.00				x	
	3b(2)	Technical / Procedural Documentation	4.00	3.00				x	
	3b(3)	Report Generation	4.00	3.00				x	
	3c	5-S / Visual Deployment	4.00	2.00			x		
	3d	Work Measurement	4.00	2.00			x		
	3e	Work Order Usage	4.00	3.00				x	
	3f	Schedule Compliance	4.00				x		
	3g	Value Stream	4.00	2.00			x		
4	-	MAINTENANCE ENGINEERING	13.51	9.46					
	4a	Knowledge of Lean Maintenance Tools	4.00	2.00			x		
	4b	Predictive Maintenance / Condition Monitor & Testing	4.00	3.00				x	
	4c	Continuous PM Evaluation & Improvement	4.00	3.00				x	
	4d	Planning & Scheduling	4.00	3.00				x	
5	-	MRO STOREROOM	13.51	4.73					
	5a	Inventory Control	4.00	3.00				x	
	5b	Storeroom Organization	4.00	2.00			x		
	5c	CMMS Integration / Work Order Use	4.00				x		
	5d	Stockouts	4.00			x			
	5e	Use of JIT Vendors	4.00	2.00			x		
6	-	TRAINING	10.81	8.11					
	6a	Skill needs / Skills availability assessed	4.00	2.00			x		
	6b	Focused Training Program	4.00	2.00			x		
	6c	Qualification / Certification Program	4.00	4.00					x
	6d	Training Effectiveness Evaluation	4.00	4.00					x
		TOTALS	100.00	55.41					

FIGURE B.5. Lean Maintenance Practices Audit.

TABLE B.2.　Predictive Maintenance Data Collection Forms—1

Vibration Analysis Test Criteria

Item	Date of Inspection	Acceptable Limit	Actual Value	Inspector Initials	PASS	FAIL	Comments
Vibration Analysis Test							
Test Instrumentation							
FFT Analyzer							
Type							
Model							
Serial Number							
Last Calibration Date							
Line Resolution Bandwidth							
Dynamic Range							
Hanning Window							
Linear Non-overlap Averaging							
Anti-aliasing Filters							
Amplitude Accuracy							
Sound Disk Thickness							
Adhesive (hard/soft)							
Vibration Readings							

(Continues)

TABLE B.2. Predictive Maintenance Data Collection Forms—1—(Continued)

Vibration Analysis Test Criteria

Item	Date of Inspection	Acceptable Limit	Actual Value	Inspector Initials	PASS	FAIL	Comments
1H							
1V							
1A							
2H							
2V							
2A							
Velocity Amplitude (in./sec.—peak)							
Running Speed Order							
Acceleration Overall Amp (g—peak)							
Vibration Signatures (H, V, A)							
Frequency (CPM)							
Balanced Condition							
Balance Wt. Type							
Results:							

TABLE B.3. Predictive Maintenance Data Collection Forms—2

Lubrication Oil Test Criteria

Item	Date of Inspection	Acceptable Limit	Actual Value	Inspector Initials	PASS	FAIL	Comments
Lubrication Oil Test							
Liquids							
Viscosity Grade (ISO Units)							
AGMA/SAE Classification							
Additives							
Grease							
Type of Base Stock							
NLGI Number							
Type/% of thickener							
Dropping Point							
Base Oil Viscosity (SUS)							
Total Acid Number							
Visual Observation (Cloudiness)							
IR Spectral Analysis—Metal							
Count							
Particle Count							
Water Content							
Viscosity							
Results:							

TABLE B.4. Predictive Maintenance Data Collection Forms—3

Alignment Criteria

Item	Date of Inspection	Acceptable Limit	Actual Value	Inspector Initials	PASS	FAIL	Comments
Alignment							
RPM							
Soft Foot Actual (in.)							
Soft Foot Tolerance							
Vert. angularity at coupling—Actual							
Vert. offset at coupling—Actual							
Vert. angularity at coupling—Actual							
Vert. offset at coupling—Actual							
Axial Shaft Play							
Shims							
Shim Type							
Shim Condition							
Number of Shims in Pack							
Thickness							
Sheaves							
True to Shaft							
Runout (in.)							
Results:							

TABLE B.5.   Predictive Maintenance Data Collection Forms—4

Breaker Timing Test Criteria

Item	Date of Inspection	Acceptable Limit	Actual Value	Inspector Initials	PASS	FAIL	Comments
Breaker Timing Test							
Voltage Applied							
C1—Phase A							
C2—Phase B							
C3—Phase C							
Results:							

TABLE B.6. Predictive Maintenance Data Collection Forms—5

Contact Resistance Test Criteria

Item	Date of Inspection	Acceptable Limit	Actual Value	Inspector Initials	PASS	FAIL	Comments
Contact Resistance Test							
DC Current Applied							
Measured Voltage							
Calculated Resistance							
Manufacturer Resistance							
Results:							

TABLE B.7.   Predictive Maintenance Data Collection Forms—6

Item	Date of Inspection	Acceptable Limit	Actual Value	Inspector Initials	PASS	FAIL	Comments
Insulation Oil Test							
Dissolved Gas Analysis							
Nitrogen (N_2)		<100-ppm					
Oxygen (O_2)		<10-ppm					
Carbon Dioxide (CO_2)		<10-ppm					
Carbon Monoxide (CO)		<100-ppm					
Methane (CH_4)		None					
Ethane (C_2H_6)		None					
Ethylene (C_2H_4)		None					
Hydrogen (H_2)		None					
Acetylene (C_2H_2)		None					
Karl Fisher (@ 20.Deg.C)		<25-ppm					
Dielectric Breakdown		<30.kV					
Strength							
Neutralization Number		<0.05-mg/g					
Interfacial Tension		<40 dynes/cm					
Color (ASTM D-1524)		<3.0					
Sediment		clear					
Power Factor		<0.05%					
Sediment/Visual Examination		clear					
Results:							

TABLE B.8. Predictive Maintenance Data Collection Forms—7

Item	Date of Inspection	Acceptable Limit	Actual Value	Inspector Initials	PASS	FAIL	Comments
Power Factor Test Criteria							
Power Factor Test							
Grounded Specimen Test—GST							
Ungrounded Specimen Test—UST							
GST with Guard							
Environment Humidity							
Environment Temperature							
Surface Cleanliness							
Phase I							
Applied Voltage							
Total Current							
Capacitive Current							

(Continues)

TABLE B.8.   Predictive Maintenance Data Collection Forms—7—(Continued)

Dissipation Factor									
Power Factor									
Normal Power Factor									
Phase II									
Applied Voltage									
Total Current									
Capacitive Current									
Dissipation Factor									
Power Factor									
Normal Power Factor									
Phase III									
Applied Voltage									
Total Current									
Capacitive Current									
Dissipation Factor									
Power Factor									
Normal Power Factor									
Results:									

Procedure for Performing a Failure Modes and Effects Analysis (FMEA)

FMEAs are generally performed using the guidance provided in MIL-STD-1629A, in spite of the fact that, in theory, Military Standards no longer exist. A functional process based on MIL-STD-1629A is outlined here.

1. Describe, in words, the process and functions of the system/equipment to be analyzed. This step is really meant to ensure that the analyst has a clear understanding of what the equipment is meant to do and how it fits into the overall production scheme. There is no need to create an eloquent thesis. Just write down short "one-liners" that describe the various functions and the overall system process.
2. As you refine and put order to the written descriptions, begin creating a diagram of the process which basically will consist of ordered blocks representing the various functions. If not previously defined, system boundaries will need to be established. It may help in some cases to sketch a pictorial representation in order to better visualize components and their functions. When completed, the block diagram will need to be completed "smooth" as it will be a permanent attachment to the FMEA Data Form as in Figure B-6.
3. On a rough copy of the FMEA Data Form, begin listing the functions as identified above.
4. Identify functional failures or failure modes. Note that a failure mode in one component can be a cause of a failure mode in another component. In some cases the iterations of failure modes and causes can be extensive. All failure modes should be identified, regardless of their probability of occurring (see Figure B-7).
5. Describe the effects of each failure mode and assign a severity ranking. Effects can be on the component, on the next step in the process (or block in the system diagram), on the end result of the process or all three. Be sure to consider safety and environmental effects as well as effects on production and product. If not previously established you will need to develop a ranking system for the severity of the effect. This is normally a 1 to 10 scale where 10 is the most severe and 1 indicates none or negligible severity.
6. Identify potential causes of each failure mode. Be sure to consider all possibilities, including poor design, extremes of operating environments, operator (or maintenance tech) error—which in turn may be due to inadequate training, documentation, or procedure errors, etc.
7. Enter the probability factor for each potential cause.
8. Identify the compensating provisions, which are either design or process controls intended to (a) prevent the cause of a failure from occurring or (b) identify the potential for the cause to occur (i.e., existing PdM procedure).
9. Determine the likelihood of detection. Detection is an assessment of the likelihood that the compensating provisions (design or process) will detect the Cause of the Failure Mode or the Failure Mode itself. If the Compensating Provisions include an existing PdM procedure, the Likelihood of Detection is the likelihood that the potential for the Cause to Occur will be detected (see Figure B-8).
10. Calculate and enter the Failure Mode Ranking. The ranking is the mathematical product of the numerical Severity, Probability, and Detection ratings. Ranking = (Severity) × (Probability) × (Detection). The ranking is used to prioritize those items requiring additional action.
11. Enter any remarks pertinent to the FMEA item.

Management Oversight and Risk Tree (MORT) Analysis

A Mini-MORT Analysis Chart is shown on page 312. This chart is a checklist of what happened (less-than-adequate spe-cific barriers and controls) and why it happened (less-than-adequate management).

To perform the MORT analysis:

1. Identify the problem associated with the occurrence and list it as the top event.
2. Identify the elements on the "what" side of the tree that describe what happened in the occurrence (what barrier or control problems existed).
3. For each barrier or control problem, identify the management elements on the "why" side of the tree that permitted the barrier control problem.
4. Describe each of the identified inadequate elements (problems) and summarize your findings.

A brief explanation of the "what" and "why" may assist in using mini-MORT for causal analyses.

When a target inadvertently comes in contact with a hazard and sustains damage, the event is an accident. A hazard is any condition, situation, or activity representing a potential for adversely affecting economic values or the health or quality of people's lives. A target can be any process, hardware, people, the environment, product quality, or schedule—anything that has economic or personal value.

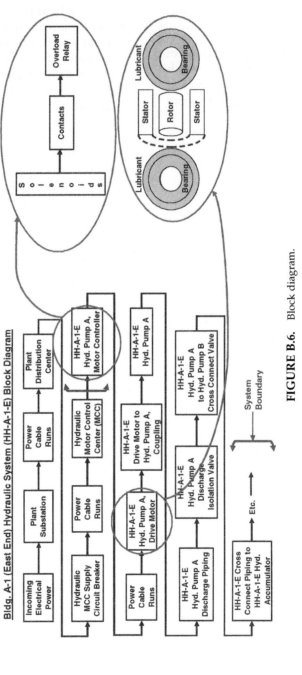

FIGURE B.6. Block diagram.

Hydraulic System HH-A-1-E

Function	Functional Failure	Failure Mode	Source of Failure
Provide hydraulic oil at specified Pressure and Demand volume (flow rate)	Total loss of pressure, volume and flow	Electric Motor Failed	(See below)
		Pump Failed	
		Major Leak	
		Blocked Line	
		Valve out of position	

Electric Motor

Function	Functional Failure	Failure Mode	Source of Failure
Drive hydraulic pump at specified power level	Motor will not turn	Insulation failure in stator	(See below)
		Insulation failure in rotor	
		Bearing Seized	
		Motor Controller Contactor Failed	
		Circuit Breaker tripped	

Motor Bearings

Failure Mode	Mechanism	Reason	Cause
Bearing seized (this includes seals, shields, lubrication system, and lock nuts)	Lubrication	Contamination	Seal failed
			Supply dirty
		Wrong Type	Procedure wrong
			Supply information wrong
		Too Little	Leak
			Human error
			Procedure error
		Too Much	Human error
			Procedure error
	Fatigue	Metallurgical	Inherent
			Excessive temperature
		Excessive Load	Mechanical imbalance
			Misalignment
			Wrong application (bearing not sized for the load)
	Etc.		

FIGURE B.7. Hydraulic system.

Detection	Rank
Almost Impossible	10
Very Remote	9
Remote	8
Very Low	7
Low	6
Moderate	5
Moderately High	4
High	3
Very High	2
Almost Certain	1

FIGURE B.8. Detection.

What prevents accidents or adverse programmatic impact events?

- Barriers that surround the hazard and/or the target and prevent contact or controls and procedures that ensure separation of the hazard from the target.
- Plans and procedures that avoid conflicting conditions and prevent programmatic impacts. In a facility, what functions implement and maintain these barriers, controls, plans, and procedures?
- Identifying the hazards, targets, and potential contacts or interactions and specifying the barriers/controls that minimize the likelihood and consequences of these contacts.
- Identifying potential conflicts/problems in areas such as operations, scheduling, or quality and specifying management policy, plans, and programs that minimize the likelihood and consequences of these adverse occurrences.
- Providing the physical barriers: designing, installation, signs/warnings, training, or procedures.
- Providing planning/scheduling, administrative controls, resources, or constraints.
- Verifying that the barriers/controls have been implemented and are being maintained by operational readiness, inspections, audits, maintenance, and configuration/change control.
- Verifying that planning, scheduling, and administrative controls have been implemented and are adequate.
- Policy and policy implementation (identification of requirements, assignment of responsibility, allocation of responsibility, accountability, vigor, and example in leadership and planning).

Definitions used with this method:

- A cause (causal factor) is any weakness or deficiency in the barrier/control functions or in the administration/management functions that implement and maintain the barriers/controls and the plans/procedures.
- A causal factor chain (sequence or series) is a logical hierarchal chain of causal factors that extends from policy and policy implementation through the verification and implementation functions to the actual problem with the barrier/control or administrative functions.
- A direct cause is a barrier/control problem that immediately preceded the occurrence and permitted the condition to exist or adverse event to occur. Since any element on the chart can be an occurrence, the next upstream condition or event on the chart is the direct cause and can be a management factor. (Management is seldom a direct cause for a real-time loss event such as injury or property damage but may very well be a direct cause for conditions.)
- A root cause is the fundamental cause, which, if corrected, will prevent recurrence of this and similar events. This is usually not a barrier/control problem but a weakness or deficiency in the identification, provision or maintenance of the barriers/controls or the administrative functions. A root cause is ordinarily control-related involving such upstream elements as management and administration. In any case, it is the original or source cause.
- A contributing cause is any cause that had some bearing on the occurrence, on the direct cause, or on the root cause but is not the direct or the root cause.

Failure Mode and Effects Analyis
(System Name)

System:
Indenture Level:
Reference Drawing:
Mission:

Date: ____ of ____
Page ____ of ____
Compiled by: ____
Approved by: ____

Identification Number	Item Functional Identification (Nomenclature)	Function	Failure Modes and Causes	Probability Factor	Failure Effects			Likelihood of Failure Detection	Compensating Provisions	Severity Class	Ranking	Remarks
					Local Effects	Next Higher Level	End Effects					

FIGURE B.9. Failure Mode and Effects Analysis.

METHOD	WHEN TO USE	ADVANTAGES	DISADVANTAGES	REMARKS
Events and Causal Factor Analysis	Use for multi-faceted problems with long or complex causal factor chain.	Provides visual display of analysis process. Identifies probable contributors to the condition.	Time-consuming and requires process to be effective.	Requires a broad perspective of the event to identify unrelated problems. Helps to identify where deviations occurred from acceptable methods.
Change Analysis	Use when cause is obscure. Especially useful in evaluating equipment failures.	Simple six-step process.	Limited value because of the danger of accepting wrong, "obvious" answer.	A singular problem technique that can be used in support of a larger investigation. All root causes may not be identified.
Barrier Analysis	Use to identify barrier and equipment failures and procedural or administrative problems.	Provides systematic approach.	Requires familiarity with process to be effective.	This process is based on the MORT Hazard/Target Concept.
MORT/Mini-MORT	Use when there is a shortage of experts to ask the right questions and whenever the problem is a recurring one. Helpful in solving programmatic problems.	Can be used with limited prior training. Provides a list of questions for specific control and management factors.	May only identify area of cause, not specific causes.	If this process fails to identify problem areas, seek additional help or use cause-and-effect analysis.
Human Performance Evaluations (HPE)	Use whenever people have been identified as being involved in the problem cause.	Thorough analysis.	None if process is closely followed.	Requires HPE training.
Kepner-Tregoe	Use for major concerns where all aspects need thorough analysis.	Highly structured approach focuses on all aspects of the occurrence and problem resolution.	More comprehensive than may be needed.	Requires Kepner-Tregoe training.

FIGURE B.10. Summary of Root Cause failure analysis methods.

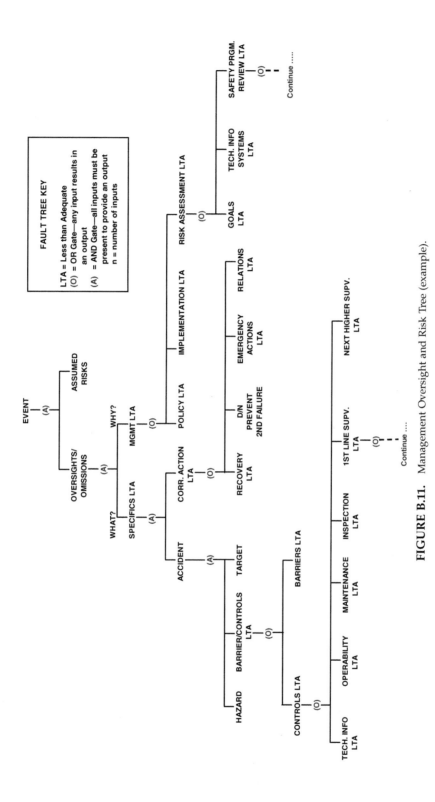

FIGURE B.11. Management Oversight and Risk Tree (example).

MORT BASED ROOT CAUSE ANALYSIS FORM

	Management																														
	Policy	Policy Implementation						Risk Assessment					Bridge Elements						Specific Factors							Task Performance					
	1	2	3	4	5	6	7	8	9	10	11	12	13	14	15	16	17	18	19	20	21	22	23	24	25	26	27	28	29	30	31
Findings or Conclusions	Policy	Line Responsibility	Staff Responsibility	Accountability	Vigor/Example	Methods/Criteria Analysis	Open	Goals	Technical Info Systems	Hazard Analysis Process	Program Review	Open	Management Services	Directives	Budget	Information Flow	Open	Open	Technical Information	Operational Readiness	Maintenance	Inspection	Supervision	Higher Supervisory Services	Open	Task Assignment	Task Safety Analysis	Procedural Incompatibility	Personnel Performance Discrep.	Open	Open

FIGURE B.12. MORT RCFA Report Form.

Answer the following (and any related items unique to your particular operation) questions in order to fill out the Change Analysis Worksheet (see Figure B-13).

WHAT?
What is the condition?
What occurred to create the condition?
What occurred prior to the condition?
What occurred following the condition?
What activity was in progress when the condition occurred?
What activity was in progress when the condition was identified?
   Operational evolution in the work space?
     Surveillance test?
     Starting/stopping equipment?
   Operational evolution outside the work space?
     Valve line-up?
     Removing equipment from service?
     Returning equipment to service?
   Maintenance activity?
     Surveillance?
     Corrective maintenance?
     Modification installation?
     Troubleshooting?
   Training activity?
What equipment was involved in the condition?
   What equipment initiated the condition?

What equipment was affected by the condition?
What equipment mitigated the condition?
What is the equipment's function?
How does it work?
How is it operated?
What failed first?
Did anything else fail due to the first problem?
What form of energy caused the equipment problem?
What are recurring activities associated with the equipment?
What corrective maintenance has been performed on the equipment?
What modifications have been made to the equipment?
What system or controls (barriers) should have prevented the condition?
What barrier(s) mitigated the consequences of the condition?
WHEN?
When did the condition occur?
What was the facility's status at the time of occurrence?
When was the condition identified?
What was the facility's status at the time of identification?
What effects did the time of day have on the condition? Did it affect:
   Information availability?

Company _____		Date _____	
Plant _____		Name of Person Performing Analysis	
		_____	
Occurrence Description _____			
Change Factor	**Difference/Change**	**Effect**	**Questions to Answer**
What (conditions, occurrence, activity, equipment)			
When (occurred, identified, plant status, schedule)			
Where (physical location, environmental conditions)			
How (work practice, omission, extraneous action, out of sequence procedure)			
Who (personnel involved, training, qualification, supervision)			

FIGURE B.13. Change Analysis Worksheet.

Personnel availability?
Ambient lighting?
Ambient temperature?
Did the condition involve shift-work personnel? If so:
    What type of shift rotation was in use?
    Where in the rotation were the personnel?
For how many continuous hours had any involved personnel been working?
WHERE?
Where did the condition occur?
What were the physical conditions in the area?
Where was the condition identified?
Was location a factor in causing the condition?
    Human factor?
        Lighting?
        Noise?
        Temperature?
        Equipment labeling?
        Radiation levels?
        Personal protective equipment required in the area?
        Radiological protective equipment required in the area?
        Accessibility?
        Indication availability?
        Other activities in the area?
        What position is required to perform tasks in the area?
    Equipment factor?
        Humidity?
        Temperature?
        Cleanliness?
HOW?
Was the condition an inappropriate action or was it caused by an inappropriate action?
    An omitted action?
    An extraneous action?
    An action performed out of sequence?
    An action performed to too small of a degree? To too large of a degree?
Was there an applicable procedure?
Was the correct procedure used?
Was the procedure followed?
    Followed in sequence?
    Followed "blindly"—without thought?
Was the procedure:
    Legible?
    Misleading?
    Confusing?
    An approved, current revision?
    Adequate to do the task?
    In compliance with other applicable codes and regulations?
Did the procedure:
    Have sufficient detail?
    Have sufficient warnings and precautions?
    Adequately identify techniques and compo-nents?
    Have steps in the proper sequence?
    Cover all involved systems?

    Require adequate work review?
WHO?
Which personnel:
    Were involved with the condition?
    Observed the condition?
    Identified the condition?
    Reported the condition?
    Corrected the condition?
    Mitigated the condition?
    Missed the condition?
What were:
    The qualifications of these personnel?
    The experience levels of these personnel?
    The work groups of these personnel?
    The attitudes of these personnel?
    Their activities at the time of involvement with the condition?
Did the personnel involved:
    Have adequate instruction?
    Have adequate supervision?
    Have adequate training?
    Have adequate knowledge?
    Communicate effectively?
    Perform correct actions?
    Worsen the condition?
    Mitigate the condition?

Barrier Analysis Description

There are many things that should be addressed during the performance of a Barrier Analysis. Note: In this usage, a barrier is from Management Oversight and Risk Tree (MORT) terminology and is something that separates an affected component from an undesirable condition/situation. The figure at the end of this description provides an example of Barrier Analysis. The questions listed below are designed to aid in determining what barrier failed, thus resulting in the occurrence.

What barriers existed between the second, third, etc. condition/situation and the second, third, etc. problems?

If there were barriers, did they perform their functions? Why?

Did the presence of any barriers mitigate or increase the occurrence severity? Why?

Were any barriers not functioning as designed? Why?

Was the barrier design adequate? Why?

Were there any barriers in the condition/situation source(s)? Did they fail? Why?

Were there any barriers on the affected component(s)? Did they fail? Why?

Were the barriers adequately maintained?

Were the barriers inspected prior to expected use?

Why were any unwanted energies present?

Is the affected system/component designed to withstand the condition/situation without the barriers? Why?

What design changes could have prevented the unwanted flow of energy? Why?

What operating changes could have prevented the unwanted flow of energy? Why?

What maintenance changes could have prevented the unwanted flow of energy? Why?

Could the unwanted energy have been deflected or evaded? Why?

What other controls are the barriers subject to? Why?

Was this event foreseen by the designers, operators, maintainers, anyone?

Is it possible to have foreseen the occurrence? Why?

Is it practical to have taken further steps to have reduced the risk of the occurrence?

Can this reasoning be extended to other similar systems/components?

Were adequate human factors considered in the design of the equipment?

What additional human factors could be added? Should be added?

Is the system/component user friendly?

Is the system/component adequately labeled for ease of operation?

Is there sufficient technical information for operating the component properly? How do you know?

Is there sufficient technical information for maintaining the component properly? How do you know?

Did the environment mitigate or increase the severity of the occurrence? Why?

What changes were made to the system/component immediately after the occurrence?

What changes are planned to be made? What might be made?

Have these changes been properly and adequately analyzed for effect?

What related changes to operations and maintenance have to be made now?

Are expected changes cost effective? Why? How do you know?

What would you have done differently to have prevented the occurrence, disregarding all economic considerations (as regards operation, maintenance, and design)?

What would you have done differently to have prevented the occurrence, considering all economic concerns (as regards operation, maintenance, and design)?

Approximating Failure Distributions

The four failure rate functions or hazard functions corresponding to the probability density functions (exponential, Weibull, lognormal and normal), are shown in Figure B-16.

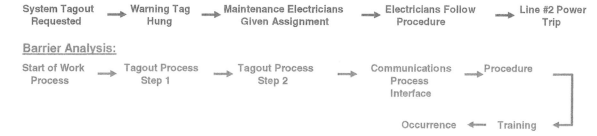

Work Task: Clean Control Relay Panel and Contacts

Occurrence: Production Line #2 Power Trip

<u>Sequence of Events:</u>

System Tagout Requested → Warning Tag Hung → Maintenance Electricians Given Assignment → Electricians Follow Procedure → Line #2 Power Trip

<u>Barrier Analysis:</u>

Start of Work Process → Tagout Process Step 1 → Tagout Process Step 2 → Communications Process Interface → Procedure → Training → Occurrence

Maint. requests de-energizing two panels so relays can be cleaned. Operations will only allow one panel at a time to be tagged out. Electrical supervisor told and agrees.	Tag hung on P689 – only P690 is still energized.	Electricians given W.O. to work, which references a Maint. Procedure, but not told of change in scope by supervisor.	Electricians go to P690 and begin procedure. Procedure has no step to verify dead power supply before starting. They open first relay and Line #2 trips.	Electricians never trained to always check power supply prior to working on electrical equipment.
Barrier Holds	**Barrier Holds**	**Barrier Fails**	**Barrier Fails**	**Barrier Fails**

FIGURE B.14. Work Task.

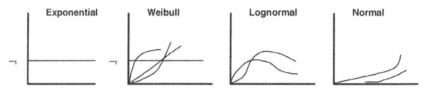

FIGURE B.15.   Failure rate functions.

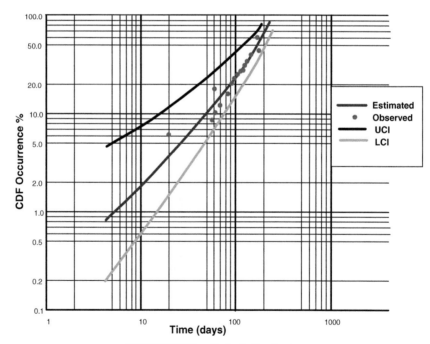

FIGURE B.16.   Weibull distribution.

Glossary

Note: (J) indicates that the term is Japanese.

Andon (J) A system of flashing lights used to indicate production status in one or more work centers.

Availability (1) Informally, the time a machine or system is available for use.

$$\text{Availability} = MTBF \sim (MTBF + MTTR)$$

From the Overall Equipment Effectiveness calculation, the actual run time of a machine or system divided by the scheduled run time. Note that Availability differs slightly from Asset Utilization (Uptime) in that scheduled run time varies between facilities and is changed by factors such as scheduled maintenance actions, logistics, or administrative delays.

Critical Failure A failure involving a loss of function or secondary damage that could have a direct adverse effect on operating safety, on mission, or have significant economic impact.

Critical Failure Mode A failure mode that has significant mission, safety or maintenance effects that warrant the selection of maintenance tasks to prevent the critical failure mode from occurring.

Current State Map Process map of existing practices. A visual method of succinctly recording the key aspects of the current structure or process in the whole or any part.

Failure A cessation of proper function or performance; the inability to meet a standard; nonperformance of what is requested or expected.

Failure Effect The consequences of failure.

Failure Mode The manner of failure. For example, the motor stops is the failure—the reason the motor failed was the motor bearing seized which is the failure mode.

Failure Modes and Effects Analysis (FMEA) Analysis used to determine what parts fail, why they usually fail, and what effect their failure has on the systems in total.

Failure Rate The mean number of failures in a given time. Often "assumed" to be $1 = (MTBF)^{-1}$.

Five S's Five activities for improving the workplace environment:

1. Seiketsu (J)—Sort (remove unnecessary items)
2. Seiri (J)—Straighten (organize)
3. Seiso (J)—Scrub (clean everything)
4. Seiton (J)—Standardize (standard routine to sort, straighten, and scrub)
5. Shitsuke (J)—Spread (expand the process to other areas)

Future State Map Value stream map of an improved process (non-value adding activities removed or minimized)

Jidoka (J) Quality at the source. Autonomation—a contraction of "autonomous automation." The concept of adding an element of human judgment to automated equipment so that the equipment becomes capable of discriminating against unacceptable quality.

Jishu kanri (J) Self-management or voluntary participation.

JIT Just-in-Time. Receiving parts, material or product precisely at the time it is needed. Avoids inventory pile-up.

Kaizen The philosophy of continual improvement, that every process can and should be continually evaluated and improved in terms of time required, resources used, resultant quality, and other aspects relevant to the process.

Kaizen Event Often referred to as Kaizen Blitz—A fast turn-around (one week or less) application of Kaizen "improvement" tools to realize quick results.

Kanban A card, sheet, or other visual device to signal readiness to previous process. (Related—visual cues: operating and maintenance visual aids for quick recognition of normal operating ranges on gauges, lubrication points, lubricant, and amount, etc.).

Karoshi Death from overwork.

Lean Enterprise Any enterprise subscribing to the reduction of waste in all business processes.

Lean Manufacturing The philosophy of continually reducing waste in all areas and in all forms; an English phrase coined to summarize Japanese manufacturing techniques (specifically, the Toyota Production System).

Mean Time Between Failures (MTBF) The mean time between failures that are repaired and returned to use.

Mean Time To Failures (MTTF) The mean time between failures that are not repaired. (Applicable to non-repairable items, e.g., light bulbs, transistors, etc.)

Mean Time To Repair (MTTR) The mean time taken to repair failures of a repairable item.

Muda (J) Waste. There are seven deadly wastes:

1. Overproduction—Excess production and early production
2. Waiting—Delays—Poor balance of work
3. Transportation—Long moves, redistributing, pickup/put-down
4. Processing—Poor process design
5. Inventory—Too much material, excess storage space required
6. Motion—Walking to get parts, tools, etc.; lost motion due to poor equipment access
7. Defects—Part defects, shelf life expiration, process errors, etc.

Mura (J) Inconsistencies (J).

Muri (J) Unreasonableness (J).

Non-Value Added Those activities within a company that do not directly contribute to satisfying end consumers' requirements. Useful to think of these as activities which consumers would not be happy to pay for.

Overall Equipment Effectiveness (OEE) A composite measure of the ability of a machine or process to carry out value-adding activity. OEE = % time machine available × % of maximum output achieved × % perfect output.

Pareto Analysis Sometimes referred to as the "80:20 rule." The tendency in many business situations for a small number of factors to account for a large proportion of events.

PDCA or PDSA Shewhart Cycle: Plan-Do-Check (or Study) Act. A process for planning, executing, evaluating, and implementing improvements.

PF Interval The amount of time (or the number of stress cycles) that elapse between the point where a potential failure (P) occurs and the point where it deteriorates into a functional failure (F). Used in determining application frequency of Predictive Maintenance (PdM) Technologies.

Poka Yoke (J) A mistake-proofing device or procedure to prevent a defect during order intake, manufacturing process, or maintenance process.

Predictive Maintenance (PdM) The use of advanced technology to assess machinery condition. The PdM data obtained allows for planning and scheduling preventive maintenance or repairs in advance of failure.

Preventive Maintenance Time- or cycle-based actions performed to prevent failure, monitor condition, or inspect for failure.

Proactive Maintenance The collection of efforts to identify, monitor and control future failure with an emphasis on the understanding and elimination of the cause of failure. Proactive maintenance activities include the development of design specifications to incorporate maintenance lessons learned and to ensure future maintainability and supportability, the development of repair specifications to eliminate underlying causes of failure, and performing root cause failure analysis to understand why inservice systems failed.

Reliability The dependability constituent or dependability characteristic of design. From MIL-STC-721C: Reliability—(1) The duration or probability of failure-free performance under stated conditions. (2) The probability that an item can perform its intended function for a specified interval under stated conditions.

Reliability-Centered Maintenance (RCM) The process that is used to determine the most effective approach to maintenance. It involves identifying actions that, when taken, will reduce the probability of failure and which are the most cost effective. It seeks the optimal mix of Condition-Based Actions, other Time- or Cycle-Based Actions, or Run-to-Failure Approach.

Return on Investment (ROI) A measure of the cost benefits derived from an investment. ROI (in %) = [(total benefits − total costs) ~ total costs] × 100.

Sensei One who provides information; a teacher, instructor.

TAKT Time Cycle Time—In Production it is the daily production number required to meet orders in hand divided into the number of working hours in the day.

Total Productive Maintenance A manufacturing-led initiative for optimizing the effectiveness of manufacturing equipment. TPM is team-based productive maintenance and involves every level and function in the organization, from top executives to the shop floor. The goal of TPM is "profitable PM." This requires you to not only prevent breakdowns and defects, but to do so in ways that are efficient and economical.

Value Adding Those activities within a company that directly contribute to satisfying end consumers, or those activities consumers would be happy to pay for.

Value Stream The specific value adding activities within a process.

Value Stream Mapping Process mapping of current state, adding value or removing waste to create future state map or ideal value stream for the process.

Vibration Analysis The dominant technique used in predictive maintenance. Uses noise or vibration created by mechanical equipment to determine the equipment's actual condition. Uses transducers to translate a vibration amplitude and frequency into electronic signals. When measurements of both amplitude and frequency are available, diagnostic methods can be used to determine both the magnitude of a problem and its probable cause. Vibration techniques most often used include broadband trending (looks at the overall machine condition), narrowband trending (looks at the condition of a specific component), and signature analysis (visual comparison of current versus normal condition). Vibration analysis most often reveals problems in machines involving mechanical imbalance, electrical imbalance, misalignment, looseness, and degenerative problems.

Wiebull Distribution A statistical representation of the probability distribution of random failures.

Index

Printed and bound by CPI Group (UK) Ltd, Croydon, CR0 4YY

08/05/2025

01864910-0001